Entropy and Information Theory

Robert M. Gray

Entropy and Information Theory

Springer Science+Business Media, LLC

Robert M. Gray
Information Systems Laboratory
Electrical Engineering Department
Stanford University
Stanford, CA 94305-2099
USA

Printed on acid-free paper.

Originally published by Springer-Verlag New York, Inc. in 1990
MyCopy version of the original edition 1990

Camera-ready text prepared by the author using LaTeX.

9 8 7 6 5 4 3 2 1

DOI 10.1007/978-1-4757-3982-4
www.springer.com/mycopy

to Tim, Lori, Julia, Peter,
Gus, Amy Elizabeth, and Alice
and in memory of Tino

Contents

Prologue

This book is devoted to the theory of probabilistic information measures and their application to coding theorems for information sources and noisy channels. The eventual goal is a general development of Shannon's mathematical theory of communication, but much of the space is devoted to the tools and methods required to prove the Shannon coding theorems. These tools form an area common to ergodic theory and information theory and comprise several quantitative notions of the information in random variables, random processes, and dynamical systems. Examples are entropy, mutual information, conditional entropy, conditional information, and discrimination or relative entropy, along with the limiting normalized versions of these quantities such as entropy rate and information rate. Much of the book is concerned with their properties, especially the long term asymptotic behavior of sample information and expected information.

The book has been strongly influenced by M. S. Pinsker's classic *Information and Information Stability of Random Variables and Processes* and by the seminal work of A. N. Kolmogorov, I. M. Gelfand, A. M. Yaglom, and R. L. Dobrushin on information measures for abstract alphabets and their convergence properties. Many of the results herein are extensions of their generalizations of Shannon's original results. The mathematical models of this treatment are more general than traditional treatments in that nonstationary and nonergodic information processes are treated. The models are somewhat less general than those of the Soviet school of information theory in the sense that standard alphabets rather than completely abstract alphabets are considered. This restriction, however, permits many stronger results as well as the extension to nonergodic processes. In addition, the assumption of standard spaces simplifies many proofs and such spaces include as examples virtually all examples of engineering interest.

The information convergence results are combined with ergodic theorems to prove general Shannon coding theorems for sources and channels. The results are not the most general known and the converses are not the strongest available, but they are sufficently general to cover most systems encountered in applications and they provide an introduction to recent extensions requiring significant additional mathematical machinery. Several of the generalizations have not previously been treated in book form. Examples of novel topics for an information theory text include asymptotic mean stationary sources, one-sided sources as well as two-sided sources, nonergodic sources, $\bar{d}$-continuous channels, and sliding block codes. Another novel aspect is the use of recent proofs of general Shannon-McMillan-Breiman theorems which do not use martingale theory: A coding proof of Ornstein and Weiss [117] is used to prove the almost everywhere convergence of sample entropy for discrete alphabet processes and a variation on the sandwich approach of Algoet and Cover [7] is used to prove the convergence of relative entropy densities for general standard alphabet processes. Both results are proved for asymptotically mean stationary processes which need not be ergodic.

This material can be considered as a sequel to my book *Probability, Random Processes, and Ergodic Properties* [51] wherein the prerequisite results on probability, standard spaces, and ordinary ergodic properties may be found. This book is self contained with the exception of common (and a few less common) results which may be found in the first book.

It is my hope that the book will interest engineers in some of the mathematical aspects and general models of the theory and mathematicians in some of the important engineering applications of performance bounds and code design for communication systems.

Information theory or the mathematical theory of communication has two primary goals: The first is the development of the fundamental theoretical limits on the achievable performance when communicating a given information source over a given communications channel using coding schemes from within a prescribed class. The second goal is the development of coding schemes that provide performance that is reasonably good in comparison with the optimal performance given by the theory. Information theory was born in a surprisingly rich state in the classic papers of Claude E. Shannon [129] [130] which contained the basic results for simple memoryless sources and channels and introduced more general communication systems models, including finite state sources and channels. The key tools used to prove the original results and many of those that followed were special cases of the ergodic theorem and a new variation of the ergodic theorem which considered sample averages of a measure of the entropy or self information in a process.

Information theory can be viewed as simply a branch of applied probability theory. Because of its dependence on ergodic theorems, however, it can also be viewed as a branch of ergodic theory, the theory of invariant transformations and transformations related to invariant transformations. In order to develop the ergodic theory example of principal interest to information theory, suppose that one has a random process, which for the moment we consider as a sample space or ensemble of possible output sequences together with a probability measure on events composed of collections of such sequences. The shift is the transformation on this space of sequences that takes a sequence and produces a new sequence by shifting the first sequence a single time unit to the left. In other words, the shift transformation is a mathematical model for the effect of time on a data sequence. If the probability of any sequence event is unchanged by shifting the event, that is, by shifting all of the sequences in the event, then the shift transformation is said to be *invariant* and the random process is said to be *stationary*. Thus the theory of stationary random processes can be considered as a subset of ergodic theory. Transformations that are not actually invariant (random processes which are not actually stationary) can be considered using similar techniques by studying transformations which are almost invariant, which are invariant in an asymptotic sense, or which are dominated or asymptotically dominated in some sense by an invariant transformation. This generality can be important as many real processes are not well modeled as being stationary. Examples are processes with transients, processes that have been parsed into blocks and coded, processes that have been encoded using variable-length codes or finite state codes and channels with arbitrary starting states.

Ergodic theory was originally developed for the study of statistical mechanics as a means of quantifying the trajectories of physical or dynamical systems. Hence, in the language of random processes, the early focus was on ergodic theorems: theorems relating the time or sample average behavior of a random process to its ensemble or expected behavior. The work of Hoph [65], von Neumann [146] and others culminated in the pointwise or almost everywhere ergodic theorem of Birkhoff [16].

In the 1940's and 1950's Shannon made use of the ergodic theorem in the simple special case of memoryless processes to characterize the optimal performance theoretically achievable when communicating information sources over constrained random media called channels. The ergodic theorem was applied in a direct fashion to study the asymptotic behavior of error frequency and time average distortion in a communication system, but a new variation was introduced by defining a mathematical measure of the entropy or information in a random process and characterizing its asymptotic behavior. These results are known as *coding theorems*. Results describing performance that is actually achievable, at least in the limit of unbounded complexity and time, are known as *positive coding theorems*. Results providing unbeatable bounds on performance are known as *converse coding theorems* or *negative coding theorems*. When the same quantity is given by both positive and negative coding theorems, one has exactly the optimal performance theoretically achievable by the given communication systems model.

While mathematical notions of information had existed before, it was Shannon who coupled the notion with the ergodic theorem and an ingenious idea known as "random coding" in order to develop the coding theorems and to thereby give operational significance to such information measures. The name "random coding" is a bit misleading since it refers to the random selection of a deterministic code and not a coding system that operates in a random or stochastic manner. The basic approach to proving positive coding theorems was to analyze the average performance over a random selection of codes. If the average is good, then there must be at least one code in the ensemble of codes with performance as good as the average. The ergodic theorem is crucial to this argument for determining such average behavior. Unfortunately, such proofs promise the existence of good codes but give little insight into their construction.

Shannon's original work focused on memoryless sources whose probability distribution did not change with time and whose outputs were drawn from a finite alphabet or the real line. In this simple case the well-known ergodic theorem immediately provided the required result concerning the asymptotic behavior of information. He observed that the basic ideas extended in a relatively straightforward manner to more complicated Markov sources. Even this generalization, however, was a far cry from the general stationary sources considered in the ergodic theorem.

To continue the story requires a few additional words about measures of information. Shannon really made use of two different but related measures. The first was entropy, an idea inherited from thermodynamics and previously proposed as a measure of the information in a random signal by Hartley [64]. Shannon defined the entropy of a discrete time discrete alphabet random process $\{X_n\}$, which we denote by $H(X)$ while deferring its definition, and made rigorous the idea that the the entropy of a process is the amount of information in the process. He did this by proving a coding theorem showing that if one wishes to code the given process into a sequence of binary symbols so that a receiver viewing the binary sequence can reconstruct the original process perfectly (or nearly so), then one needs at least $H(X)$ binary symbols or bits (converse theorem) and one can accomplish the task with very close to $H(X)$ bits (positive theorem). This coding theorem is known as the *noiseless source coding theorem*.

The second notion of information used by Shannon was mutual information. Entropy is really a notion of self information–the information provided by a random process about itself. Mutual information is a measure of the information contained in one process about another process. While entropy is sufficient to study the reproduction of a single process through a noiseless environment, more often one has two or more distinct random processes, e.g., one random process representing an information source and another representing the output of a communication medium wherein the coded source has been corrupted by another random process called noise. In such cases observations are made on one process in order to make decisions on another. Suppose that $\{X_n, Y_n\}$ is a random process with a discrete alphabet, that is, taking on values in a discrete set. The coordinate random processes $\{X_n\}$ and $\{Y_n\}$ might correspond, for example, to the input and output of a communication system. Shannon introduced the notion of the average mutual information between the two processes:

$$I(X,Y) = H(X) + H(Y) - H(X,Y), \tag{1}$$

the sum of the two self entropies minus the entropy of the pair. This proved to be the relevant quantity in coding theorems involving more than one distinct random process: the channel coding theorem describing reliable communication through a noisy channel, and the general source coding theorem describing the coding of a source for a user subject to a fidelity criterion. The first theorem focuses on error detection and correction and the second on analog-to-digital conversion and data compression. Special cases of both of these coding theorems were given in Shannon's original work.

Average mutual information can also be defined in terms of *conditional entropy* (or *equivocation*) $H(X|Y) = H(X,Y) - H(Y)$ and hence

$$I(X,Y) = H(X) - H(X|Y) = H(Y) - H(X|Y). \tag{2}$$

In this form the mutual information can be interpreted as the information contained in one process minus the information contained in the process when the other process is known. While elementary texts on information theory abound with such intuitive descriptions of information measures, we will minimize such discussion because of the potential pitfall of using the interpretations to apply such measures to problems where they are not appropriate. (See, e.g., P. Elias' "Information theory, photosynthesis, and religion" in his "Two famous papers" [36].) Information measures are important because coding theorems exist imbuing them with operational significance and not because of intuitively pleasing aspects of their definitions.

We focus on the definition (1) of mutual information since it does not require any explanation of what conditional entropy means and since it has a more symmetric form than the conditional definitions. It turns out that

$H(X,X) = H(X)$ (the entropy of a random variable is not changed by repeating it) and hence from (1)

$$I(X,X) = H(X) \tag{3}$$

so that entropy can be considered as a special case of average mutual information.

To return to the story, Shannon's work spawned the new field of information theory and also had a profound effect on the older field of ergodic theory.

Information theorists, both mathematicians and engineers, extended Shannon's basic approach to ever more general models of information sources, coding structures, and performance measures. The fundamental ergodic theorem for entropy was extended to the same generality as the ordinary ergodic theorems by McMillan [103] and Breiman [19] and the result is now known as the Shannon-McMillan-Breiman theorem. (Other names are the asymptotic equipartition theorem or AEP, the ergodic theorem of information theory, and the entropy theorem.) A variety of detailed proofs of the basic coding theorems and stronger versions of the theorems for memoryless, Markov, and other special cases of random processes were developed, notable examples being the work of Feinstein [38] [39] and Wolfowitz (see, e.g., Wolfowitz [151].) The ideas of measures of information, channels, codes, and communications systems were rigorously extended to more general random processes with abstract alphabets and discrete and continuous time by Khinchine [72], [73] and by Kolmogorov and his colleagues, especially Gelfand, Yaglom, Dobrushin, and Pinsker [45], [90], [87], [32], [125]. (See, for example, "Kolmogorov's contributions to information theory and algorithmic complexity" [23].) In almost all of the early Soviet work, it was average mutual information that played the fundamental role. It was the more natural quantity when more than one process were being considered. In addition, the notion of entropy was not useful when dealing with processes with continuous alphabets since it is virtually always infinite in such cases. A generalization of the idea of entropy called *discrimination* was developed by Kullback (see, e.g., Kullback [92]) and was further studied by the Soviet school. This form of information measure is now more commonly referred to as relative entropy or cross entropy (or Kullback-Leibler number) and it is better interpreted as a measure of similarity between probability distributions than as a measure of information between random variables. Many results for mutual information and entropy can be viewed as special cases of results for relative entropy and the formula for relative entropy arises naturally in some proofs.

It is the mathematical aspects of information theory and hence the descendants of the above results that are the focus of this book, but the developments in the engineering community have had as significant an impact on the foundations of information theory as they have had on applications. Simpler proofs of the basic coding theorems were developed for special cases and, as a natural offshoot, the rate of convergence to the optimal performance bounds characterized in a variety of important cases. See, e.g., the texts by Gallager [43], Berger [11], and Csiszàr and Körner [26]. Numerous practicable coding techniques were developed which provided performance reasonably close to the optimum in many cases: from the simple linear error correcting and detecting codes of Slepian [137] to the huge variety of algebraic codes currently being implemented (see, e.g., [13], [148],[95], [97], [18]) and the various forms of convolutional, tree, and trellis codes for error correction and data compression (see, e.g., [145], [69]). Clustering techniques have been used to develop good nonlinear codes (called "vector quantizers") for data compression applications such as speech and image coding [49], [46], [99], [69], [118]. These clustering and trellis search techniques have been combined to form single codes that combine the data compression and reliable communication operations into a single coding system [8].

The engineering side of information theory through the middle 1970's has been well chronicled by two IEEE collections: *Key Papers in the Development of Information Theory,* edited by D. Slepian [138], and *Key Papers in the Development of Coding Theory,* edited by E. Berlekamp [14]. In addition there have been several survey papers describing the history of information theory during each decade of its existence published in the *IEEE Transactions on Information Theory.*

The influence on ergodic theory of Shannon's work was equally great but in a different direction. After the development of quite general ergodic theorems, one of the principal issues of ergodic theory was the isomorphism problem, the characterization of conditions under which two dynamical systems are really the same in the sense that each could be obtained from the other in an invertible way by coding. Here, however, the coding was not of the variety considered by Shannon: Shannon considered block codes, codes that parsed the data into nonoverlapping blocks or windows of finite length and separately mapped each input block into an output block. The more natural construct in ergodic theory can be called a sliding block code: Here the encoder views a block of possibly infinite length and produces a single symbol of the output sequence using some mapping (or code or filter). The input sequence is then shifted one time unit to the left, and the same mapping applied to produce the next output symbol, and so on. This is a smoother operation than the block coding structure since the outputs are produced based on overlapping windows of data instead of on a completely different set of data each time. Unlike the Shannon codes, these codes will produce stationary output processes if given stationary input processes. It should be mentioned that examples of such sliding block codes often occurred in the information theory literature: time-invariant convolutional codes or, simply, time-invariant linear filters are sliding block codes. It is perhaps odd that virtually all of the theory for such codes in the information theory literature was developed by effectively considering the sliding block codes as very long block codes. Recently sliding block codes have proved a useful structure for the design of noiseless codes for constrained alphabet channels such as magnetic recording devices, and techniques from symbolic dynamics have been applied to the design of such codes. See, for example [3], [100].

Shannon's noiseless source coding theorem suggested a solution to the isomorphism problem: If we assume for the moment that one of the two processes is binary, then perfect coding of a process into a binary process and back into the original process requires that the original process and the binary process have the same entropy. Thus a natural conjecture is that two processes are isomorphic if and only if they have the same entropy. A major difficulty was the fact that two different kinds of coding were being considered: stationary sliding block codes with zero error by the ergodic theorists and either fixed length block codes with small error or variable length (and hence nonstationary) block codes with zero error by the Shannon theorists. While it was plausible that the former codes might be developed as some sort of limit of the latter, this proved to be an extremely difficult problem. It was Kolmogorov [88], [89] who first reasoned along these lines and proved that in fact equal entropy (appropriately defined) was a necessary condition for isomorphism.

Kolmogorov's seminal work initiated a new branch of ergodic theory devoted to the study of entropy of dynamical systems and its application to the isomorphism problem. Most of the original work was done by Soviet mathematicians; notable papers are those by Sinai [134] [135] (in ergodic theory entropy is also known as the Kolmogorov-Sinai invariant), Pinsker [125], and Rohlin and Sinai [127]. An actual construction of a perfectly noiseless sliding block code for a special case was provided by Meshalkin [104]. While much insight was gained into the behavior of entropy and progress was made on several simplified versions of the isomorphism problem, it was several years before Ornstein [114] proved a result that has since come to be known as the Kolmogorov-Ornstein isomorphism theorem.

Ornstein showed that if one focused on a class of random processes which we shall call B-processes, then two processes are indeed isomorphic if and only if they have the same entropy. B-processes have several equivalent definitions, perhaps the simplest is that they are processes which can be obtained by encoding a memoryless process using a sliding block code.

This class remains the most general class known for which the isomorphism conjecture holds. In the course of his proof, Ornstein developed intricate connections between block coding and sliding block coding. He used Shannonlike techniques on the block codes, then imbedded the block codes into sliding block codes, and then used the stationary structure of the sliding block codes to advantage in limiting arguments to obtain the required zero error codes. Several other useful techniques and results were introduced in the proof: notions of the distance between processes and relations between the goodness of approximation and the difference of entropy. Ornstein expanded these results into a book [116] and gave a tutorial discussion in the premier issue of the *Annals of Probability* [115]. Several correspondence items by other ergodic theorists discussing the paper accompanied the article.

The origins of this book lie in the tools developed by Ornstein for the proof of the isomorphism theorem rather than with the result itself. During the early 1970's I first become interested in ergodic theory because of joint work with Lee D. Davisson on source coding theorems for stationary nonergodic processes. The ergodic decomposition theorem discussed in Ornstein [115] provided a needed missing link and led to an intense campaign on my part to learn the fundamentals of ergodic theory and perhaps find other useful tools. This effort was greatly eased by Paul Shields' book *The Theory of Bernoulli Shifts* [131] and by discussions with Paul on topics in both ergodic theory and information theory. This in turn led to a variety of other applications of ergodic theoretic techniques and results to information theory, mostly in the area of source coding theory: proving source coding theorems for sliding block codes and using process distance measures to prove universal source coding theorems and to provide new characterizations of Shannon distortion-rate functions. The work was done with Dave Neuhoff, like me then an apprentice ergodic theorist, and Paul Shields.

With the departure of Dave and Paul from Stanford, my increasing interest led me to discussions with Don Ornstein on possible applications of his techniques to channel coding problems. The interchange often consisted of my describing a problem, his generation of possible avenues of solution, and then my going off to work for a few weeks to understand his suggestions and work them through.

One problem resisted our best efforts–how to synchronize block codes over channels with memory, a prerequisite for constructing sliding block codes for such channels. In 1975 I had the good fortune to meet and talk with Roland Dobrushin at the 1975 IEEE/USSR Workshop on Information Theory in Moscow. He observed that some of his techniques for handling synchronization in memoryless channels should immediately generalize to our case and therefore should provide the missing link. The key elements were all there, but it took seven years for the paper by Ornstein, Dobrushin and me to evolve and appear [59].

Early in the course of the channel coding paper, I decided that having the solution to the sliding block channel coding result in sight was sufficient excuse to write a book on the overlap of ergodic theory and information theory. The intent was to develop the tools of ergodic theory of potential use to information theory and to demonstrate their use by proving Shannon coding theorems for the most general known information sources, channels, and code structures. Progress on the book was disappointingly slow, however, for a number of reasons. As delays mounted, I saw many of the general coding theorems extended and improved by others (often by J. C. Kieffer) and new applications of ergodic theory to information theory developed, such as the channel modeling work of Neuhoff and Shields [110], [113], [112], [111] and design methods for sliding block codes for input restricted noiseless channels by Adler, Coppersmith, and Hasner [3] and Marcus [100]. Although I continued to work in some aspects of the area, especially with nonstationary and nonergodic processes and processes with standard alphabets, the area remained for me a relatively minor one and I had little time to write. Work and writing came in bursts during sabbaticals and occasional advanced topic seminars. I abandoned the idea of providing the most general possible coding theorems and decided instead to settle for coding theorems that were sufficiently general to cover most applications and which possessed proofs I liked and could understand. The mantle of the most general theorems will go to a book in progress by J.C. Kieffer [85]. That book shares many topics with this one, but the approaches and viewpoints and many of the results treated are quite different. At the risk of generalizing, the books will reflect our differing backgrounds: mine as an engineer by training and a would-be mathematician, and his as a mathematician by training migrating to an engineering school. The proofs of the principal results often differ in significant ways and the two books contain a variety of different minor results developed as tools along the way. This book is perhaps more "old fashioned" in that the proofs often retain the spirit of the original "classical" proofs, while Kieffer has developed a variety of new and powerful techniques to obtain the most general known results. I have also taken more detours along the way in order to catalog various properties of entropy and other information measures that I found interesting in their own right, even though they were not always necessary for proving the coding theorems. Only one third of this book is actually devoted to Shannon source and channel coding theorems; the remainder can be viewed as a monograph on information measures and their properties, especially their ergodic properties.

Because of delays in the original project, the book was split into two smaller books and the first, ***Probability, Random Processes, and Ergodic Properties***, was published by Springer-Verlag in 1988 [50]. It treats advanced probability and random processes with an emphasis on processes with standard alphabets, on nonergodic and nonstationary processes, and on necessary and sufficient conditions for the convergence of long term sample averages. Asymptotically mean stationary sources and the ergodic decomposition are there treated in depth and recent simplified proofs of the ergodic theorem due to Ornstein and Weiss [117] and others were incorporated. That book provides the background material and introduction to this book, the split naturally falling before the introduction of entropy. The first chapter of this book reviews some of the basic notation of the first one in information theoretic terms, but results are often simply quoted as needed from the first book without any attempt to derive them. The two books together are self-contained in that all supporting results from probability theory and ergodic theory needed here may be found in the first book. This book is self-contained so far as its information theory content, but it should be considered as an advanced text on the subject and not as an introductory treatise to the reader only wishing an intuitive overview.

Here the Shannon-McMillan-Breiman theorem is proved using the coding approach of Ornstein and Weiss [117] (see also Shield's tutorial paper [132]) and hence the treatments of ordinary ergodic theorems in the first book and the ergodic theorems for information measures in this book are consistent. The extension of the Shannon-McMillan-Breiman theorem to densities is proved using the "sandwich" approach of Algoet and Cover [7], which depends strongly on the usual pointwise or Birkhoff ergodic theorem: sample entropy is asymptotically sandwiched between two functions whose limits can be determined from the ergodic theorem. These results are the most general yet published in book form and differ from traditional developments in that martingale theory is not required in the proofs.

A few words are in order regarding topics that are not contained in this book. I have not included multiuser information theory for two reasons: First, after including the material that I wanted most, there was no room left. Second, my experience in the area is slight and I believe this topic can be better handled by others. Results as general as the single user systems described here have not yet been developed. Good surveys of the multiuser area may be found in El Gamal and Cover [44], Van der Meulen [142], and Berger [12].

Traditional noiseless coding theorems and actual codes such as the Huffman codes are not considered in depth because quite good treatments exist in the literature, e.g., [43], [1], [102]. The corresponding ergodic theory result–the Kolmogorov-Ornstein isomorphism theorem–is also not proved, because its proof is difficult and the result is not needed for the Shannon coding theorems. Many techniques used in its proof, however, are used here for similar and other purposes.

The actual computation of channel capacity and distortion rate functions has not been included because existing treatments [43], [17], [11], [52] are quite adequate.

This book does not treat code design techniques. Algebraic coding is well developed in existing texts on the subject [13], [148], [95], [18]. Allen Gersho and I are currently writing a book on the theory and design of nonlinear coding techniques such as vector quantizers and trellis codes for analog-to-digital conversion and for source coding (data compression) and combined source and channel coding applications [47]. A less mathematical treatment of rate-distortion theory along with other source coding topics not treated here (including asymptotic, or high rate, quantization theory and uniform quantizer noise theory) may be found in my book [52].

Universal codes, codes which work well for an unknown source, and variable rate codes, codes producing a variable number of bits for each input vector, are not considered. The interested reader is referred to [109] [96] [77] [78] [28] and the references therein.

A recent active research area that has made good use of the ideas of relative entropy to characterize exponential growth is that of large deviations theory[143][31]. These techniques have been used to provide new proofs of the basic source coding theorems[22]. These topics are not treated here.

Lastly, J. C. Kieffer has recently developed a powerful new ergodic theorem that can be used to prove both traditional ergodic theorems and the extended Shannon-McMillan-Brieman theorem [83]. He has used this theorem to prove new strong (almost everywhere) versions of the souce coding theorem and its converse, that is, results showing that sample average distortion is with probability one no smaller than the distortion-rate function and that there exist codes with sample average distortion arbitrarily close to the distortion-rate function [84] [82]. These results should have a profound impact on the future development of the theoretical tools and results of information theory. Their imminent publication provide a strong motivation for the completion of this monograph, which is devoted to the traditional methods. Tradition has its place, however, and the methods and results treated here should retaihn much of their role at the core of the theory of entropy and information. It is hoped that this collection of topics and methods will find a niche in the literature.

Acknowledgments

The research in information theory that yielded many of the results and some of the new proofs for old results in this book was supported by the National Science Foundation. Portions of the research and much of the early writing were supported by a fellowship from the John Simon Guggenheim Memorial Foundation. The book was originally written using the eqn and troff utilities on several UNIX systems and was subsequently translated into LaTeX on both UNIX and Apple Macintosh systems. All of these computer systems were supported by the Industrial Affiliates Program of the Stanford University Information Systems Laboratory. Much helpful advice on the mysteries of LaTeX was provided by Richard Roy and Mark Goldburg.

The book benefited greatly from comments from numerous students and colleagues over many years; most notably Paul Shields, Paul Algoet, Ender Ayanoglu, Lee Davisson, John Kieffer, Dave Neuhoff, Don Ornstein, Bob Fontana, Jim Dunham, Farivar Saadat, Michael Sabin, Andrew Barron, Phil Chou, Tom Lookabaugh, Andrew Nobel, and Bradley Dickinson.

Robert M. Gray
La Honda, California
April 1990

Chapter 1

Information Sources

1.1 Introduction

An *information source* or *source* is a mathematical model for a physical entity that produces a succession of symbols called "outputs" in a random manner. The symbols produced may be real numbers such as voltage measurements from a transducer, binary numbers as in computer data, two dimensional intensity fields as in a sequence of images, continuous or discontinuous waveforms, and so on. The space containing all of the possible output symbols is called the *alphabet* of the source and a source is essentially an assignment of a probability measure to events consisting of sets of sequences of symbols from the alphabet. It is useful, however, to explicitly treat the notion of time as a transformation of sequences produced by the source. Thus in addition to the common random process model we shall also consider modeling sources by dynamical systems as considered in ergodic theory.

The material in this chapter is a distillation of [50] and is intended to establish notation.

1.2 Probability Spaces and Random Variables

A measurable space $(\Omega, \mathcal{B})$ is a pair consisting of a sample space Ω together with a σ-field $\mathcal{B}$ of subsets of Ω (also called the event space). A σ-field or σ-algebra $\mathcal{B}$ is a nonempty collection of subsets of Ω with the following properties:

$$\Omega \in \mathcal{B}. \tag{1.2.1}$$

$$\text{If } F \in \mathcal{B}, \text{ then } F^c = \{\omega : \omega \notin F\} \in \mathcal{B}. \tag{1.2.2}$$

$$\text{If } F_i \in \mathcal{B}; i = 1, 2, \cdots, \text{ then } \bigcup_i F_i \in \mathcal{B}. \tag{1.2.3}$$

From de Morgan's "laws" of elementary set theory it follows that also

$$\bigcap_{i=1}^{\infty} F_i = (\bigcup_{i=1}^{\infty} F_i^c)^c \in \mathcal{B}.$$

An event space is a collection of subsets of a sample space (called events by virtue of belonging to the event space) such that any countable sequence of set theoretic operations (union, intersection, complementation) on events produces other events. Note that there are two extremes: the largest possible σ-field of Ω is the collection of all subsets of Ω (sometimes called the *power set*) and the smallest possible σ-field is $\{\Omega, \emptyset\}$, the entire space together with the null set $\emptyset = \Omega^c$ (called the *trivial space*).

If instead of the closure under countable unions required by (1.2.3), we only require that the collection of subsets be closed under finite unions, then we say that the collection of subsets is a *field*.

While the concept of a field is simpler to work with, a σ-field possesses the additional important property that it contains all of the limits of sequences of sets in the collection. That is, if F_n, $n = 1, 2, \cdots$ is an increasing sequence of sets in a σ-field, that is, if $F_{n-1} \subset F_n$ and if $F = \bigcup_{n=1}^{\infty} F_n$ (in which case we write $F_n \uparrow F$ or $\lim_{n\to\infty} F_n = F$), then also F is contained in the σ-field. In a similar fashion we can define decreasing sequences of sets: If F_n decreases to F in the sense that $F_{n+1} \subset F_n$ and $F = \bigcap_{n=1}^{\infty} F_n$, then we write $F_n \downarrow F$. If $F_n \in \mathcal{B}$ for all n, then $F \in \mathcal{B}$.

A *probability space* $(\Omega, \mathcal{B}, P)$ is a triple consisting of a sample space Ω , a σ-field $\mathcal{B}$ of subsets of Ω , and a probability measure P which assigns a real number $P(F)$ to every member F of the σ-field $\mathcal{B}$ so that the following conditions are satisfied:

- *Nonnegativity*:

$$P(F) \geq 0, \text{ all } F \in \mathcal{B}; \tag{1.2.4}$$

- *Normalization*:

$$P(\Omega) = 1; \tag{1.2.5}$$

- *Countable Additivity*:

If $F_i \in \mathcal{B}, i = 1, 2, \cdots$ are disjoint, then

$$P(\bigcup_{i=1}^{\infty} F_i) = \sum_{i=1}^{\infty} P(F_i). \tag{1.2.6}$$

A set function P satisfying only (1.2.4) and (1.2.6) but not necessarily (1.2.5) is called a *measure* and the triple $(\Omega, \mathcal{B}, P)$ is called a *measure space*. Since the probability measure is defined on a σ-field, such countable unions of subsets of Ω in the σ-field are also events in the σ-field.

A standard result of basic probability theory is that if $G_n \downarrow \emptyset$ (the empty or null set), that is, if $G_{n+1} \subset G_n$ for all n and $\bigcap_{n=1}^{\infty} G_n = \emptyset$, then we have

- *Continuity at $\emptyset$*:

$$\lim_{n\to\infty} P(G_n) = 0. \tag{1.2.7}$$

similarly it follows that we have

- *Continuity from Below*:

$$\text{If } F_n \uparrow F, \text{ then } \lim_{n\to\infty} P(F_n) = P(F), \tag{1.2.8}$$

and

- *Continuity from Above*:

$$\text{If } F_n \downarrow F, \text{ then } \lim_{n\to\infty} P(F_n) = P(F). \tag{1.2.9}$$

Given a measurable space $(\Omega, \mathcal{B})$, a collection $\mathcal{G}$ of members of $\mathcal{B}$ is said to *generate* $\mathcal{B}$ and we write $\sigma(\mathcal{G}) = \mathcal{B}$ if $\mathcal{B}$ is the smallest σ-field that contains $\mathcal{G}$; that is, if a σ-field contains all of the members of $\mathcal{G}$, then it must also contain all of the members of $\mathcal{B}$. The following is a fundamental approximation theorem of probability theory. A proof may be found in Corollary 1.5.3 of [50]. The result is most easily stated in terms of the symmetric difference Δ defined by

$$F \Delta G \equiv (F \bigcap G^c) \bigcup (F^c \bigcup G).$$

Theorem 1.2.1: Given a probability space $(\Omega, \mathcal{B}, P)$ and a generating field $\mathcal{F}$, that is, $\mathcal{F}$ is a field and $\mathcal{B} = \sigma(\mathcal{F})$, then given $F \in \mathcal{B}$ and $\epsilon > 0$, there exists an $F_0 \in \mathcal{F}$ such that $P(F \Delta F_0) \leq \epsilon$.

Let $(A, \mathcal{B}_A)$ denote another measurable space. A *random variable* or *measurable function* defined on $(\Omega, \mathcal{B})$ and taking values in (A,$\mathcal{B}_A$) is a mapping or function $f : \Omega \to A$ with the property that

$$\text{if } F \in \mathcal{B}_A, \text{ then } f^{-1}(F) = \{\omega : f(\omega) \in F\} \in \mathcal{B}. \tag{1.2.10}$$

The name "random variable" is commonly associated with the special case where A is the real line and $\mathcal{B}$ the Borel field, the smallest σ-field containing all the intervals. Occasionally a more general sounding name such as "random object" is used for a measurable function to implicitly include random variables (A the real line), random vectors (A a Euclidean space), and random processes (A a sequence or waveform space). We will use the terms "random variable" in the more general sense.

A random variable is just a function or mapping with the property that inverse images of "output events" determined by the random variable are events in the original measurable space. This simple property ensures that the output of the random variable will inherit its own probability measure. For example, with the probability measure P_f defined by

$$P_f(B) = P(f^{-1}(B)) = P(\omega : f(\omega) \in B); \; B \in \mathcal{B}_A,$$

$(A, \mathcal{B}_A, P_f)$ becomes a probability space since measurability of f and elementary set theory ensure that P_f is indeed a probability measure. The induced probability measure P_f is called the *distribution* of the random variable f. The measurable space $(A,\mathcal{B}_A)$ or, simply, the sample space A, is called the alphabet of the random variable f. We shall occasionally also use the notation Pf^{-1} which is a mnemonic for the relation $Pf^{-1}(F) = P(f^{-1}(F))$ and which is less awkward when f itself is a function with a complicated name, e.g., $\Pi_{\mathcal{I} \to \mathcal{M}}$.

If the alphabet A of a random variable f is not clear from context, then we shall refer to f as an *A-valued random variable*. If f is a measurable function from $(\Omega, \mathcal{B})$ to $(A, \mathcal{B}_A)$, we will say that f is $\mathcal{B}/\mathcal{B}_A$-measurable if the σ-fields might not be clear from context.

Given a probability space $(\Omega, \mathcal{B}, P)$, a collection of subsets $\mathcal{G}$ is a sub-σ-field if it is a σ-field and all its members are in $\mathcal{B}$. A random variable $f : \Omega \to A$ is said to be measurable with respect to a sub-σ-field $\mathcal{G}$ if $f^{-1}(H) \in \mathcal{G}$ for all $H \in \mathcal{B}_A$.

Given a probability space $(\Omega, \mathcal{B}, P)$ and a sub-σ-field $\mathcal{G}$, for any event $H \in \mathcal{B}$ the conditional probability $m(H|\mathcal{G})$ is defined as any function, say g, which satisfies the two properties

$$g \text{ is measurable with respect to } \mathcal{G} \tag{1.2.11}$$

$$\int_G g h dP = m(G \bigcap H); \quad \text{all } G \in \mathcal{G}. \tag{1.2.12}$$

An important special case of conditional probability occurs when studying the distributions of random variables defined on an underlying probability space. Suppose that $X : \Omega \to A_X$ and $Y : \Omega \to A_Y$ are two random variables defined on $(\Omega, \mathcal{B}, P)$ with alphabets A_X and A_Y and σ-fields $\mathcal{B}_{A_X}$ and $\mathcal{B}_{A_Y}$, respectively. Let P_{XY} denote the induced distribution on $(A_X \times A_Y, \mathcal{B}_{A_X} \times \mathcal{B}_{A_Y})$, that is, $P_{XY}(F \times G) = P(X \in F, Y \in G)$

$= P(X^{-1}(F)\bigcap Y^{-1}(G))$. Let $\sigma(Y)$ denote the sub-σ-field of $\mathcal{B}$ generated by Y, that is, $Y^{-1}(\mathcal{B}_{A_Y})$. Since the conditional probability $P(F|\sigma(Y))$ is real-valued and measurable with respect to $\sigma(Y)$, it can be written as $g(Y(\omega))$, $\omega \in \Omega$, for some function $g(y)$. (See, for example, Lemma 5.2.1 of [50].) Define $P(F|y) = g(y)$. For a fixed $F \in \mathcal{B}_{A_X}$ define the *conditional distribution* of F given $Y = y$ by

$$P_{X|Y}(F|y) = P(X^{-1}(F)|y);\ y \in \mathcal{B}_{A_Y}.$$

From the properties of conditional probability,

$$P_{XY}(F \times G) = \int_G P_{X|Y}(F|y)dP_Y(y); F \in \mathcal{B}_{A_X}, G \in \mathcal{B}_{A_Y}. \tag{1.2.13}$$

It is tempting to think that for a fixed y, the set function defined by $P_{X|Y}(F|y)$; $F \in \mathcal{B}_{A_X}$ is actually a probability measure. This is not the case in general. When it does hold for a conditional probability measure, the conditional probability measure is said to be *regular*. As will be emphasized later, this text will focus on standard alphabets for which regular conditional probabilites always exist.

1.3 Random Processes and Dynamical Systems

We now consider two mathematical models for a source: A random process and a dynamical system. The first is the familiar one in elementary courses, a source is just a random process or sequence of random variables. The second model is possibly less familiar; a random process can also be constructed from an abstract dynamical system consisting of a probability space together with a transformation on the space. The two models are connected by considering a time shift to be a transformation.

A *discrete time random process* or for our purposes simply a *random process* is a sequence of random variables $\{X_n\}_{n \in \mathcal{T}}$ or $\{X_n; n \in \mathcal{T}\}$, where $\mathcal{T}$ is an index set, defined on a common probability space $(\Omega, \mathcal{B}, P)$. We define a *source* as a random process, although we could also use the alternative definition of a dynamical system to be introduced shortly. We usually assume that all of the random variables share a common alphabet, say A. The two most common index sets of interest are the set of all integers $\mathcal{Z} = \{\cdots, -2, -1, 0, 1, 2, \cdots\}$, in which case the random process is referred to as a *two-sided* random process, and the set of all nonnegative integers $\mathcal{Z}_+ = \{0, 1, 2, \cdots\}$, in which case the random process is said to be *one-sided*. One-sided random processes will often prove to be far more difficult in theory, but they provide better models for physical random processes that must be "turned on" at some time or which have transient behavior.

Observe that since the alphabet A is general, we could also model continuous time random processes in the above fashion by letting A consist of a family of waveforms defined on an interval, e.g., the random variable X_n could in fact be a continuous time waveform $X(t)$ for $t \in [nT, (n+1)T)$, where T is some fixed positive real number.

The above definition does not specify any structural properties of the index set $\mathcal{T}$. In particular, it does not exclude the possibility that $\mathcal{T}$ be a finite set, in which case "random vector" would be a better name than "random process." In fact, the two cases of $\mathcal{T} = \mathcal{Z}$ and $\mathcal{T} = \mathcal{Z}_+$ will be the only important examples for our purposes. Nonetheless, the general notation of $\mathcal{T}$ will be retained in order to avoid having to state separate results for these two cases.

An abstract dynamical system consists of a probability space $(\Omega, \mathcal{B}, P)$ together with a measurable transformation $T : \Omega \to \Omega$ of Ω into itself. Measurability means that if $F \in \mathcal{B}$, then also $T^{-1}F = \{\omega : T\omega \in F\} \in \mathcal{B}$. The quadruple $(\Omega,\mathcal{B},P,T)$ is called a *dynamical system* in ergodic theory. The interested reader can find excellent introductions to classical ergodic theory and dynamical system theory in the books of Halmos [62] and Sinai [136]. More complete treatments may be found in [15], [131], [124], [30], [147], [116], [42]. The term "dynamical systems" comes from the focus of the theory on the long term "dynamics" or "dynamical behavior" of repeated applications of the transformation T on the underlying measure space.

An alternative to modeling a random process as a sequence or family of random variables defined on a common probability space is to consider a single random variable together with a transformation defined on the underlying probability space. The outputs of the random process will then be values of the random variable taken on transformed points in the original space. The transformation will usually be related to shifting in time and hence this viewpoint will focus on the action of time itself. Suppose now that T is a measurable mapping of points of the sample space Ω into itself. It is easy to see that the cascade or composition of measurable functions is also measurable. Hence the transformation T^n defined as $T^2\omega = T(T\omega)$ and so on ($T^n\omega = T(T^{n-1}\omega)$) is a measurable function for all positive integers n. If f is an A-valued random variable defined on $(\Omega, \mathcal{B})$, then the functions $fT^n : \Omega \to A$ defined by $fT^n(\omega) = f(T^n\omega)$ for $\omega \in \Omega$ will also be random variables for all n in $\mathcal{Z}_+$. Thus a dynamical system together with a random variable or measurable function f defines a one-sided random process $\{X_n\}_{n\in\mathcal{Z}_+}$ by $X_n(\omega) = f(T^n\omega)$. If it should be true that T is invertible, that is, T is one-to-one and its inverse T^{-1} is measurable, then one can define a two-sided random process by $X_n(\omega) = f(T^n\omega)$, all n in $\mathcal{Z}$.

The most common dynamical system for modeling random processes is that consisting of a sequence space Ω containing all one- or two-sided A-valued sequences together with the shift transformation T, that is, the transformation that maps a sequence $\{x_n\}$ into the sequence $\{x_{n+1}\}$ wherein each coordinate has been shifted to the left by one time unit. Thus, for example, let $\Omega = A^{\mathcal{Z}_+} = \{\text{all } x = (x_0, x_1, \cdots) \text{ with } x_i \in A \text{ for all } i\}$ and define $T : \Omega \to \Omega$ by $T(x_0, x_1, x_2, \cdots) = (x_1, x_2, x_3, \cdots)$. T is called the *shift* or *left shift* transformation on the one-sided sequence space. The shift for two-sided spaces is defined similarly.

The different models provide equivalent models for a given process: one emphasizing the sequence of outputs and the other emphasising the action of a transformation on the underlying space in producing these outputs. In order to demonstrate in what sense the models are equivalent for given random processes, we next turn to the notion of the distribution of a random process.

1.4 Distributions

While in principle all probabilistic quantities associated with a random process can be determined from the underlying probability space, it is often more convenient to deal with the induced probability measures or distributions on the space of possible outputs of the random process. In particular, this allows us to compare different random processes without regard to the underlying probability spaces and thereby permits us to reasonably equate two random processes if their outputs have the same probabilistic structure, even if the underlying probability spaces are quite different.

We have already seen that each random variable X_n of the random process $\{X_n\}$ inherits a distribution because it is measurable. To describe a process, however, we need more than simply probability measures on output values of separate single random variables; we require probability measures on collections of random variables, that is, on sequences of outputs. In order to place probability measures on sequences of outputs of a random process, we first must construct the appropriate measurable spaces. A convenient technique for accomplishing this is to consider product spaces, spaces for sequences formed by concatenating spaces for individual outputs.

Let $\mathcal{T}$ denote any finite or infinite set of integers. In particular, $\mathcal{T} = \mathcal{Z}(n) = \{0, 1, 2, \cdots, n-1\}$, $\mathcal{T} = \mathcal{Z}$, or $\mathcal{T} = \mathcal{Z}_+$. Define $x^{\mathcal{T}} = \{x_i\}_{i \in \mathcal{T}}$. For example, $x^{\mathcal{Z}} = (\cdots, x_{-1}, x_0, x_1, \cdots)$ is a two-sided infinite sequence. When $\mathcal{T} = \mathcal{Z}(n)$ we abbreviate $x^{\mathcal{Z}(n)}$ to simply x^n . Given alphabets $A_i, i \in \mathcal{T}$, define the cartesian product space

$$\mathop{\times}_{i \in \mathcal{T}} A_i = \{ \text{ all } x^{\mathcal{T}} : x_i, \in A_i \text{ all } i \text{ in } \mathcal{T}\}.$$

In most cases all of the A_i will be replicas of a single alphabet A and the above product will be denoted simply by $A^{\mathcal{T}}$. Thus, for example, $A^{\{m,m+1,\cdots,n\}}$ is the space of all possible outputs of the process from time m to time n; $A^{\mathcal{Z}}$ is the sequence space of all possible outputs of a two-sided process. We shall abbreviate the notation for the space $A^{\mathcal{Z}(n)}$, the space of all n dimensional vectors with coordinates in A, by A^n.

To obtain useful σ-fields of the above product spaces, we introduce the idea of a rectangle in a product space. A *rectangle* in $A^{\mathcal{T}}$ taking values in the coordinate σ-fields $\mathcal{B}_i$, $i \in \mathcal{J}$, is defined as any set of the form

$$B = \{x^{\mathcal{T}} \in A^{\mathcal{T}} : x_i \in B_i; \text{ all } i \text{ in } \mathcal{J}\}, \tag{1.4.1}$$

where $\mathcal{J}$ is a finite subset of the index set $\mathcal{T}$ and $B_i \in \mathcal{B}_i$ for all $i \in \mathcal{J}$. (Hence rectangles are sometimes referred to as finite dimensional rectangles.) A rectangle as in (1.4.1) can be written as a finite intersection of one-dimensional rectangles as

$$B = \bigcap_{i \in \mathcal{J}} \{x^{\mathcal{T}} \in A^{\mathcal{T}} : x_i \in B_i\} = \bigcap_{i \in \mathcal{J}} X_i^{-1}(B_i) \tag{1.4.2}$$

where here we consider X_i as the coordinate functions $X_i : A^{\mathcal{T}} \to A$ defined by $X_i(x^{\mathcal{T}}) = x_i$.

As rectangles in $A^{\mathcal{T}}$ are clearly fundamental events, they should be members of any useful σ-field of subsets of $A^{\mathcal{T}}$. Define the product σ-field $\mathcal{B}_A{}^{\mathcal{T}}$ as the smallest σ-field containing all of the rectangles, that is, the collection of sets that contains the clearly important class of rectangles and the minimum amount of other stuff required to make the collection a σ-field. To be more precise, given an index set $\mathcal{T}$ of integers, let $RECT(\mathcal{B}_i, i \in \mathcal{T})$ denote the set of all rectangles in $A^{\mathcal{T}}$ taking coordinate values in sets in $\mathcal{B}_i, i \in \mathcal{T}$. We then define the product σ-field of $A^{\mathcal{T}}$ by

$$\mathcal{B}_A{}^{\mathcal{T}} = \sigma(RECT(\mathcal{B}_i, i \in \mathcal{T})). \tag{1.4.3}$$

Consider an index set $\mathcal{T}$ and an A-valued random process $\{X_n\}_{n \in \mathcal{T}}$ defined on an underlying probability space $(\Omega, \mathcal{B}, P)$. Given any index set $\mathcal{J} \subset \mathcal{T}$, measurability of the individual random variables X_n implies that of the random vectors $X^{\mathcal{J}} = \{X_n; n \in \mathcal{J}\}$. Thus the measurable space $(A^{\mathcal{J}}, \mathcal{B}_A{}^{\mathcal{J}})$ inherits a probability measure from the underlying space through the random variables $X^{\mathcal{J}}$. Thus in particular the measurable space

$(A^{\mathcal{T}}, \mathcal{B}_A{}^{\mathcal{T}})$ inherits a probability measure from the underlying probability space and thereby determines a new probability space $(A^{\mathcal{T}}, \mathcal{B}_A{}^{\mathcal{T}}, P_{X^{\mathcal{T}}})$, where the induced probability measure is defined by

$$P_{X^{\mathcal{T}}}(F) = P((X^{\mathcal{T}})^{-1}(F)) = P(\omega : X^{\mathcal{T}}(\omega) \in F);\ F \in \mathcal{B}_A{}^{\mathcal{T}}. \quad (1.4.4)$$

Such probability measures induced on the outputs of random variables are referred to as *distributions* for the random variables, exactly as in the simpler case first treated. When $\mathcal{T} = \{m, m+1, \cdots, m+n-1\}$, e.g., when we are treating $X_m^n = (X_n, \cdots, X_{m+n-1})$ taking values in A^n, the distribution is referred to as an n-dimensional or nth order distribution and it describes the behavior of an n-dimensional random variable. If $\mathcal{T}$ is the entire process index set, e.g., if $\mathcal{T} = \mathcal{Z}$ for a two-sided process or $\mathcal{T} = \mathcal{Z}_+$ for a one-sided process, then the induced probability measure is defined to be the distribution of the process. Thus, for example, a probability space $(\Omega, \mathcal{B}, P)$ together with a doubly infinite sequence of random variables $\{X_n\}_{n \in \mathcal{Z}}$ induces a new probability space $(A^{\mathcal{Z}}, \mathcal{B}_A{}^{\mathcal{Z}}, P_{X^{\mathcal{Z}}})$ and $P_{X^{\mathcal{Z}}}$ is the distribution of the process. For simplicity, let us now denote the process distribution simply by m. We shall call the probability space $(A^{\mathcal{T}}, \mathcal{B}_A{}^{\mathcal{T}}, m)$ induced in this way by a random process $\{X_n\}_{n \in \mathcal{Z}}$ the output space or sequence space of the random process.

Since the sequence space $(A^{\mathcal{T}}, \mathcal{B}_A{}^{\mathcal{T}}, m)$ of a random process $\{X_n\}_{n \in \mathcal{Z}}$ is a probability space, we can define random variables and hence also random processes on this space. One simple and useful such definition is that of a sampling or coordinate or projection function defined as follows: Given a product space $A^{\mathcal{T}}$, define the sampling functions $\Pi_n : A^{\mathcal{T}} \to A$ by

$$\Pi_n(x^{\mathcal{T}}) = x_n, x^{\mathcal{T}} \in A^{\mathcal{T}};\ n \in \mathcal{T}. \quad (1.4.5)$$

The sampling function is named Π since it is also a projection. Observe that the distribution of the random process $\{\Pi_n\}_{n \in \mathcal{T}}$ defined on the probability space $(A^{\mathcal{T}}, \mathcal{B}_A{}^{\mathcal{T}}, m)$ is exactly the same as the distribution of the random process $\{X_n\}_{n \in \mathcal{T}}$ defined on the probability space $(\Omega, \mathcal{B}, P)$. In fact, so far they are the same process since the $\{\Pi_n\}$ simply read off the values of the $\{X_n\}$.

What happens, however, if we no longer build the Π_n on the X_n, that is, we no longer first select ω from Ω according to P, then form the sequence $x^{\mathcal{T}} = X^{\mathcal{T}}(\omega) = \{X_n(\omega)\}_{n \in \mathcal{T}}$, and then define $\Pi_n(x^{\mathcal{T}}) = X_n(\omega)$? Instead we directly choose an x in $A^{\mathcal{T}}$ using the probability measure m and then view the sequence of coordinate values. In other words, we are considering two completely separate experiments, one described by the probability space $(\Omega, \mathcal{B}, P)$ and the random variables $\{X_n\}$ and the other described by the probability space $(A^{\mathcal{T}}, \mathcal{B}_A{}^{\mathcal{T}}, m)$ and the random variables $\{\Pi_n\}$.

In these two separate experiments, the actual sequences selected may be completely different. Yet intuitively the processes should be the "same" in the sense that their statistical structures are identical, that is, they have the same distribution. We make this intuition formal by defining two processes to be *equivalent* if their process distributions are identical, that is, if the probability measures on the output sequence spaces are the same, regardless of the functional form of the random variables of the underlying probability spaces. In the same way, we consider two random variables to be equivalent if their distributions are identical.

We have described above two equivalent processes or two equivalent models for the same random process, one defined as a sequence of random variables on a perhaps very complicated underlying probability space, the other defined as a probability measure directly on the measurable space of possible output sequences. The second model will be referred to as a *directly given* random process.

Which model is "better" depends on the application. For example, a directly given model for a random process may focus on the random process itself and not its origin and hence may be simpler to deal with. If the random process is then coded or measurements are taken on the random process, then it may be better to model the encoded random process in terms of random variables defined on the original random process and not as a directly given random process. This model will then focus on the input process and the coding operation. We shall let convenience determine the most appropriate model.

We can now describe yet another model for the above random process, that is, another means of describing a random process with the same distribution. This time the model is in terms of a dynamical system. Given the probability space $(A^{\mathcal{T}}, \mathcal{B}_A{}^{\mathcal{T}}, m)$, define the (left) shift transformation $T : A^{\mathcal{T}} \to A^{\mathcal{T}}$ by

$$T(x^{\mathcal{T}}) = T(\{x_n\}_{n\in\mathcal{T}}) = y^{\mathcal{T}} = \{y_n\}_{n\in\mathcal{T}},$$

where

$$y_n = x_{n+1}, n \in \mathcal{T}.$$

Thus the nth coordinate of $y^{\mathcal{T}}$ is simply the $(n+1)$st coordinate of $x^{\mathcal{T}}$. (We assume that $\mathcal{T}$ is closed under addition and hence if n and 1 are in $\mathcal{T}$, then so is $(n+1)$.) If the alphabet of such a shift is not clear from context, we will occasionally denote the shift by T_A or $T_{A^{\mathcal{T}}}$. The shift can easily be shown to be measurable.

Consider next the dynamical system $(A^{\mathcal{T}}, \mathcal{B}_A{}^{\mathcal{T}}, P, T)$ and the random process formed by combining the dynamical system with the zero time sampling function Π_0 (we assume that 0 is a member of $\mathcal{T}$). If we define

$Y_n(x) = \Pi_0(T^n x)$ for $x = x^{\mathcal{T}} \in A^{\mathcal{T}}$, or, in abbreviated form, $Y_n = \Pi_0 T^n$, then the random process $\{Y_n\}_{n\in\mathcal{T}}$ is equivalent to the processes developed above. Thus we have developed three different, but equivalent, means of producing the same random process. Each will be seen to have its uses.

The above development shows that a dynamical system is a more fundamental entity than a random process since we can always construct an equivalent model for a random process in terms of a dynamical system–use the directly given representation, shift transformation, and zero time sampling function.

The shift transformation on a sequence space introduced above is the most important transformation that we shall encounter. It is not, however, the only important transformation. When dealing with transformations we will usually use the notation T to reflect the fact that it is often related to the action of a simple left shift of a sequence, yet it should be kept in mind that occasionally other operators will be considered and the theory to be developed will remain valid, even if T is not required to be a simple time shift. For example, we will also consider block shifts.

Most texts on ergodic theory deal with the case of an invertible transformation, that is, where T is a one-to-one transformation and the inverse mapping T^{-1} is measurable. This is the case for the shift on $A^{\mathbf{Z}}$, the two-sided shift. It is not the case, however, for the one-sided shift defined on $A^{\mathbf{Z}_+}$ and hence we will avoid use of this assumption. We will, however, often point out in the discussion what simplifications or special properties arise for invertible transformations.

Since random processes are considered equivalent if their distributions are the same, we shall adopt the notation $[A, m, X]$ for a random process $\{X_n; n \in \mathcal{T}\}$ with alphabet A and process distribution m, the index set $\mathcal{T}$ usually being clear from context. We will occasionally abbreviate this to the more common notation $[A, m]$, but it is often convenient to note the name of the output random variables as there may be several, e.g., a random process may have an input X and output Y. By "the associated probability space" of a random process $[A, m, X]$ we shall mean the sequence probability space $(A^{\mathcal{T}}, \mathcal{B}_A{}^{\mathcal{T}}, m)$. It will often be convenient to consider the random process as a directly given random process, that is, to view X_n as the coordinate functions Π_n on the sequence space $A^{\mathcal{T}}$ rather than as being defined on some other abstract space. This will not always be the case, however, as often processes will be formed by coding or communicating other random processes. Context should render such bookkeeping details clear.

1.5 Standard Alphabets

A measurable space $(A, \mathcal{B}_A)$ is a *standard space* if there exists a sequence of finite fields $\mathcal{F}_n$; $n = 1, 2, \cdots$ with the following properties:

(1) $\mathcal{F}_n \subset \mathcal{F}_{n+1}$ (the fields are increasing).

(2) $\mathcal{B}_A$ is the smallest σ-field containing all of the $\mathcal{F}_n$ (the $\mathcal{F}_n$ *generate* $\mathcal{B}_A$ or $\mathcal{B}_A = \sigma(\bigcup_{n=1}^{\infty} \mathcal{F}_n)$).

(3) An event $G_n \in \mathcal{F}_n$ is called an *atom* of the field if it is nonempty and and its only subsets which are also field members are itself and the empty set. If $G_n \in \mathcal{F}_n$; $n = 1, 2, \cdots$ are atoms and $G_{n+1} \subset G_n$ for all n, then

$$\bigcap_{n=1}^{\infty} G_n \neq \emptyset.$$

Standard spaces are important for several reasons: First, they are a general class of spaces for which two of the key results of probability hold: (1) the Kolmogorov extension theorem showing that a random process is completely described by its finite order distributions, and (2) the existence of regular conditional probability measures. Thus, in particular, the conditional probability measure $P_{X|Y}(F|y)$ of (1.2.13) is regular if the alphabets A_X and A_Y are standard and hence for each fixed $y \in A_Y$ the set function $P_{X|Y}(F|y)$; $F \in \mathcal{B}_{A_X}$ is a probability measure. In this case we can interpret $P_{X|Y}(F|y)$ as $P(X \in F|Y = y)$. Second, the ergodic decomposition theorem of ergodic theory holds for such spaces. Third, the class is sufficiently general to include virtually all examples arising in applications, e.g., discrete spaces, the real line, Euclidean vector spaces, Polish spaces (complete separable metric spaces), etc. The reader is referred to [50] and the references cited therein for a detailed development of these properties and examples of standard spaces.

Standard spaces are not the most general space for which the Kolmogorov extension theorem, the existence of conditional probability, and the ergodic decomposition theorem hold. These results also hold for perfect spaces which include standard spaces as a special case. (See, e.g., [128],[139],[126], [98].) We limit discussion to standard spaces, however, as they are easier to characterize and work with and they are sufficiently general to handle most cases encountered in applications. Although standard spaces are not the most general for which the required probability theory results hold, they are the most general for which all finitely additive normalized measures extend to countably additive probability measures, a property which greatly eases the proof of many of the desired results.

Throughout this book we shall assume that the alphabet A of the information source is a standard space.

1.6 Expectation

Let $(\Omega, \mathcal{B}, m)$ be a probability space, e.g., the probability space of a directly given random process with alphabet A, $(A^{\mathcal{T}}, {\mathcal{B}_A}^{\mathcal{T}}, m)$. A real-valued random variable $f : \Omega \to \mathbf{R}$ will also be called a *measurement* since it is often formed by taking a mapping or function of some other set of more general random variables, e.g., the outputs of some random process which might not have real-valued outputs. Measurements made on such processes, however, will always be assumed to be real.

Suppose next we have a measurement f whose range space or *alphabet* $f(\Omega) \subset \mathbf{R}$ of possible values is finite. Then f is called a *discrete random variable* or *discrete measurement* or *digital measurement* or, in the common mathematical terminology, a *simple function*.

Given a discrete measurement f, suppose that its range space is $f(\Omega) = \{b_i, i = 1, \cdots, N\}$, where the b_i are distinct. Define the sets $F_i = f^{-1}(b_i) = \{x : f(x) = b_i\}$, $i = 1, \cdots, N$. Since f is measurable, the F_i are all members of $\mathcal{B}$. Since the b_i are distinct, the F_i are disjoint. Since every input point in Ω must map into some b_i, the union of the F_i equals Ω. Thus the collection $\{F_i; i = 1, 2, \cdots, N\}$ forms a partition of Ω. We have therefore shown that any discrete measurement f can be expressed in the form

$$f(x) = \sum_{i=1}^{M} b_i 1_{F_i}(x), \tag{1.6.1}$$

where $b_i \in \mathbf{R}$, the $F_i \in \mathcal{B}$ form a partition of Ω, and 1_{F_i} is the indicator function of F_i, $i = 1, \cdots, M$. Every simple function has a unique representation in this form with distinct b_i and $\{F_i\}$ a partition.

The *expectation* or *ensemble average* or *probabilistic average* or *mean* of a discrete measurement $f : \Omega \to \mathbf{R}$ as in (1.6.1) with respect to a probability measure m is defined by

$$E_m f = \sum_{i=0}^{M} b_i m(F_i). \tag{1.6.2}$$

An immediate consequence of the definition of expectation is the simple but useful fact that for any event F in the original probability space,

$$E_m 1_F = m(F),$$

that is, probabilities can be found from expectations of indicator functions.

Again let $(\Omega, \mathcal{B}, m)$ be a probability space and $f : \Omega \to \mathcal{R}$ a measurement, that is, a real-valued random variable or measurable real-valued function. Define the sequence of *quantizers* $q_n : \mathcal{R} \to \mathcal{R}$, $n = 1, 2, \cdots$, as follows:

$$q_n(r) = \begin{cases} n & n \leq r \\ (k-1)2^{-n} & (k-1)2^{-n} \leq r < k2^{-n};\ k = 1, 2, \cdots, n2^n \\ -(k-1)2^{-n} & -k2^{-n} \leq r < -(k-1)2^{-n};\ k = 1, 2, \cdots, n2^n \\ -n & r < -n\ . \end{cases}$$

We now define expectation for general measurements in two steps. If $f \geq 0$, then define

$$E_m f = \lim_{n \to \infty} E_m(q_n(f)). \tag{1.6.3}$$

Since the q_n are discrete measurements on f, the $q_n(f)$ are discrete measurements on Ω ($q_n(f)(x) = q_n(f(x))$ is a simple function) and hence the individual expectations are well defined. Since the $q_n(f)$ are nondecreasing, so are the $E_m(q_n(f))$ and this sequence must either converge to a finite limit or grow without bound, in which case we say it converges to ∞. In both cases the expectation $E_m f$ is well defined, although it may be infinite.

If f is an arbitrary real random variable, define its positive and negative parts $f^+(x) = \max(f(x), 0)$ and $f^-(x) = -\min(f(x), 0)$ so that $f(x) = f^+(x) - f^-(x)$ and set

$$E_m f = E_m f^+ - E_m f^- \tag{1.6.4}$$

provided this does not have the form $+\infty - \infty$, in which case the expectation does not exist. It can be shown that the expectation can also be evaluated for nonnegative measurements by the formula

$$E_m f = \sup_{\text{discrete } g:\ g \leq f} E_m g.$$

The expectation is also called an *integral* and is denoted by any of the following:

$$E_m f = \int f dm = \int f(x) dm(x) = \int f(x) m(dx).$$

The subscript m denoting the measure with respect to which the expectation is taken will occasionally be omitted if it is clear from context.

A measurement f is said to be *integrable* or *m-integrable* if $E_m f$ exists and is finite. A function is integrable if and only if its absolute value is integrable. Define $L^1(m)$ to be the space of all m-integrable functions. Given any m-integrable f and an event B, define

$$\int_B f dm = \int f(x) 1_B(x)\, dm(x).$$

Two random variables f and g are said to be equal m-almost-everywhere or equal m-a.e. or equal with m-probability one if $m(f = g) = m(\{x : f(x) = g(x)\}) = 1$. The m- is dropped if it is clear from context.

Given a probability space $(\Omega, \mathcal{B}, m)$, suppose that $\mathcal{G}$ is a sub-σ-field of $\mathcal{B}$, that is, it is a σ-field of subsets of Ω and all those subsets are in $\mathcal{B}$ ($\mathcal{G} \subset \mathcal{B}$). Let $f : \Omega \to \mathcal{R}$ be an integrable measurement. Then the *conditional expectation* $E(f|\mathcal{G})$ is described as any function, say $h(\omega)$, that satisfies the following two properties:

$$h(\omega) \text{ is measurable with respect to } \mathcal{G} \tag{1.6.5}$$

$$\int_G h\, dm = \int_G f\, dm; \text{ all } G \in \mathcal{G}. \tag{1.6.6}$$

If a regular conditional probability distribution given $\mathcal{G}$ exists, e.g., if the space is standard, then one has a constructive definition of conditional expectation: $E(f|\mathcal{G})(\omega)$ is simply the expectation of f with respect to the conditional probability measure $m(.|\mathcal{G})(\omega)$. Applying this to the example of two random variables X and Y with standard alphabets described in Section 1.2 we have from (1.6.6) that for integrable $f : A_X \times A_Y \to \mathcal{R}$

$$E(f) = \int f(x,y) dP_{XY}(x,y) = \int (\int f(x,y) dP_{X|Y}(x|y)) dP_Y(y). \tag{1.6.7}$$

In particular, for fixed y, $f(x,y)$ is an integrable (and measurable) function of x.

Equation (1.6.7) provides a generalization of (1.2.13) from rectangles to arbitrary events. For an arbitrary $F \in \mathcal{B}_{A_X \times A_Y}$ we have that

$$P_{XY}(F) = \int \int (1_F(x,y) dP_{X|Y}(x|y)) dP_Y(y) = \int P_{X|Y}(F_y|y) dP_Y(y), \tag{1.6.8}$$

where $F_y = \{x : (x,y) \in F\}$ is called the *section* of F at y. If F is measurable, then so is F_y for all y.
The inner integral is just

$$\int_{x:(x,y)\in F} dP_{X|Y}(x|y) = P_{X|Y}(F_y|y),$$

where the set $F_y = \{x : (x,y) \in F\}$ is called the *section* of F at y. Since $1_F(x,y)$ is measurable with respect to x for each fixed y, $F_y \in \mathcal{B}_{A_X}$.

1.7 Asymptotic Mean Stationarity

A dynamical system (or the associated source) $(\Omega, \mathcal{B}, P, T)$ is said to be *stationary* if

$$P(T^{-1}G) = P(G)$$

for all $G \in \mathcal{B}$. It is said to be ***asymptotically mean stationary*** or, simply, AMS if the limit

$$\bar{P}(G) = \lim_{n\to\infty} \frac{1}{n} \sum_{k=0}^{n-1} P(T^{-k}G) \tag{1.7.1}$$

exists for all $G \in \mathcal{B}$. The following theorems summarize several important properties of AMS sources. Details may be found in Chapter 6 of [50].

Theorem 1.7.1: If a dynamical system $(\Omega, \mathcal{B}, P, T)$ is AMS, then $\bar{P}$ defined in (1.7.1) is a probability measure and $(\Omega, \mathcal{B}, \bar{P}, T)$ is stationary. ($\bar{P}$ is called the ***stationary mean*** of P.) If an event G is invariant in the sense that $T^{-1}G = G$, then

$$P(G) = \bar{P}(G).$$

If a random variable g is invariant in the sense that $g(Tx) = g(x)$ with P probability 1, then

$$E_P g = E_{\bar{P}} g.$$

The stationary mean $\bar{P}$ ***asymptotically dominates*** P in the sense that if $\bar{P}(G) = 0$, then

$$\limsup_{n\to\infty} P(T^{-n}G) = 0.$$

Theorem 1.7.2: Given an AMS source $\{X_n\}$ let $\sigma(X_n, X_{n+1}, \cdots)$ denote the σ-field generated by the random variables $X_n, \cdots$, that is, the smallest σ-field with respect to which all these random variables are measurable. Define the *tail σ-field* $\mathcal{F}_\infty$ by

$$\mathcal{F}_\infty = \bigcap_{n=0}^{\infty} \sigma(X_n, \cdots).$$

If $G \in \mathcal{F}_\infty$ and $\bar{P}(G) = 0$, then also $P(G) = 0$.

The tail σ-field can be thought of as events that are determinable by looking only at samples of the sequence in the arbitrarily distant future. The theorem states that the stationary mean *dominates* the original measure on such tail events in the sense that zero probability under the stationary mean implies zero probability under the original source.

1.8 Ergodic Properties

Two of the basic results of ergodic theory that will be called upon extensively are the pointwise or almost-everywhere ergodic theorem and the ergodic decomposition theorem We quote these results along with some relevant notation for reference. Detailed developments may be found in Chapters 6-8 of [50]. The ergodic theorem states that AMS dynamical systems (and hence also sources) have convergent sample averages, and it characterizes the limits.

Theorem 1.8.1: If a dynamical system $(\Omega, \mathcal{B}, m, T)$ is AMS with stationary mean $\bar{m}$ and if $f \in L^1(\bar{m})$, then with probability one under m and $\bar{m}$

$$\lim_{n\to\infty} \frac{1}{n} \sum_{i=0}^{n-1} fT^i = E_{\bar{m}}(f|\mathcal{I}),$$

where $\mathcal{I}$ is the sub-σ-field of invariant events, that is, events G for which $T^{-1}G = G$.

The basic idea of the ergodic decomposition is that any stationary source which is not ergodic can be represented as a mixture of stationary ergodic components or subsources.

Theorem 1.8.2: Given the standard sequence space $(\Omega, \mathcal{B})$ with shift T as previously, there exists a family of stationary ergodic measures $\{p_x;\ x \in \Omega\}$, called the *ergodic decomposition*, with the following properties:

(a) $p_{Tx} = p_x$.

(b) For any stationary measure m,

$$m(G) = \int p_x(G)dm(x); \text{ all } G \in \mathcal{B}.$$

(c) For any $g \in L^1(m)$

$$\int g dm = \int (\int g dp_x) dm(x).$$

It is important to note that the same collection of stationary ergodic components works for any stationary measure m. This is the strong form of the ergodic decomposition.

The final result of this section is a variation on the ergodic decomposition that will be useful. To describe the result, we need to digress briefly to introduce a metric on spaces of probability measures. A thorough development can be found in Chapter 8 of [50]. We have a standard sequence measurable space $(\Omega, \mathcal{B})$ and hence we can generate the σ-field $\mathcal{B}$ by a countable field $\mathcal{F} = \{F_n; n = 1, 2, \cdots\}$. Given such a countable generating field, a ***distributional distance*** between two probability measures p and m on $(\Omega, \mathcal{B})$ is defined by

$$d(p, m) = \sum_{n=1}^{\infty} 2^{-n} |p(F_n) - m(F_n)|.$$

Any choice of a countable generating field yields a distributional distance. Such a distance or metric yields a measurable space of probability measures as follows: Let Λ denote the space of all probability measures on the original measurable space $(\Omega, \mathcal{B})$. Let $\mathcal{B}(\Lambda)$ denote the σ-field of subsets of Λ generated by all open spheres using the distributional distance, that is, all sets of the form $\{p : d(p, m) \leq \epsilon\}$ for some $m \in \Lambda$ and some $\epsilon > 0$. We can now consider properties of functions that carry sequences in our original space into probability measures. The following is Theorem 8.5.1 of [50].

Theorem 1.8.3: Fix a standard measurable space $(\Omega, \mathcal{B})$ and a transformation $T : \Omega \to \Omega$. Then there are a standard measurable space $(\Lambda, \mathcal{L})$, a family of stationary ergodic measures $\{m_\lambda; \lambda \in \Lambda\}$ on $(\Omega, \mathcal{B})$, and a measurable mapping $\psi : \Omega \to \Lambda$ such that

(a) ψ is invariant ($\psi(Tx) = \psi(x)$ all x);

(b) if m is a stationary measure on $(\Omega, \mathcal{B})$ and P_ψ is the induced distribution; that is, $P_\psi(G) = m(\psi^{-1}(G))$ for $G \in \mathbf{\Lambda}$ (which is well defined from (a)), then

$$m(F) = \int dm(x) m_{\psi(x)}(F) = \int dP_\psi(\lambda) m_\lambda(F), \text{ all } F \in \mathcal{B},$$

and if $f \in L^1(m)$, then so is $\int f dm_\lambda$ P_ψ-a.e. and

$$E_m f = \int dm(x) E_{m_{\psi(x)}} f = \int dP_\psi(\lambda) E_{m_\lambda} f.$$

Finally, for any event F, $m_\psi(F) = m(F|\psi)$, that is, given the ergodic decomposition and a stationary measure m , the ergodic component λ is a version of the conditional probability under m given $\psi = \lambda$.

The following corollary to the ergodic decomposition is Lemma 8.6.2 of [50]. It states that the conditional probability of a future event given the entire past is unchanged by knowing the ergodic component in effect. This is because the infinite past determines the ergodic component in effect.

Corollary 1.8.1: Suppose that $\{X_n\}$ is a two-sided stationary process with distribution m and that $\{m_\lambda; \lambda \in \Lambda\}$ is the ergodic decomposition and ψ the ergodic component function. Then the mapping ψ is measurable with respect to $\sigma(X_{-1}, X_{-2}, \cdots)$ and

$$m((X_0, X_1, \cdots) \in F | X_{-1}, X_{-2}, \cdots)$$

$$= m_\psi((X_0, X_1, \cdots) \in F | X_{-1}, X_{-2}, \cdots); \; m - \text{a.e.}$$

Chapter 2

Entropy and Information

2.1 Introduction

The development of the idea of entropy of random variables and processes by Claude Shannon provided the beginnings of information theory and of the modern age of ergodic theory. We shall see that entropy and related information measures provide useful descriptions of the long term behavior of random processes and that this behavior is a key factor in developing the coding theorems of information theory. We now introduce the various notions of entropy for random variables, vectors, processes, and dynamical systems and we develop many of the fundamental properties of entropy.

In this chapter we emphasize the case of finite alphabet random processes for simplicity, reflecting the historical development of the subject. Occasionally we consider more general cases when it will ease later developments.

2.2 Entropy and Entropy Rate

There are several ways to introduce the notion of entropy and entropy rate. We take some care at the beginning in order to avoid redefining things later. We also try to use definitions resembling the usual definitions of elementary information theory where possible. Let $(\Omega, \mathcal{B}, P, T)$ be a dynamical system. Let f be a finite alphabet measurement (a simple function) defined on Ω and define the one-sided random process $f_n = fT^n$; $n = 0, 1, 2, \cdots$. This process can be viewed as a coding of the original space, that is, one produces successive coded values by transforming (e.g., shifting) the points of the space, each time producing an output symbol using the same rule or mapping. In the usual way we can construct an equivalent directly given model of this process. Let $A = \{a_1, a_2, \cdots, a_{||A||}\}$ denote the finite alphabet of f and let $(A^{\mathcal{Z}_+}, \mathcal{B}_A^{\mathcal{Z}_+})$ be the resulting one-sided sequence space, where $\mathcal{B}_A$ is the power set. We abbreviate the notation for this sequence space to $(A^\infty, \mathcal{B}_A^\infty)$. Let T_A denote the shift on this space and let X denote the time zero sampling or coordinate function and define $X_n(x) = X(T_A^n x) = x_n$. Let m denote the process distribution induced by the original space and the fT^n, i.e., $m = P_{\bar{f}} = P\bar{f}^{-1}$ where $\bar{f}(\omega) = (f(\omega), f(T\omega), f(T^2\omega), \cdots)$.

Observe that by construction, shifting the input point yields an output sequence that is also shifted, that is,

$$\bar{f}(T\omega) = T_A \bar{f}(\omega).$$

Sequence-valued measurements of this form are called ***stationary*** or ***invariant*** codings (or ***time invariant*** or ***shift invariant*** codings in the case of the shift) since the coding commutes with the transformations.

The entropy and entropy rates of a finite alphabet measurement depend only on the process distributions and hence are usually more easily stated in terms of the induced directly given model and the process distribution. For the moment, however, we point out that the definition can be stated in terms of either system. Later we will see that the entropy of the underlying system is defined as a supremum of the entropy rates of all finite alphabet codings of the system.

The ***entropy*** of a discrete alphabet random variable f defined on the probability space $(\Omega, \mathcal{B}, P)$ is defined by

$$H_P(f) = -\sum_{a \in A} P(f = a) \ln P(f = a). \tag{2.2.1}$$

We define 0ln0 to be 0 in the above formula. We shall often use logarithms to the base 2 instead of natural logarithms. The units for entropy are "nats" when the natural logarithm is used and "bits" for base 2 logarithms. The natural logarithms are usually more convenient for mathematics while the base 2 logarithms provide more intuitive descriptions. The subscript P can be omitted if the measure is clear from context. Be forewarned that the measure will often not be clear from context since more than one measure may be under consideration and hence the subscripts will be required. A discrete alphabet random variable f has a probability mass function (pmf), say p_f, defined by $p_f(a) = P(f = a) = P(\{\omega : f(\omega) = a\})$ and hence we can also write

$$H(f) = -\sum_{a \in A} p_f(a) \ln p_f(a).$$

It is often convenient to consider the entropy not as a function of the particular outputs of f but as a function of the partition that f induces on Ω. In particular, suppose that the alphabet of f is $A = \{a_1, a_2, \cdots, a_{||A||}\}$ and define the partition $\mathcal{Q} = \{Q_i; i = 1, 2, \cdots, ||A||\}$ by $Q_i = \{\omega : f(\omega) = a_i\} = f^{-1}(\{a_i\})$. In other words, $\mathcal{Q}$ consists of disjoint sets which group the points in Ω together according to what output the measurement f produces. We can consider the entropy as a function of the partition and write

$$H_P(\mathcal{Q}) = -\sum_{i=1}^{||A||} P(Q_i) \ln P(Q_i). \tag{2.2.2}$$

Clearly different mappings with different alphabets can have the same entropy if they induce the same partition. Both notations will be used according to the desired emphasis. We have not yet defined entropy for random variables that do not have discrete alphabets; we shall do that later.

Return to the notation emphasizing the mapping f rather than the partition. Defining the random variable $P(f)$ by $P(f)(\omega) = P(\lambda : f(\lambda) = f(\omega))$ we can also write the entropy as

$$H_P(f) = E_P(-\ln P(f)).$$

Using the equivalent directly given model we have immediately that

$$H_P(f) = H_P(\mathcal{Q}) = H_m(X_0) = E_m(-\ln m(X_0)). \tag{2.2.3}$$

At this point one might ask why we are carrying the baggage of notations for entropy in both the original space and in the sequence space. If we were dealing with only one measurement f (or X_n), we could confine interest to the simpler directly-given form. More generally, however, we will be interested in different measurements or codings on a common system. In this case we will require the notation using the original system. Hence for the moment we keep both forms, but we shall often focus on the second where possible and the first only when necessary.

The *nth order entropy of a discrete alphabet measurement f with respect to T* is defined as

$$H_P^{(n)}(f) = n^{-1} H_P(f^n)$$

where $f^n = (f, fT, fT^2, \cdots, fT^{n-1})$ or, equivalently, we define the discrete alphabet random process $X_n(\omega) = f(T^n\omega)$, then

$$f^n = X^n = X_0, X_1, \cdots, X_{n-1}.$$

As previously, this is given by

$$H_m^{(n)}(X) = n^{-1} H_m(X^n) = n^{-1} E_m(-\ln m(X^n)).$$

This is also called the entropy (per-coordinate or per-sample) of the random vector f^n or X^n. We can also use the partition notation here. The partition corresponding to f^n has a particular form: Suppose that we have two partitions, $\mathcal{Q} = \{Q_i\}$ and $\mathcal{P} = \{P_i\}$. Define their *join* $\mathcal{Q}\bigvee\mathcal{P}$ as the partition containing all nonempty intersection sets of the form $Q_i \bigcap P_j$. Define also $T^{-1}\mathcal{Q}$ as the partition containing the atoms $T^{-1}Q_i$. Then f^n induces the partition

$$\bigvee_{i=0}^{n-1} T^{-i}\mathcal{Q}$$

and we can write

$$H_P^{(n)}(f) = H_P^{(n)}(\mathcal{Q}) = n^{-1}H_P(\bigvee_{i=0}^{n-1} T^{-i}\mathcal{Q}).$$

As before, which notation is preferable depends on whether we wish to emphasize the mapping f or the partition $\mathcal{Q}$.

The *entropy rate* or *mean entropy* of a discrete alphabet measurement f with respect to the transformation T is defined by

$$\bar{H}_P(f) = \limsup_{n\to\infty} H_P^{(n)}(f)$$

$$= \bar{H}_P(\mathcal{Q}) = \limsup_{n\to\infty} H_P^{(n)}(\mathcal{Q}) = \bar{H}_m(X) = \limsup_{n\to\infty} H_m^{(n)}(X).$$

Given a dynamical system $(\Omega, \mathcal{B}, P, T)$, the *entropy* $H(P,T)$ of the system (or of the measure with respect to the transformation) is defined by

$$H(P,T) = \sup_f \bar{H}_P(f) = \sup_{\mathcal{Q}} \bar{H}_P(\mathcal{Q}),$$

where the supremum is over all finite alphabet measurements (or codings) or, equivalently, over all finite measurable partitions of Ω. (We emphasize that this means alphabets of size M for all finite values of M.) The entropy of a system is also called the *Kolmogorov-Sinai invariant* of the system because of the generalization by Kolmogorov [88] and Sinai [134] of Shannon's entropy rate concept to dynamical systems and the demonstration that equal entropy was a necessary condition for two dynamical systems to be isomorphic.

Suppose that we have a dynamical system corresponding to a finite alphabet random process $\{X_n\}$, then one possible finite alphabet measurement on the process is $f(x) = x_0$, that is, the time 0 output. In this case clearly $\bar{H}_P(f) = \bar{H}_P(X)$ and hence, since the system entropy is defined as the supremum over *all* simple measurements,

$$H(P,T) \geq \bar{H}_P(X). \tag{2.2.4}$$

We shall later see that (2.2.4) holds with equality for finite alphabet random processes and provides a generalization of entropy rate for processes that do not have finite alphabets.

2.3 Basic Properties of Entropy

For simplicity we focus on the entropy rate of a directly given finite alphabet random process $\{X_n\}$. We also will emphasize stationary measures, but we will try to clarify those results that require stationarity and those that are more general.

Let A be a finite set. Let $\Omega = A^{\mathbf{Z}_+}$ and let $\mathcal{B}$ be the sigma-field of subsets of Ω generated by the rectangles. Since A is finite, $(A, \mathcal{B}_A)$ is standard, where $\mathcal{B}_A$ is the power set of A. Thus $(\Omega, \mathcal{B})$ is also standard by Lemma 2.4.1 of [50]. In fact, from the proof that cartesian products of standard spaces are standard, we can take as a basis for $\mathcal{B}$ the fields $\mathcal{F}_n$ generated by the finite dimensional rectangles having the form $\{x : X^n(x) = x^n = a^n\}$ for all $a^n \in A^n$ and all positive integers n. (Members of this class of rectangles are called *thin cylinders*.) The union of all such fields, say $\mathcal{F}$, is then a generating field.

Many of the basic properties of entropy follow from the following simple inequality.

Lemma 2.3.1: Given two probability mass functions $\{p_i\}$ and $\{q_i\}$, that is, two countable or finite sequences of nonnegative numbers that sum to one, then

$$\sum_i p_i \ln \frac{p_i}{q_i} \geq 0$$

with equality if and only if $q_i = p_i$, all i.

Proof: The lemma follows easily from the elementary inequality for real numbers

$$\ln x \leq x - 1 \tag{2.3.1}$$

(with equality if and only if $x = 1$) since

$$\sum_i p_i \ln \frac{q_i}{p_i} \leq \sum_i p_i (\frac{q_i}{p_i} - 1) = \sum_i q_i - \sum_i p_i = 0$$

with equality if and only if $q_i/p_i = 1$ all i. Alternatively, the inequality follows from Jensen's inequality [63] since ln is a convex $\bigcap$ function:

$$\sum_i p_i \ln \frac{q_i}{p_i} \leq \ln \left(\sum_i p_i \frac{q_i}{p_i} \right) = 0$$

with equality if and only if $q_i/p_i = 1$, all i. □

The quantity used in the lemma is of such fundamental importance that we pause to introduce another notion of information and to recast the inequality in terms of it. As with entropy, the definition for the moment is only for finite alphabet random variables. Also as with entropy, there are a variety of ways to define it. Suppose that we have an underlying measurable space $(\Omega, \mathcal{B})$ and two measures on this space, say P and M, and we have a random variable f with finite alphabet A defined on the space and that $\mathcal{Q}$ is the induced partition $\{f^{-1}(a); a \in A\}$. Let P_f and M_f be the induced distributions and let p and m be the corresponding probability mass functions, e.g., $p(a) = P_f(\{a\}) = P(f = a)$. Define the *relative entropy* of a measurement f with measure P with respect to the measure M by

$$H_{P||M}(f) = H_{P||M}(\mathcal{Q}) = \sum_{a \in A} p(a) \ln \frac{p(a)}{m(a)} = \sum_{i=1}^{||A||} P(Q_i) \ln \frac{P(Q_i)}{M(Q_i)}.$$

Observe that this only makes sense if $p(a)$ is 0 whenever $m(a)$ is, that is, if P_f is absolutely continuous with respect to M_f or $M_f >> P_f$. Define $H_{P||M}(f) = \infty$ if P_f is not absolutely continuous with respect to M_f. The measure M is referred to as the *reference measure*. Relative entropies will play an increasingly important role as general alphabets are considered. In the early chapters the emphasis will be on ordinary entropy with similar properties for relative entropies following almost as an afterthought. When considering more abstract (nonfinite) alphabets later on, relative entropies will prove indispensible.

Analogous to entropy, given a random process $\{X_n\}$ described by two process distributions p and m, if it is true that

$$m_{X^n} >> p_{X^n};\ n = 1, 2, \cdots,$$

then we can define for each n the *nth order relative entropy* $n^{-1} H_{p||m}(X^n)$ and the *relative entropy rate*

$$\bar{H}_{p||m}(X) \equiv \limsup_{n \to \infty} \frac{1}{n} H_{p||m}(X^n).$$

When dealing with relative entropies it is often the measures that are important and not the random variable or partition. We introduce a special notation which emphasizes this fact. Given a probability space $(\Omega, \mathcal{B}, P)$, with Ω a finite space, and another measure M on the same space, we define the *divergence of P with respect to M* as the relative entropy of the identity mapping with respect to the two measures:

$$D(P||M) = \sum_{\omega \in \Omega} P(\omega) \ln \frac{P(\omega)}{M(\omega)}.$$

Thus, for example, given a finite alphabet measurement f on an arbitrary probability space $(\Omega, \mathcal{B}, P)$, if M is another measure on $(\Omega, \mathcal{B})$ then

$$H_{P||M}(f) = D(P_f||M_f).$$

Similarly,

$$H_{p||m}(X^n) = D(P_{X^n}||M_{X^n}),$$

where P_{X^n} and M_{X^n} are the distributions for X^n induced by process measures p and m, respectively. The theory and properties of relative entropy are therefore determined by those for divergence.

There are many names and notations for relative entropy and divergence throughout the literature. The idea was introduced by Kullback for applications of information theory to statistics (see, e.g., Kullback [92] and the references therein) and was used to develop information theoretic results by Perez [120] [122] [121], Dobrushin [32], and Pinsker [125]. Various names in common use for this quantity are discrimination, discrimination information, Kullback-Leibler number, directed divergence, and cross entropy.

The lemma can be summarized simply in terms of divergence as in the following theorem, which is commonly referred to as the divergence inequality.

Theorem 2.3.1: Given any two probability measures P and M on a common finite alphabet probability space, then

$$D(P||M) \geq 0 \tag{2.3.2}$$

with equality if and only if $P = M$.

In this form the result is known as the *divergence inequality*. The fact that the divergence of one probability measure with respect to another is nonnegative and zero only when the two measures are the same suggest the interpretation of divergence as a "distance" between the two probability measures, that is, a measure of how different the two measures are. It is not a true distance or metric in the usual sense since it is not a symmetric function of the two measures and it does not satisfy the triangle inequality. The interpretation is, however, quite useful for adding insight into results characterizing the behavior of divergence and it will later be seen to have implications for ordinary distance measures between probability measures.

The divergence plays a basic role in the family of information measures all of the information measures that we will encounter–entropy, relative entropy, mutual information, and the conditional forms of these information measures–can be expressed as a divergence.

There are three ways to view entropy as a special case of divergence. The first is to permit M to be a general measure instead of requiring it to be a probability measure and have total mass 1. In this case entropy is minus the divergence if M is the counting measure, i.e., assigns measure 1 to every point in the discrete alphabet. If M is not a probability measure, then the divergence inequality (2.3.2) need not hold. Second, if the alphabet of f is A_f and has $||A_f||$ elements, then letting M be a uniform pmf assigning probability $1/||A||$ to all symbols in A yields

$$D(P||M) = \ln ||A_f|| - H_P(f) \geq 0$$

and hence the entropy is the log of the alphabet size minus the divergence with respect to the uniform distribution. Third, we can also consider entropy a special case of divergence while still requiring that M be a probability measure by using product measures and a bit of a trick. Say we have two measures P and Q on a common probability space $(\Omega, \mathcal{B})$. Define two measures on the product space $(\Omega \times \Omega, \mathcal{B}(\Omega \times \Omega))$ as follows: Let $P \times Q$ denote the usual product measure, that is, the measure specified by its values on rectangles as $P \times Q(F \times G) = P(F)Q(G)$. Thus, for example, if P and Q are discrete distributions with pmf's p and q, then the pmf for $P \times Q$ is just $p(a)q(b)$. Let P' denote the "diagonal" measure defined by its values on rectangles as $P'(F \times G) = P(F \bigcap G)$. In the discrete case P' has pmf $p\prime(a,b) = p(a)$ if $a = b$ and 0 otherwise. Then

$$H_P(f) = D(P'||P \times P).$$

Note that if we let X and Y be the coordinate random variables on our product space, then both P' and $P \times P$ give the same marginal probabilities to X and Y, that is, $P_X = P_Y = P$. P' is an extreme distribution on (X, Y) in the sense that with probability one $X = Y$; the two coordinates are deterministically dependent on one another. $P \times P$, however, is the opposite extreme in that it makes the two random variables X and Y independent of one another. Thus the entropy of a distribution P can be viewed as the relative entropy between these two extreme joint distributions having marginals P.

We now return to the general development for entropy. For the moment fix a probability measure m on a measurable space $(\Omega, \mathcal{B})$ and let X and Y be two finite alphabet random variables defined on that space. Let A_X and A_Y denote the corresponding alphabets. Let P_{XY}, P_X, and P_Y denote the distributions of (X,Y), X, and Y, respectively.

First observe that since $P_X(a) \leq 1$, all a, $-\ln P_X(a)$ is positive and hence

$$H(X) = -\sum_{a \in A} P_X(a) \ln P_X(a) \geq 0. \tag{2.3.3}$$

From (2.3.2) with M uniform as in the second interpretation of entropy above, if X is a random variable with alphabet A_X, then

$$H(X) \leq \ln ||A_X||.$$

Since for any $a \in A_X$ and $b \in A_Y$ we have that $P_X(a) \geq P_{XY}(a,b)$, it follows that

$$H(X,Y) = -\sum_{a,b} P_{XY}(a,b) \ln P_{XY}(a,b)$$

$$\geq -\sum_{a,b} P_{XY}(a,b) \ln P_X(a) = H(X).$$

Using Lemma 2.3.1 we have that since P_{XY} and $P_X P_Y$ are probability mass functions,

$$H(X,Y) - (H(X) + H(Y)) = \sum_{a,b} P_{XY}(a,b) \ln \frac{P_X(a)P_Y(b)}{P_{XY}(a,b)} \leq 0.$$

This proves the following result:

Lemma 2.3.2: Given two discrete alphabet random variables X and Y defined on a common probability space, we have

$$0 \leq H(X) \tag{2.3.4}$$

and

$$\max(H(X), H(Y)) \leq H(X,Y) \leq H(X) + H(Y) \tag{2.3.5}$$

where the right hand inequality holds with equality if and only if X and Y are independent. If the alphabet of X has $||A_X||$ symbols, then

$$H_X(X) \leq \ln ||A_X||. \tag{2.3.6}$$

There is another proof of the left hand inequality in (2.3.5) that uses an inequality for relative entropy that will be useful later when considering codes. The following lemma gives the inequality. First we introduce a definition. A partition $\mathcal{R}$ is said to *refine* a partion $\mathcal{Q}$ if every atom in $\mathcal{Q}$ is a union of atoms of $\mathcal{R}$, in which case we write $\mathcal{Q} < \mathcal{R}$.

Lemma 2.3.3: Suppose that P and M are two measures defined on a common measurable space $(\Omega, \mathcal{B})$ and that we are given a finite partitions $\mathcal{Q} < \mathcal{R}$. Then

$$H_{P||M}(\mathcal{Q}) \leq H_{P||M}(\mathcal{R})$$

and

$$H_P(\mathcal{Q}) \leq H_P(\mathcal{R})$$

Comments: The lemma can also be stated in terms of random variables and mappings in an intuitive way: Suppose that U is a random variable with finite alphabet A and $f : A \to B$ is a mapping from A into another finite alphabet B. Then the composite random variable $f(U)$ defined by $f(U)(\omega) = f(U(\omega))$ is also a finite random variable. If U induces a partition $\mathcal{R}$ and $f(U)$ a partition $\mathcal{Q}$, then $\mathcal{Q} < \mathcal{R}$ (since knowing the value of U implies the value of $f(U)$). Thus the lemma immediately gives the following corollary.

Corollary 2.3.1 If $M >> P$ are two measures describing a random variable U with alphabet A and if $f : A \to B$, then

$$H_{P||M}(f(U)) \leq H_{P||M}(U)$$

and

$$H_P(f(U)) \leq H_P(U).$$

Since $D(P_f||M_f) = H_{P||M}(f)$, we have also the following corollary which we state for future reference.

Corollary 2.3.2: Suppose that P and M are two probability measures on a discrete space and that f is a random variable defined on that space, then

$$D(P_f||M_f) \leq D(P||M).$$

The lemma, discussion, and corollaries can all be interpreted as saying that taking a measurement on a finite alphabet random variable lowers the entropy and the relative entropy of that random variable. By choosing U as (X, Y) and $f(X, Y) = X$ or Y, the lemma yields the promised inequality of the previous lemma.

Proof of Lemma: If $H_{P||M}(\mathcal{R}) = +\infty$, the result is immediate. If $H_{P||M}(\mathcal{Q}) = +\infty$, that is, if there exists at least one Q_j such that $M(Q_j) = 0$ but $P(Q_j) \neq 0$, then there exists an $R_i \subset Q_j$ such that $M(R_i) =$

0 and $P(R_i) > 0$ and hence $H_{P||M}(\mathcal{R}) = +\infty$. Lastly assume that both $H_{P||M}(\mathcal{R})$ and $H_{P||M}(\mathcal{Q})$ are finite and consider the difference

$$H_{P||M}(\mathcal{R}) - H_{P||M}(\mathcal{Q}) =$$

$$\sum_i P(R_i) \ln \frac{P(R_i)}{M(R_i)} - \sum_j P(Q_j) \ln \frac{P(Q_j)}{M(Q_j)} =$$

$$\sum_j [\sum_{i:R_i \subset Q_j} P(R_i) \ln \frac{P(R_i)}{M(R_i)} - P(Q_j) \ln \frac{P(Q_j)}{M(Q_j)}].$$

We shall show that each of the bracketed terms is nonnegative, which will prove the first inequality. Fix j. If $P(Q_j)$ is 0 we are done since then also $P(R_i)$ is 0 for all i in the inner sum since these R_i all belong to Q_j. If $P(Q_j)$ is not 0, we can divide by it to rewrite the bracketed term as

$$P(Q_j) \left(\sum_{i:R_i \subset Q_j} \frac{P(R_i)}{P(Q_j)} \ln \frac{P(R_i)/P(Q_j)}{M(R_i)/M(Q_j)} \right),$$

where we also used the fact that $M(Q_j)$ cannot be 0 since then $P(Q_j)$ would also have to be zero. Since $R_i \subset Q_j$, $P(R_i)/P(Q_j) = P(R_i \bigcap Q_j)/P(Q_j)$ $= P(R_i|Q_j)$ is an elementary conditional probability. Applying a similar argument to M and dividing by $P(Q_j)$, the above expression becomes

$$\sum_{i:R_i \subset Q_j} P(R_i|Q_j) \ln \frac{P(R_i|Q_j)}{M(R_i|Q_j)}$$

which is nonnegative from Lemma 2.3.1, which proves the first inequality. The second inequality follows similarly: Consider the difference

$$H_P(\mathcal{R}) - H_P(\mathcal{Q}) = \sum_j [\sum_{i:R_i \subset Q_j} P(R_i) \ln \frac{P(Q_j)}{P(R_i)}] =$$

$$\sum_j P(Q_j)[- \sum_{i:R_i \subset Q_j} P(R_i|Q_j) \ln P(R_i|Q_j)]$$

and the result follows since the bracketed term is nonnegative since it is an entropy for each value of j(Lemma 2.3.2). □

The next result provides useful inequalities for entropy considered as a function of the underlying distribution. In particular, it shows that entropy is a concave (or convex $\bigcap$) function of the underlying distribution. Define the binary entropy function (the entropy of a binary random variable with probability mass function $(\lambda, 1-\lambda)$) by

$$h_2(\lambda) = -\lambda \ln \lambda - (1-\lambda)\ln(1-\lambda).$$

Lemma 2.3.4: Let m and p denote two distributions for a discrete alphabet random variable X and let $\lambda \in (0,1)$. Then for any $\lambda \in (0,1)$

$$\lambda H_m(X) + (1-\lambda)H_p(X) \le H_{\lambda m+(1-\lambda)p}(X)$$

$$\le \lambda H_m(X) + (1-\lambda)H_p(X) + h_2(\lambda). \tag{2.3.7}$$

Proof: We do a little extra here to save work in a later result. Define the quantities

$$I = -\sum_x m(x)\ln(\lambda m(x) + (1-\lambda)p(x))$$

and

$$J = H_{\lambda m+(1-\lambda)p}(X) = -\lambda \sum_x m(x)\ln(\lambda m(x) + (1-\lambda)p(x))$$

$$-(1-\lambda)\sum_x p(x)\ln(\lambda m(x) + (1-\lambda)p(x)).$$

First observe that

$$\lambda m(x) + (1-\lambda)p(x) \ge \lambda m(x)$$

and therefore applying this bound to both m and p

$$I \le -\ln\lambda - \sum_x m(x)\ln m(x) = -\ln\lambda + H_m(X),$$

$$J \le -\lambda \sum_x m(x)\ln m(x) - (1-\lambda)\sum_x p(x)\ln p(x) + h_2(\lambda)$$

$$= \lambda H_m(X) + (1-\lambda)H_p(X) + h_2(\lambda). \tag{2.3.8}$$

To obtain the lower bounds of the lemma observe that

$$I = -\sum_x m(x)\ln m(x)(\lambda + (1-\lambda)\frac{p(x)}{m(x)})$$

$$= -\sum_x m(x) \ln m(x) - \sum_x m(x) \ln(\lambda + (1-\lambda)\frac{p(x)}{m(x)}).$$

Using (2.3.1) the rightmost term is bound below by

$$-\sum_x m(x)((\lambda + (1-\lambda)\frac{p(x)}{m(x)} - 1)$$

$$= -\lambda - 1 + \lambda \sum_{a \in A} p(X = a) + 1 = 0.$$

Thus for all n

$$I \geq -\sum_x m(x) \ln m(x) = H_m(X). \tag{2.3.9}$$

and hence also

$$J \geq -\lambda \sum_x m(x) \ln m(x) - (1-\lambda) \sum_x p(x) \ln p(x)$$

$$= \lambda H_m(X) + (1-\lambda) H_p(X). \quad \Box$$

The next result presents an interesting connection between combinatorics and binomial sums with a particular entropy. We require the familiar definition of the binomial coefficient:

$$\binom{n}{k} = \frac{n!}{k!(n-k)!}.$$

Lemma 2.3.5: Given $\delta \in (0, \frac{1}{2}]$ and a positive integer M, we have

$$\sum_{i \leq \delta M} \binom{M}{i} \leq e^{M h_2(\delta)}. \tag{2.3.10}$$

If $0 < \delta \leq p \leq 1$, then

$$\sum_{i \leq \delta M} \binom{M}{i} p^i (1-p)^{M-i} \leq e^{-M h_2(\delta || p)}, \tag{2.3.11}$$

where

$$h_2(\delta || p) = \delta \ln \frac{\delta}{p} + (1-\delta) \ln \frac{1-\delta}{1-p}.$$

Proof: We have after some simple algebra that

$$e^{-h_2(\delta)M} = \delta^{\delta M}(1-\delta)^{(1-\delta)M}.$$

If $\delta < 1/2$, then $\delta^k(1-\delta)^{M-k}$ increases as k decreases (since we are having more large terms and fewer small terms in the product) and hence if $i \leq M\delta$,

$$\delta^{\delta M}(1-\delta)^{(1-\delta)M} \leq \delta^i(1-\delta)^{M-i}.$$

Thus we have the inequalities

$$1 = \sum_{i=0}^{M} \binom{M}{i} \delta^i(1-\delta)^{M-i} \geq \sum_{i \leq \delta M} \binom{M}{i} \delta^i(1-\delta)^{M-i}$$

$$\geq e^{-h_2(\delta)M} \sum_{i \leq \delta M} \binom{M}{i}$$

which completes the proof of (2.3.10). In a similar fashion we have that

$$e^{Mh_2(\delta||p)} = (\frac{\delta}{p})^{\delta M}(\frac{1-\delta}{1-p})^{(1-\delta)M}.$$

Since $\delta \leq p$, we have as in the first argument that for $i \leq M\delta$

$$(\frac{\delta}{p})^{\delta M}(\frac{1-\delta}{1-p})^{(1-\delta)M} \leq (\frac{\delta}{p})^i(\frac{1-\delta}{1-p})^{M-i}$$

and therefore after some algebra we have that if $i \leq M\delta$ then

$$p^i(1-p)^{M-i} \leq \delta^i(1-\delta)^{M-i}e^{-Mh_2(\delta||p)}$$

and hence

$$\sum_{i \leq \delta M} \binom{M}{i} p^i(1-p)^{M-i} \leq e^{-Mh_2(\delta||p)} \sum_{i \leq \delta M} \binom{M}{i} \delta^i(1-\delta)^{M-i}$$

$$\leq e^{-nh_2(\delta||p)} \sum_{i=0}^{M} \binom{M}{i} \delta^i(1-\delta)^{M-i} = e^{-Mh_2(\delta||p)},$$

which proves (2.3.11). □

The following is a technical but useful property of sample entropies. The proof follows Billingsley [15].

Lemma 2.3.6: Given a finite alphabet process $\{X_n\}$ (not necessarily stationary) with distribution m, let $X_k^n = (X_k, X_{k+1}, \cdots, X_{k+n-1})$ denote the random vectors giving a block of samples of dimension n starting at time k. Then the random variables $n^{-1} \ln m(X_k^n)$ are m-uniformly integrable (uniform in k and n).

Proof: For each nonnegative integer r define the sets

$$E_r(k,n) = \{x : -\frac{1}{n} \ln m(x_k^n) \in [r, r+1)\}$$

and hence if $x \in E_r(k,n)$ then

$$r \leq -\frac{1}{n} \ln m(x_k^n) < r+1$$

or

$$e^{-nr} \geq m(x_k^n) > e^{-n(r+1)}.$$

Thus for any r

$$\int_{E_r(k,n)} (-\frac{1}{n} \ln m(X_k^n))\, dm < (r+1) m(E_r(k,n))$$

$$= (r+1) \sum_{x_k^n \in E_r(k,n)} m(x_k^n) \leq (r+1) \sum_{x_k^n} e^{-nr}$$

$$= (r+1) e^{-nr} ||A||^n \leq (r+1) e^{-nr},$$

where the final step follows since there are at most $||A||^n$ possible n-tuples corresponding to thin cylinders in $E_r(k,n)$ and by construction each has probability less than e^{-nr}.

To prove uniform integrability we must show uniform convergence to 0 as $r \to \infty$ of the integral

$$\gamma_r(k,n) = \int_{x: -\frac{1}{n} \ln m(x_k^n) \geq r} (-\frac{1}{n} \ln m(X_k^n))\, dm$$

$$= \sum_{i=0}^{\infty} \int_{E_{r+i}(k,n)} (-\frac{1}{n} \ln m(X_k^n))\, dm \leq \sum_{i=0}^{\infty} (r+i+1) e^{-n(r+i)} ||A||^n$$

$$\leq \sum_{i=0}^{\infty} (r+i+1) e^{-n(r+i-\ln ||A||)}.$$

Taking r large enough so that $r > \ln ||A||$, then the exponential term is bound above by the special case $n = 1$ and we have the bound

$$\gamma_r(k,n) \leq \sum_{i=0}^{\infty}(r+i+1)e^{-(r+i-\ln ||A||)}$$

a bound which is finite and independent of k and n. The sum can easily be shown to go to zero as $r \to \infty$ using standard summation formulas. (The exponential terms shrink faster than the linear terms grow.) □

Variational Description of Divergence

Divergence has a variational characterization that is a fundamental property for its applications to large deviations theory [143] [31]. Although this theory will not be treated here, the basic result of this section provides an alternative description of divergence and hence of relative entropy that has intrinsic interest. The basic result is originally due to Donsker and Varadhan [34].

Suppose now that P and M are two probability measures on a common discrete probability space, say $(\Omega, \mathcal{B})$. Given any real-valued random variable Φ defined on the probability space, we will be interested in the quantity

$$E_M e^{\Phi}. \tag{2.3.12}$$

which is called the *cumulant generating function* of Φ with respect to M and is related to the characteristic function of the random variable Φ as well as to the moment generating function and the operational transform of the random variable. The following theorem provides a variational description of divergence in terms of the cumulant generating function.

Theorem 2.3.2:

$$D(P||M) = \sup_{\Phi} \left(E_P\Phi - \ln(E_M(e^{\Phi}))\right). \tag{2.3.13}$$

Proof: First consider the random variable Φ defined by

$$\Phi(\omega) = \ln(P(\omega)/M(\omega))$$

and observe that

$$E_P\Phi - \ln(E_M(e^{\Phi})) = \sum_{\omega} P(\omega) \ln \frac{P(\omega)}{M(\omega)} - \ln(\sum_{\omega} M(\omega)\frac{P(\omega)}{M(\omega)})$$

$$= D(P||M) - \ln 1 = D(P||M).$$

This proves that the supremum over all Φ is no smaller than the divergence.

To prove the other half observe that for any bounded random variable Φ,

$$E_P\Phi - \ln E_M(e^{\Phi}) = E_P\left(\ln \frac{e^{\Phi}}{E_M(e^{\Phi})}\right) = \sum_{\omega} P(\omega)\left(\ln \frac{M^{\Phi}(\omega)}{M(\omega)}\right),$$

where the probability measure M^{Φ} is defined by

$$M^{\Phi}(\omega) = \frac{M(\omega)e^{\Phi(\omega)}}{\sum_x M(x)e^{\Phi(x)}}.$$

We now have for any Φ that

$$\begin{aligned} &D(P\|Q) - (E_P\Phi - \ln(E_M(e^{\Phi}))) \\ &= \sum_{\omega} P(\omega)\left(\ln \frac{P(\omega)}{M(\omega)}\right) - \sum_{\omega} P(\omega)\left(\ln \frac{M^{\Phi}(\omega)}{M(\omega)}\right) \\ &= \sum_{\omega} P(\omega)\left(\ln \frac{P(\omega)}{M^{\Phi}(\omega)}\right) \geq 0 \end{aligned}$$

using the divergence inequality. Since this is true for any Φ, it is also true for the supremum over Φ and the theorem is proved. □

2.4 Entropy Rate

Again let $\{X_n; n = 0, 1, \cdots\}$ denote a finite alphabet random process and apply Lemma 2.3.2 to vectors and obtain

$$H(X_0, X_1, \cdots, X_{n-1}) \leq$$

$$H(X_0, X_1, \cdots, X_{m-1}) + H(X_m, X_{m+1}, \cdots, X_{n-1});\ 0 < m < n. \tag{2.4.1}$$

Define as usual the random vectors $X_k^n = (X_k, X_{k+1}, \cdots, X_{k+n-1})$, that is, X_k^n is a vector of dimension n consisting of the samples of X from k to $k+n-1$. If the underlying measure is stationary, then the distributions of the random vectors X_k^n do not depend on k. Hence if we define the sequence $h(n) = H(X^n) = H(X_0, \cdots, X_{n-1})$, then the above equation becomes

$$h(k+n) \leq h(k) + h(n);\ \text{all } k, n > 0.$$

Thus $h(n)$ is a subadditive sequence as treated in Section 7.5 of [50]. A basic property of subadditive sequences is that the limit $h(n)/n$ as $n \to \infty$

exists and equals the infimum of $h(n)/n$ over n. (See, e.g., Lemma 7.5.1 of [50].) This immediately yields the following result.

Lemma 2.4.1: If the distribution m of a finite alphabet random process $\{X_n\}$ is stationary, then

$$\bar{H}_m(X) \equiv \lim_{n\to\infty} \frac{1}{n} H_m(X^n) = \inf_{n\geq 1} \frac{1}{n} H_m(X^n).$$

Thus the limit exists and equals the infimum.

The next two properties of entropy rate are primarily of interest because they imply a third property, the ergodic decomposition of entropy rate, which will be described in Theorem 2.4.1. They are also of some independent interest. The first result is a continuity result for entropy rate when considered as a function or functional on the underlying process distribution. The second property demonstrates that entropy rate is actually an affine functional (both convex $\bigcup$ and convex $\bigcap$) of the underlying distribution, even though finite order entropy was only convex $\bigcap$ and not affine.

We apply the distributional distance described in Section 1.8 to the standard sequence measurable space $(\Omega, \mathcal{B}) = (A^{\mathcal{Z}_+}, \mathcal{B}_A^{\mathcal{Z}_+})$ with a σ-field generated by the countable field $\mathcal{F} = \{F_n;\ n = 1, 2, \cdots\}$ generated by all thin rectangles.

Corollary 2.4.1: The entropy rate $\bar{H}_m(X)$ of a discrete alphabet random process considered as a functional of stationary measures is upper semicontinuous; that is, if probability measures m and m_n, $n = 1, 2, \cdots$ have the property that $d(m, m_n) \to 0$ as $n \to \infty$, then

$$\bar{H}_m(X) \geq \limsup_{n\to\infty} \bar{H}_{m_n}(X).$$

Proof: For each fixed n

$$H_m(X^n) = -\sum_{a^n\in A^n} m(X^n = a^n) \ln m(X^n = a^n)$$

is a continuous function of m since for the distance to go to zero, the probabilities of all thin rectangles must go to zero and the entropy is the sum of continuous real-valued functions of the probabilities of thin rectangles. Thus we have from Lemma 2.4.1 that if $d(m_k, m) \to 0$, then

$$\bar{H}_m(X) = \inf_n \frac{1}{n} H_m(X^n) = \inf_n \frac{1}{n} \lim_{k\to\infty} H_{m_k}(X^n)$$

$$\geq \limsup_{k\to\infty}\left(\inf_n \frac{1}{n}H_{m_k}(X^n)\right) = \limsup_{k\to\infty}\bar{H}_{m_k}(X). \ \Box$$

The next lemma uses Lemma 2.3.4 to show that entropy rates are affine functions of the underlying probability measures.

Lemma 2.4.2: Let m and p denote two distributions for a discrete alphabet random process $\{X_n\}$. Then for any $\lambda \in (0,1)$,

$$\lambda H_m(X^n) + (1-\lambda)H_p(X^n) \leq H_{\lambda m+(1-\lambda)p}(X^n)$$

$$\leq \lambda H_m(X^n) + (1-\lambda)H_p(X^n) + h_2(\lambda), \tag{2.4.2}$$

and

$$\limsup_{n\to\infty}(-\int dm(x)\frac{1}{n}\ln(\lambda m(X^n(x)) + (1-\lambda)p(X^n(x))))$$

$$= \limsup_{n\to\infty} - \int dm(x)\frac{1}{n}\ln m(X^n(x)) = \bar{H}_m(X). \tag{2.4.3}$$

If m and p are stationary then

$$\bar{H}_{\lambda m+(1-\lambda)p}(X) = \lambda\bar{H}_m(X) + (1-\lambda)\bar{H}_p(X) \tag{2.4.4}$$

and hence the entropy rate of a stationary discrete alphabet random process is an affine function of the process distribution. □

Comment: Eq. (2.4.2) is simply Lemma 2.3.4 applied to the random vectors X^n stated in terms of the process distributions. Eq. (2.4.3) states that if we look at the limit of the normalized log of a mixture of a pair of measures when one of the measures governs the process, then the limit of the expectation does not depend on the other measure at all and is simply the entropy rate of the driving source. Thus in a sense the sequences produced by a measure are able to select the true measure from a mixture.

Proof: Eq. (2.4.2) is just Lemma 2.3.4. Dividing by n and taking the limit as $n \to \infty$ proves that entropy rate is affine. Similarly, take the limit supremum in expressions (2.3.8) and (2.3.9) and the lemma is proved. □

We are now prepared to prove one of the fundamental properties of entropy rate, the fact that it has an ergodic decomposition formula similar to property (c) of Theorem 1.8.2 when it is considered as a functional on the underlying distribution. In other words, the entropy rate of a stationary source is given by an integral of the entropy rates of the stationary ergodic components. This is a far more complicated result than property (c) of the ordinary ergodic decomposition because the entropy rate depends on the distribution; it is not a simple function of the underlying sequence. The result is due to Jacobs [68].

Theorem 2.4.1: The Ergodic Decomposition of Entropy Rate Let $(A^{\mathcal{Z}_+}, \mathcal{B}(A)^{\mathcal{Z}_+}, m, T)$ be a stationary dynamical system corresponding to a stationary finite alphabet source $\{X_n\}$. Let $\{p_x\}$ denote the ergodic decomposition of m. If $\bar{H}_{p_x}(X)$ is m-integrable, then

$$\bar{H}_m(X) = \int dm(x) \bar{H}_{p_x}(X).$$

Proof: The theorem follows immediately from Corollary 2.4.1 and Lemma 2.4.2 and the ergodic decomposition of semi-continuous affine funtionals as in Theorem 8.9.1 of [50]. □

Relative Entropy Rate

The properties of relative entropy rate are more difficult to demonstrate. In particular, the obvious analog to (2.4.1) does not hold for relative entropy rate without the requirement that the reference measure by memoryless, and hence one cannot immediately infer that the relative entropy rate is given by a limit for stationary sources. The following lemma provides a condition under which the relative entropy rate is given by a limit. The condition, that the dominating measure be a kth order (or k-step) Markov source will occur repeatedly when dealing with relative entropy rates. A source is kth order Markov or k-step Markov (or simply Markov if k is clear from context) if for any n and any $N \geq k$

$$P(X_n = x_n | X_{n-1} = x_{n-1}, \cdots, X_{n-N} = x_{n-N})$$
$$= P(X_n = x_n | X_{n-1} = x_{n-1}, \cdots, X_{n-k} = x_{n-k});$$

that is, conditional probabilities given the infinite past depend only on the most recent k symbols. A 0-step Markov source is a memoryless source. A Markov source is said to have ***stationary transitions*** if the above conditional probabilities do not depend on n, that is, if for any n

$$P(X_n = x_n | X_{n-1} = x_{n-1}, \cdots, X_{n-N} = x_{n-N})$$
$$= P(X_k = x_n | X_{k-1} = x_{n-1}, \cdots, X_0 = x_{n-k}).$$

Lemma 2.4.3 If p is a stationary process and m is a k-step Markov process with stationary transitions, then

$$\bar{H}_{p||m}(X) = \lim_{n\to\infty} \frac{1}{n} H_{p||m}(X^n) = -\bar{H}_p(X) - E_p[\ln m(X_k | X^k)],$$

where $E_p[\ln m(X_k|X^k)]$ is an abbreviation for

$$E_p[\ln m(X_k|X^k)] = \sum_{x^{k+1}} p_{X^{k+1}}(x^{k+1}) \ln m_{X_k|X^k}(x_k|x^k).$$

Proof: If for any n it is not true that $m_{X^n} >> p_{X^n}$, then $H_{p||m}(X^n) = \infty$ for that and all larger n and both sides of the formula are infinite, hence we assume that all of the finite dimensional distributions satisfy the absolute continuity relation. Since m is Markov,

$$m_{X^n}(x^n) = \prod_{l=k}^{n-1} m_{X_l|X^l}(x_l|x^l) m_{X^k}(x^k).$$

Thus

$$\frac{1}{n} H_{p||m}(X^n) = -\frac{1}{n} H_p(X^n) - \frac{1}{n} \sum_{x^n} p_{X^n}(x^n) \ln m_{X^n}(x^n)$$

$$= -\frac{1}{n} H_p(X^n) - \frac{1}{n} \sum_{x^k} p_{X^k}(x^k) \ln m_{X^k}(x^k)$$

$$-\frac{n-k}{n} \sum_{x^{k+1}} p_{X^{k+1}}(x^{k+1}) \ln m_{X_k|X^k}(x_k|x^k).$$

Taking limits then yields

$$\bar{H}_{p||m}(X) = -\bar{H}_p - \sum_{x^{k+1}} p_{X^{k+1}}(x^{k+1}) \ln m_{X_k|X^k}(x_k|x^k),$$

where the sum is well defined because if $m_{X_k|X^k}(x_k|x^k) = 0$, then so must $p_{X^{k+1}}(x^{k+1}) = 0$ from absolute continuity. □

Combining the previous lemma with the ergodic decomposition of entropy rate yields the following corollary.

Corollary 2.4.2: The Ergodic Decomposition of Relative Entropy Rate Let $(A^{\mathcal{Z}_+}, \mathcal{B}(A)^{\mathcal{Z}_+}, p, T)$ be a stationary dynamical system corresponding to a stationary finite alphabet source $\{X_n\}$. Let m be a kth order Markov process for which $m_{X^n} >> p_{X^n}$ for all n. Let $\{p_x\}$ denote the ergodic decomposition of p. If $\bar{H}_{p_x||m}(X)$ is p-integrable, then

$$\bar{H}_{p||m}(X) = \int dp(x) \bar{H}_{p_x||m}(X).$$

2.5 Conditional Entropy and Information

We now turn to other notions of information. While we could do without these if we confined interest to finite alphabet processes, they will be essential for later generalizations and provide additional intuition and results even in the finite alphabet case. We begin by adding a second finite alphabet measurement to the setup of the previous sections. To conform more to information theory tradition, we consider the measurements as finite alphabet random variables X and Y rather than f and g. This has the advantage of releasing f and g for use as functions defined on the random variables: $f(X)$ and $g(Y)$. Let $(\Omega, \mathcal{B}, P, T)$ be a dynamical system. Let X and Y be finite alphabet measurements defined on Ω with alphabets A_X and A_Y. Define the *conditional entropy* of X given Y by

$$H(X|Y) \equiv H(X,Y) - H(Y).$$

The name conditional entropy comes from the fact that

$$H(X|Y) = -\sum_{x,y} P(X = a, Y = b) \ln P(X = a|Y = b)$$

$$= -\sum_{x,y} p_{X,Y}(x,y) \ln p_{X|Y}(x|y),$$

where $p_{X,Y}(x,y)$ is the joint pmf for (X,Y) and $p_{X|Y}(x|y) = p_{X,Y}(x,y)/p_Y(y)$ is the conditional pmf. Defining

$$H(X|Y = y) = -\sum_{x} p_{X|Y}(x|y) \ln p_{X|Y}(x|y)$$

we can also write

$$H(X|Y) = \sum_{y} p_Y(y) H(X|Y = y).$$

Thus conditional entropy is an average of entropies with respect to conditional pmf's. We have immediately from Lemma 2.3.2 and the definition of conditional entropy that

$$0 \le H(X|Y) \le H(X). \tag{2.5.1}$$

The inequalities could also be written in terms of the partitions induced by X and Y. Recall that according to Lemma 2.3.2 the right hand inequality will be an equality if and only if X and Y are independent.

Define the *average mutual information* between X and Y by

$$I(X;Y) = H(X) + H(Y) - H(X,Y)$$

$$= H(X) - H(X|Y) = H(Y) - H(Y|X).$$

In terms of distributions and pmf's we have that

$$I(X;Y) = \sum_{x,y} P(X = x, Y = y) \ln \frac{P(X = x, Y = y)}{P(X = x)P(Y = y)}$$

$$= \sum_{x,y} p_{X,Y}(x,y) \ln \frac{p_{X,Y}(x,y)}{p_X(x)p_Y(y)} = \sum_{x,y} p_{X,Y}(x,y) \ln \frac{p_{X|Y}(x|y)}{p_X(x)}$$

$$= \sum_{x,y} p_{X,Y}(x,y) \ln \frac{p_{Y|X}(y|x)}{p_Y(y)}.$$

Note also that mutual information can be expressed as a divergence by

$$I(X;Y) = D(P_{X,Y} || P_X \times P_Y),$$

where $P_X \times P_Y$ is the product measure on X, Y, that is, a probability measure which gives X and Y the same marginal distributions as P_{XY}, but under which X and Y are independent. Entropy is a special case of mutual information since

$$H(X) = I(X;X).$$

We can collect several of the properties of entropy and relative entropy and produce corresponding properties of mutual information. We state these in the form using measurements, but they can equally well be expressed in terms of partitions.

Lemma 2.5.1: Suppose that X and Y are two finite alphabet random variables defined on a common probability space. Then

$$0 \leq I(X;Y) \leq \min(H(X), H(Y)).$$

Suppose that $f : A_X \rightarrow A$ and $g : A_Y \rightarrow B$ are two measurements. Then

$$I(f(X); g(Y)) \leq I(X;Y).$$

Proof: The first result follows immediately from the properties of entropy. The second follows from Lemma 2.3.3 applied to the measurement (f, g) since mutual information is a special case of relative entropy. □

The next lemma collects some additional, similar properties.

Lemma 2.5.2: Given the assumptions of the previous lemma,

$$H(f(X)|X) = 0,$$

$$H(X, f(X)) = H(X),$$

$$H(X) = H(f(X)) + H(X|f(X),$$

$$I(X; f(X)) = H(f(X)),$$

$$H(X|g(Y)) \geq H(X|Y),$$

$$I(f(X); g(Y)) \leq I(X; Y),$$

$$H(X|Y) = H(X, f(X, Y))|Y),$$

and, if Z is a third finite alphabet random variable defined on the same probability space,

$$H(X|Y) \geq H(X|Y, Z).$$

Comments: The first relation has the interpretation that given a random variable, there is no additional information in a measurement made on the random variable. The second and third relationships follow from the first and the definitions. The third relation is a form of chain rule and it implies that given a measurement on a random variable, the entropy of the random variable is given by that of the measurement plus the conditional entropy of the random variable given the measurement. This provides an alternative proof of the second result of Lemma 2.3.3. The fifth relation says that conditioning on a measurement of a random variable is less informative than conditioning on the random variable itself. The sixth relation states that coding reduces mutual information as well as entropy. The seventh relation is a conditional extension of the second. The eighth relation says that conditional entropy is nonincreasing when conditioning on more information.

Proof: Since $g(X)$ is a deterministic function of X, the conditional pmf is trivial (a Kronecker delta) and hence $H(g(X)|X = x)$ is 0 for all x, hence the first relation holds. The second and third relations follow from the first and the definition of conditional entropy. The fourth relation follows from the first since $I(X;Y) = H(Y) - H(Y|X)$. The fifth relation follows from the previous lemma since

$$H(X) - H(X|g(Y)) = I(X;g(Y)) \leq I(X;Y) = H(X) - H(X|Y).$$

The sixth relation follows from Corollary 2.3.2 and the fact that

$$I(X;Y) = D(P_{X,Y}||P_X \times P_Y).$$

The seventh relation follows since

$$H(X, f(X,Y))|Y) = H(X, f(X,Y)), Y) - H(Y)$$

$$= H(X,Y) - H(Y) = H(X|Y).$$

The final relation follows from the second by replacing Y by Y, Z and setting $g(Y,Z) = Y$. □

In a similar fashion we can consider conditional relative entropies. Suppose now that M and P are two probability measures on a common space, that X and Y are two random variables defined on that space, and that $M_{XY} >> P_{XY}$ (and hence also $M_X >> P_Y$). Analagous to the definition of the conditional entropy we can define

$$H_{P||M}(X|Y) \equiv H_{P||M}(X,Y) - H_{P||M}(Y).$$

Some algebra shows that this is equivalent to

$$H_{P||M}(X|Y) = \sum_{x,y} p_{X,Y}(x,y) \ln \frac{p_{X|Y}(x|y)}{m_{X|Y}(x|y)}$$

$$= \sum_x p_X(x) \left(p_{X|Y}(x|y) \ln \frac{p_{X|Y}(x|y)}{m_{X|Y}(x|y)} \right). \tag{2.5.2}$$

This can be written as

$$H_{P||M}(X|Y) = \sum_y p_Y(y) D(p_{X|Y}(\cdot|y)||m_{X|Y}(\cdot|y)),$$

an average of divergences of conditional pmf's, each of which is well defined because of the original absolute continuity of the joint measure. Manipulations similar to those for entropy can now be used to prove the following properties of conditional relative entropies.

Lemma 2.5.3 Given two probability measures M and P on a common space, and two random variables X and Y defined on that space with the property that $M_{XY} >> P_{XY}$, then the following properties hold:

$$H_{P||M}(f(X)|X) = 0,$$

$$H_{P||M}(X, f(X)) = H_{P||M}(X),$$

$$H_{P||M}(X) = H_{P||M}(f(X)) + H_{P||M}(X|f(X)), \tag{2.5.3}$$

If $M_{XY} = M_X \times M_Y$ (that is, if the pmfs satisfy $m_{X,Y}(x, y) = m_X(x)m_Y(y)$), then

$$H_{P||M}(X, Y) \geq H_{P||M}(X) + H_{P||M}(Y)$$

and

$$H_{P||M}(X|Y) \geq H_{P||M}(X).$$

Eq. (2.5.3) is a chain rule for relative entropy which provides as a corollary an immediate proof of Lemma 2.3.3. The final two inequalities resemble inequalities for entropy (with a sign reversal), but they do not hold for all reference measures.

The above lemmas along with Lemma 2.3.3 show that all of the information measures thus far considered are reduced by taking measurements or by coding. This property is the key to generalizing these quantities to nondiscrete alphabets.

We saw in Lemma 2.3.4 that entropy was a convex $\bigcap$ function of the underlying distribution. The following lemma provides similar properties of mutual information considered as a function of either a marginal or a conditional distribution.

Lemma 2.5.4: Let μ denote a pmf on a discrete space A_X, $\mu(x) = \Pr(X = x)$, and let q be a conditional pmf, $q(y|x) = \Pr(Y = y|X = x)$. Let μq denote the resulting joint pmf $\mu q(x, y) = \mu(x)q(y|x)$. Let $I_{\mu q} = I_{\mu q}(X;Y)$ be the average mutual information. Then $I_{\mu q}$ is a convex $\bigcup$ function of q; that is, given two conditional pmf's q_1 and q_2, a $\lambda \in [0, 1]$, and $\bar{q} = \lambda q_1 + (1 - \lambda)q_2$, then

$$I_{\mu\bar{q}} \leq \lambda I_{\mu q_1} + (1 - \lambda)I_{\mu q_2},$$

and $I_{\mu q}$ is a convex $\bigcap$ function of μ, that is, given two pmf's μ_1 and μ_2, $\lambda \in [0,1]$, and $\bar{\mu} = \lambda\mu_1 + (1-\lambda)\mu_2$,

$$I_{\bar{\mu}q} \geq \lambda I_{\mu_1 q} + (1-\lambda) I_{\mu_2 q}.$$

Proof: Let r (respectively, r_1, r_2, $\bar{r}$) denote the pmf for Y resulting from q (respectively q_1, q_2, $\bar{q}$), that is, $r(y) = \Pr(Y = y) = \sum_x \mu(x) q(y|x)$. From (2.3.1)

$$I_{\mu\bar{q}} = \lambda \sum_{x,y} \mu(x) q_1(x,y) \log\left(\frac{\mu(x)\bar{q}(x,y)}{\mu(x)\bar{r}(y)} \frac{\mu(x) r_1(y)}{\mu(x) q_1(x,y)} \frac{\mu(x) q_1(x,y)}{\mu(x) r_1(y)}\right)$$

$$+(1-\lambda) \sum_{x,y} \mu(x) q_2(x,y) \log\left(\frac{\mu(x)\bar{q}(x,y)}{\mu(x)\bar{r}(y)} \frac{\mu(x) r_2(y)}{\mu(x) q_2(x,y)} \frac{\mu(x) q_2(x,y)}{\mu(x) r_2(y)}\right)$$

$$\leq \lambda I_{\mu q_1} + \lambda \sum_{x,y} \mu q_1(x,y) \left(\frac{\mu(x)\bar{q}(x,y)}{\mu(x)\bar{r}(y)} \frac{\mu(x) r_1(y)}{\mu(x) q_1(x,y)} - 1\right)$$

$$+(1-\lambda) I_{\mu q_2} + (1-\lambda) \sum_{x,y} \mu(x) q_2(x,y) \left(\frac{\mu(x)\bar{q}(x,y)}{\mu(x)\bar{r}(y)} \frac{\mu(x) r_2(y)}{\mu(x) q_2(x,y)} - 1\right)$$

$$= \lambda I_{\mu q_1} + (1-\lambda) I_{\mu q_2} + \lambda(-1 + \sum_{x,y} \frac{\mu\bar{q}(x,y)}{\bar{r}(y)} r_1(y))$$

$$+(1-\lambda)(-1 + \sum_{x,y} \frac{\mu(x)\bar{q}(x,y)}{\bar{r}(y)} r_2(y)) = \lambda I_{\mu q_1} + (1-\lambda) I_{\mu q_2}.$$

Similarly, let $\bar{\mu} = \lambda\mu_1 + (1-\lambda)\mu_2$ and let r_1, r_2, and $\bar{r}$ denote the induced output pmf's. Then

$$I_{\bar{\mu}q} = \lambda \sum_{x,y} \mu_1(x) q(y|x) \log\left(\frac{q(y|x)}{\bar{r}(y)} \frac{r_1(y)}{q(y|x)} \frac{q(y|x)}{r_1(y)}\right)$$

$$+(1-\lambda) \sum_{x,y} \mu_2(x) q(y|x) \log\left(\frac{q(y|x)}{\bar{r}(y)} \frac{r_2(y)}{q(y|x)} \frac{q(y|x)}{r_2(y)}\right)$$

$$= \lambda I_{\mu_1 q} + (1-\lambda) I_{\mu_2 q} - \lambda \sum_{x,y} \mu_1(x) q(y|x) \log \frac{\bar{r}(y)}{r_1(y)}$$

$$-(1-\lambda) \sum_{x,y} \mu_2(x) q(y|x) \log \frac{\bar{r}(y)}{r_2(y)} \geq \lambda I_{\mu_1 q} + (1-\lambda) I_{\mu_2 q}$$

from another application of (2.3.1). □

We consider one other notion of information: Given three finite alphabet random variables X, Y, Z, define the *conditional mutual information* between X and Y given Z by

$$I(X;Y|Z) = D(P_{XYZ}||P_{X\times Y|Z}) \tag{2.5.4}$$

where $P_{X\times Y|Z}$ is the distribution defined by its values on rectangles as

$$P_{X\times Y|Z}(F\times G\times D) = \sum_{z\in D} P(X\in F|Z=z)P(Y\in G|Z=z)P(Z=z). \tag{2.5.5}$$

$P_{X\times Y|Z}$ has the same conditional distributions for X given Z and for Y given Z as does P_{XYZ}, but now X and Y are conditionally independent given Z. Alternatively, the conditional distribution for X, Y given Z under the distribution $P_{X\times Y|Z}$ is the product distribution $P_X|Z \times P_Y|Z$. Thus

$$I(X;Y|Z) = \sum_{x,y,z} p_{XYZ}(x,y,z)\ln\frac{p_{XYZ}(x,y,z)}{p_{X|Z}(x|z)p_{Y|Z}(y|z)p_Z(z)}$$

$$= \sum_{x,y,z} p_{XYZ}(x,y,z)\ln\frac{p_{XY|Z}(x,y|z)}{p_{X|Z}(x|z)p_{Y|Z}(y|z)}. \tag{2.5.6}$$

Since

$$\frac{p_{XYZ}}{p_{X|Z}p_{Y|Z}p_Z} = \frac{p_{XYZ}}{p_X p_{YZ}}\times\frac{p_X}{p_{X|Z}} = \frac{p_{XYZ}}{p_{XZ}p_Y}\times\frac{p_Y}{p_{Y|Z}}$$

we have the first statement in the following lemma.

Lemma 2.5.4:

$$I(X;Y|Z) + I(Y;Z) = I(Y;(X,Z)), \tag{2.5.7}$$

$$I(X;Y|Z) \geq 0, \tag{2.5.8}$$

with equality if and only if X and Y are conditionally independent given Z, that is, $p_{XY|Z} = p_{X|Z}p_{Y|Z}$. Given finite valued measurements f and g,

$$I(f(X);g(Y)|Z) \leq I(X;Y|Z).$$

Proof: The second inequality follows from the divergence inequality (2.3.2) with $P = P_{XYZ}$ and $M = P_{X\times Y|Z}$, i.e., the pmf's p_{XYZ} and $p_{X|Z}p_{Y|Z}p_Z$. The third inequality follows from Lemma 2.3.3 or its corollary applied to the same measures. □

Comments: Eq. (2.5.7) is called *Kolmogorov's formula.* If X and Y are conditionally independent given Z in the above sense, then we also have that $p_{X|YZ} = p_{XY|Z}/p_{Y|Z} = p_{X|Z}$, in which case we say that $Y \to Z \to X$ is a *Markov chain* and note that given Z, X does not depend on Y. (Note that if $Y \to Z \to X$ is a Markov chain, then so is $X \to Z \to Y$.) Thus the conditional mutual information is 0 if and only if the variables form a Markov chain with the conditioning variable in the middle. One might be tempted to infer from Lemma 2.3.3 that given finite valued measurements f, g, and r

$$I(f(X); g(Y)|r(Z)) \overset{(?)}{\leq} I(X;Y|Z).$$

This does not follow, however, since it is not true that if $\mathcal{Q}$ is the partition corresponding to the three quantizers, then $D(P_{f(X),g(Y),r(Z)}||P_{f(X)\times g(Y)|r(Z)})$ is $H_{P_{X,Y,Z}||P_{X\times Y|Z}}(f(X), g(Y), r(Z))$ because of the way that $P_{X\times Y|Z}$ is constructed; e.g., the fact that X and Y are conditionally independent given Z implies that $f(X)$ and $g(Y)$ are conditionally independent given Z, but it does not imply that $f(X)$ and $g(Y)$ are conditionally independent given $r(Z)$. Alternatively, if M is $P_{X\times Z|Y}$, then it is not true that $P_{f(X)\times g(Y)|r(Z)}$ equals $M(fgr)^{-1}$. Note that if this inequality were true, choosing $r(z)$ to be trivial (say 1 for all z) would result in $I(X;Y|Z) \geq I(X;Y|r(Z)) = I(X;Y)$. This cannot be true in general since, for example, choosing Z as (X,Y) would give $I(X;Y|Z) = 0$. Thus one must be careful when applying Lemma 2.3.3 if the measures and random variables are related as they are in the case of conditional mutual information.

We close this section with an easy corollary of the previous lemma and of the definition of conditional entropy. Results of this type are referred to as *chain rules* for information and entropy.

Corollary 2.5.1: Given finite alphabet random variables Y, X_1, X_2, $\cdots$, X_n,

$$H(X_1, X_2, \cdots, X_n) = \sum_{i=1}^{n} H(X_i|X_1, \cdots, X_{i-1})$$

$$H_{p||m}(X_1, X_2, \cdots, X_n) = \sum_{i=1}^{n} H_{p||m}(X_i|X_1, \cdots, X_{i-1})$$

$$I(Y;(X_1, X_2, \cdots, X_n)) = \sum_{i=1}^{n} I(Y; X_i|X_1, \cdots, X_{i-1}).$$

2.6 Entropy Rate Revisited

The chain rule of Corollary 2.5.1 provides a means of computing entropy rates for stationary processes. We have that

$$\frac{1}{n}H(X^n) = \frac{1}{n}\sum_{i=0}^{n-1} H(X_i|X^{i-1}).$$

First suppose that the source is a stationary kth order Markov process, that is, for any $m > k$

$$\Pr(X_n = x_n|X_i = x_i;\ i = 0, 1, \cdots, n-1)$$

$$= \Pr(X_n = x_n|X_i = x_i;\ i = n-k, \cdots, n-1).$$

For such a process we have for all $n \geq k$ that

$$H(X_n|X^n) = H(X_n|X_{n-k}^k) = H(X_k|X^k),$$

where $X_i^m = X_i, \cdots, X_{i+m-1}$. Thus taking the limit as $n \to \infty$ of the nth order entropy, all but a finite number of terms in the sum are identical and hence the Cesàro (or arithmetic) mean is given by the conditional expectation. We have therefore proved the following lemma.

Lemma 2.6.1: If $\{X_n\}$ is a stationary kth order Markov source, then

$$\bar{H}(X) = H(X_k|X^k).$$

If we have a two-sided stationary process $\{X_n\}$, then all of the previous definitions for entropies of vectors extend in an obvious fashion and a generalization of the Markov result follows if we use stationarity and the chain rule to write

$$\frac{1}{n}H(X^n) = \frac{1}{n}\sum_{i=0}^{n-1} H(X_0|X_{-1}, \cdots, X_{-i}).$$

Since conditional entropy is nonincreasing with more conditioning variables ((2.5.1) or Lemma 2.5.2), $H(X_0|X_{-1}, \cdots, X_{-i})$ has a limit. Again using the fact that a Cesàro mean of terms all converging to a common limit also converges to the same limit we have the following result.

Lemma 2.6.2: If $\{X_n\}$ is a two-sided stationary source, then

$$\bar{H}(X) = \lim_{n\to\infty} H(X_0|X_{-1}, \cdots, X_{-n}).$$

It is tempting to identify the above limit as the conditional entropy given the infinite past, $H(X_0|X_{-1},\cdots)$. Since the conditioning variable is a sequence and does not have a finite alphabet, such a conditional entropy is not included in any of the definitions yet introduced. We shall later demonstrate that this interpretation is indeed valid when the notion of conditional entropy has been suitably generalized.

The natural generalization of Lemma 2.6.2 to relative entropy rates unfortunately does not work because conditional relative entropies are not in general monotonic with increased conditioning and hence the chain rule does not immediately yield a limiting argument analogous to that for entropy. The argument does work if the reference measure is a kth order Markov, as considered in the following lemma.

Lemma 2.6.3: If $\{X_n\}$ is a source described by process distributions p and m and if p is stationary and m is kth order Markov with stationary transitions, then for $n \geq k$ $H_{p||m}(X_0|X_{-1},\cdots,X_{-n})$ is nondecreasing in n and

$$\bar{H}_{p||m}(X) = \lim_{n\to\infty} H_{p||m}(X_0|X_{-1},\cdots,X_{-n})$$

$$= -\bar{H}_p(X) - E_p[\ln m(X_k|X^k)].$$

Proof: For $n \geq k$ we have that

$$H_{p||m}(X_0|X_{-1},\cdots,X_{-n})$$

$$= -H_p(X_0|X_{-1},\cdots,X_{-n}) - \sum_{x^{k+1}} p_{X^{k+1}}(x^{k+1}) \ln m_{X_k|X^k}(x_k|x^k).$$

Since the conditional entropy is nonincreasing with n and the remaining term does not depend on n, the combination is nondecreasing with n. The remainder of the proof then parallels the entropy rate result. □

It is important to note that the relative entropy analogs to entropy properties often require kth order Markov assumptions on the reference measure (but not on the original measure).

Markov Approximations

Recall that the relative entropy rate $\bar{H}_{p||m}(X)$ can be thought of as a distance between the process with distribution p and that with distribution m and that the rate is given by a limit if the reference measure m is Markov. A particular Markov measure relevant to p is the distribution $p^{(k)}$ which is the *kth order Markov approximation to p* in the sense that it is a kth order Markov source and it has the same kth order transition probabilities as p.

To be more precise, the process distribution $p^{(k)}$ is specified by its finite dimensional distributions

$$p^{(k)}_{X^k}(x^k) = p_{X^k}(x^k)$$

$$p^{(k)}_{X^n}(x^n) = p_{X^k}(x^k) \prod_{l=k}^{n-1} p_{X_l|X^k_{l-k}}(x_l|x^k_{l-k});\ n = k, k+1, \cdots$$

so that

$$p^{(k)}_{X_k|X^k} = p_{X_k|X^k}.$$

It is natural to ask how good this approximation is, especially in the limit, that is, to study the behavior of the relative entropy rate $\bar{H}_{p||p^{(k)}}(X)$ as $k \to \infty$.

Theorem 2.6.2: Given a stationary process p, let $p^{(k)}$ denote the kth order Markov approximations to p. Then

$$\lim_{k\to\infty} \bar{H}_{p||p^{(k)}}(X) = \inf_k \bar{H}_{p||p^{(k)}}(X) = 0.$$

Thus the Markov approximations are asymptotically accurate in the sense that the relative entropy rate between the source and approximation can be made arbitrarily small (zero if the original source itself happens to be Markov).

Proof: As in the proof of Lemma 2.6.3 we can write for $n \geq k$ that

$$H_{p||p^{(k)}}(X_0|X_{-1}, \cdots, X_{-n})$$

$$= -H_p(X_0|X_{-1}, \cdots, X_{-n}) - \sum_{x^{k+1}} p_{X^{k+1}}(x^{k+1}) \ln p_{X_k|X^k}(x_k|x^k)$$

$$= H_p(X_0|X_{-1}, \cdots, X_{-k}) - H_p(X_0|X_{-1}, \cdots, X_{-n}).$$

Note that this implies that $p^{(k)}_{X^n} >> p_{X^n}$ for all n since the entropies are finite. This automatic domination of the finite dimensional distributions of a measure by those of its Markov approximation will not hold in the general case to be encountered later, it is specific to the finite alphabet case. Taking the limit as $n \to \infty$ gives

$$\bar{H}_{p||p^{(k)}}(X) = \lim_{n\to\infty} H_{p||p^{(k)}}(X_0|X_{-1}, \cdots, X_{-n})$$

$$= H_p(X_0|X_{-1}, \cdots, X_{-k}) - \bar{H}_p(X).$$

The corollary then follows immediately from Lemma 2.6.2. □

Markov approximations will play a fundamental role when considering relative entropies for general (nonfinite alphabet) processes. The basic result above will generalize to that case, but the proof will be much more involved.

2.7 Relative Entropy Densities

Many of the convergence results to come will be given and stated in terms of relative entropy densities. In this section we present a simple but important result describing the asymptotic behavior of relative entropy densities. Although the result of this section is only for finite alphabet processes, it is stated and proved in a manner that will extend naturally to more general processes later on. The result will play a fundamental role in the basic ergodic theorems to come.

Throughout this section we will assume that M and P are two process distributions describing a random process $\{X_n\}$. Denote as before the sample vector $X^n = (X_0, X_1, \cdots, X_{n-1})$, that is, the vector beginning at time 0 having length n. The distributions on X^n induced by M and P will be denoted by M_n and P_n, respectively. The corresponding pmf's are m_{X^n} and p_{X^n}. The key assumption in this section is that for all n if $m_{X^n}(x^n) = 0$, then also $p_{X^n}(x^n) = 0$, that is,

$$M_n >> P_n \text{ for all } n. \tag{2.7.1}$$

If this is the case, we can define the relative entropy density

$$h_n(x) \equiv \ln \frac{p_{X^n}(x^n)}{m_{X^n}(x^n)} = \ln f_n(x), \tag{2.7.2}$$

where

$$f_n(x) \equiv \begin{cases} \frac{p_{X^n}(x^n)}{m_{X^n}(x^n)} & \text{if } m_{X^n}(x^n) \neq 0 \\ 0 & \text{otherwise} \end{cases} \tag{2.7.3}$$

Observe that the relative entropy is found by integrating the relative entropy density:

$$\begin{aligned} H_{P||M}(X^n) = D(P_n||M_n) &= \sum_{x^n} p_{X^n}(x^n) \ln \frac{p_{X^n}(x^n)}{m_{X^n}(x^n)} \\ &= \int \ln \frac{p_{X^n}(X^n)}{m_{X^n}(X^n)}\, dP \end{aligned} \tag{2.7.4}$$

Thus, for example, if we assume that

$$H_{P||M}(X^n) < \infty, \text{ all } n, \tag{2.7.5}$$

then (2.7.1) holds.

The following lemma will prove to be useful when comparing the asymptotic behavior of relative entropy densities for different probability measures. It is the first almost everywhere result for relative entropy densities that we consider. It is somewhat narrow in the sense that it only compares limiting densities to zero and not to expectations. We shall later see that essentially the same argument implies the same result for the general case (Theorem 5.4.1), only the interim steps involving pmf's need be dropped. Note that the lemma requires neither stationarity nor asymptotic mean stationarity.

Lemma 2.7.1: Given a finite alphabet process $\{X_n\}$ with process measures P, M satisfying (2.7.1), Then

$$\limsup_{n\to\infty} \frac{1}{n} h_n \leq 0,\ M - \text{a.e.} \tag{2.7.6}$$

and

$$\liminf_{n\to\infty} \frac{1}{n} h_n \geq 0,\ P - \text{a.e..} \tag{2.7.7}$$

If in addition $M >> P$, then

$$\lim_{n\to\infty} \frac{1}{n} h_n = 0,\ P - \text{a.e..} \tag{2.7.8}$$

Proof: First consider the probability

$$M(\frac{1}{n}h_n \geq \epsilon) = M(f_n \geq e^{n\epsilon}) \leq \frac{E_M(f_n)}{e^{n\epsilon}},$$

where the final inequality is Markov's inequality. But

$$E_M(f_n) = \int dM f_n = \sum_{x^n:\, m_{X^n}(x^n)\neq 0} m_{X^n}(x^n) \frac{p_{X^n}(x^n)}{m_{X^n}(x^n)}$$

$$= \sum_{x^n:\, m_{X^n}(x^n)\neq 0} p_{X^n}(x^n) \leq 1$$

and therefore

$$M(\frac{1}{n}h_n \geq \epsilon) \leq 2^{-n\epsilon}$$

and hence

$$\sum_{n=1}^{\infty} M(\frac{1}{n}h_n > \epsilon) \leq \sum_{n=1}^{\infty} e^{-n\epsilon} < \infty.$$

From the Borel-Cantelli Lemma (e.g., Lemma 4.6.3 of [50]) this implies that $M(n^{-1}h_n \geq \epsilon \text{ i.o.}) = 0$ which implies the first equation of the lemma.

Next consider

$$P(-\frac{1}{n}h_n > \epsilon) = \sum_{x^n:-\frac{1}{n}\ln p_{X^n}(x^n)/m_{X^n}(x^n) > \epsilon} p_{X^n}(x^n)$$

$$= \sum_{x^n: -\frac{1}{n}\ln p_{X^n}(x^n)/m_{X^n}(x^n) > \epsilon \text{ and } m_{X^n}(x^n) \neq 0} p_{X^n}(x^n)$$

where the last statement follows since if $m_{X^n}(x^n) = 0$, then also $p_{X^n}(x^n) = 0$ and hence nothing would be contributed to the sum. In other words, terms violating this condition add zero to the sum and hence adding this condition to the sum does not change the sum's value. Thus

$$P(-\frac{1}{n}h_n > \epsilon) = \sum_{x^n: -\frac{1}{n}\ln p_{X^n}(x^n)/m_{X^n}(x^n) > \epsilon \text{ and } m_{X^n}(x^n) \neq 0} p_{X^n}(x^n)/m_{X^n}(x^n) m_{X^n}(x^n)$$

$$= \int_{f_n < e^{-n\epsilon}} dM f_n \leq \int_{f_n < e^{-n\epsilon}} dM e^{-n\epsilon}$$

$$= e^{-n\epsilon} M(f_n < e^{-n\epsilon}) \leq e^{-n\epsilon}.$$

Thus as before we have that $P(n^{-1}h_n > \epsilon) \leq e^{-n\epsilon}$ and hence that $P(n^{-1}h_n \leq -\epsilon \text{ i.o.}) = 0$ which proves the second claim. If also $M >> P$, then the first equation of the lemma is also true P-a.e., which when coupled with the second equation proves the third. □

Chapter 3

The Entropy Ergodic Theorem

3.1 Introduction

The goal of this chapter is to prove an ergodic theorem for sample entropy of finite alphabet random processes. The result is sometimes called the ergodic theorem of information theory or the asymptotic equipartion theorem, but it is best known as the Shannon-McMillan-Breiman theorem. It provides a common foundation to many of the results of both ergodic theory and information theory. Shannon [129] first developed the result for convergence in probability for stationary ergodic Markov sources. McMillan [103] proved L^1 convergence for stationary ergodic sources and Breiman [19] [20] proved almost everywhere convergence for stationary and ergodic sources. Billingsley [15] extended the result to stationary nonergodic sources. Jacobs [67] [66] extended it to processes dominated by a stationary measure and hence to two-sided AMS processes. Gray and Kieffer [54] extended it to processes asymptotically dominated by a stationary measure and hence to all AMS processes. The generalizations to AMS processes build on the Billingsley theorem for the stationary mean. Following generalizations of the definitions of entropy and information, corresponding generalizations of the entropy ergodic theorem will be considered in Chapter 8.

Breiman's and Billingsley's approach requires the martingale convergence theorem and embeds the possibly one-sided stationary process into a two-sided process. Ornstein and Weiss [117] recently developed a proof for the stationary and ergodic case that does not require any martingale theory and considers only positive time and hence does not require any embedding into two-sided processes. The technique was described for both the ordinary ergodic theorem and the entropy ergodic theorem by Shields [132]. In addition, it uses a form of coding argument that is both more direct and more information theoretic in flavor than the traditional martingale proofs. We here follow the Ornstein and Weiss approach for the stationary ergodic result. We also use some modifications similar to those of Katznelson and Weiss for the proof of the ergodic theorem. We then generalize the result first to nonergodic processes using the "sandwich" technique of Algoet and Cover [7] and then to AMS processes using a variation on a result of [54].

We next state the theorem to serve as a guide through the various steps. We also prove the result for the simple special case of a Markov source, for which the result follows from the usual ergodic theorem.

We consider a directly given finite alphabet source $\{X_n\}$ described by a distribution m on the sequence measurable space $(\Omega, \mathcal{B})$. Define as previously $X_k^n = (X_k, X_{k+1}, \cdots, X_{k+n-1})$. The subscript is omitted when it is zero. For any random variable Y defined on the sequence space (such as X_k^n) we define the random variable $m(Y)$ by $m(Y)(x) = m(Y = Y(x))$.

Theorem 3.1.1: The Entropy Ergodic Theorem

Given a finite alphabet AMS source $\{X_n\}$ with process distribution m and stationary mean $\bar{m}$, let $\{\bar{m}_x; x \in \Omega\}$ be the ergodic decomposition of the stationary mean $\bar{m}$. Then

$$\lim_{n\to\infty} \frac{-\ln m(X^n)}{n} = h;\ m - \text{a.e. and in } L^1(m), \tag{3.1.1}$$

where $h(x)$ is the invariant function defined by

$$h(x) = \bar{H}_{\bar{m}_x}(X). \tag{3.1.2}$$

Furthermore,

$$E_m h = \lim_{n\to\infty} \frac{1}{n} H_m(X^n) = \bar{H}_m(X); \tag{3.1.3}$$

that is, the entropy rate of an AMS process is given by the limit, and

$$\bar{H}_{\bar{m}}(X) = \bar{H}_m(X). \tag{3.1.4}$$

Comments: The theorem states that the sample entropy using the AMS measure m converges to the entropy rate of the underlying ergodic component of the stationary mean. Thus, for example, if m is itself stationary and ergodic, then the sample entropy converges to the entropy rate of the process m-a.e. and in $L^1(m)$. The $L^1(m)$ convergence follows immediately from the almost everywhere convergence and the fact that sample entropy is uniformly integrable (Lemma 2.3.6). L^1 convergence in turn immediately implies the left-hand equality of (3.1.3). Since the limit exists, it is the entropy rate. The final equality states that the entropy rates of an AMS process and its stationary mean are the same. This result follows from (3.1.2)-(3.1.3) by the following argument: We have that $\bar{H}_m(X) = E_m h$ and $\bar{H}_{\bar{m}}(X) = E_{\bar{m}} h$, but h is invariant and hence the two expectations are equal (see, e.g., Lemma 6.3.1 of [50]). Thus we need only prove almost everywhere convergence in (3.1.1) to prove the theorem.

In this section we limit ourselves to the following special case of the theorem that can be proved using the ordinary ergodic theorem without any new techniques.

Lemma 3.1.1: Given a finite alphabet stationary kth order Markov source $\{X_n\}$, then there is an invariant function h such that

$$\lim_{n\to\infty} \frac{-\ln m(X^n)}{n} = h;\ m-\text{a.e. and in } L^1(m),$$

where h is defined by

$$h(x) = -E_{\bar{m}_x} \ln m(X_k|X^k), \tag{3.1.5}$$

where $\{\bar{m}_x\}$ is the ergodic decomposition of the stationary mean $\bar{m}$. Furthermore,

$$h(x) = \bar{H}_{\bar{m}_x}(X) = H_{\bar{m}_x}(X_k|X^k). \tag{3.1.6}$$

Proof of Lemma: We have that

$$-\frac{1}{n}\ln m(X^n) = -\frac{1}{n}\sum_{i=0}^{n-1} \ln m(X_i|X^i).$$

Since the process is kth order Markov with stationary transition probabilites, for $i > k$ we have that

$$m(X_i|X^i) = m(X_i|X_{i-k},\cdots,X_{i-1}) = m(X_k|X^k)T^{i-k}.$$

The terms $-\ln m(X_i|X^i)$, $i = 0,1,\cdots,k-1$ have finite expectation and hence are finite m-a.e. so that the ergodic theorem can be applied to deduce

$$\frac{-\ln m(X^n)(x)}{n} = -\frac{1}{n}\sum_{i=0}^{k-1}\ln m(X_i|X^i)(x) - \frac{1}{n}\sum_{i=k}^{n-1}\ln m(X_k|X^k)(T^{i-k}x)$$

$$= -\frac{1}{n}\sum_{i=0}^{k-1}\ln m(X_i|X^i)(x) - \frac{1}{n}\sum_{i=0}^{n-k-1}\ln m(X_k|X^k)(T^i x)$$

$$\underset{n\to\infty}{\longrightarrow} E_{\bar{m}_x}(-\ln m(X_k|X^k)),$$

proving the first statement of the lemma. It follows from the ergodic decomposition of Markov sources (see Lemma 8.6.3) of [50]) that with probability 1, $\bar{m}_x(X_k|X^k) = m(X_k|\psi(x), X^k) = m(X_k|X^k)$, where ψ is the ergodic component function. This completes the proof. □

We prove the theorem in three steps: The first step considers stationary and ergodic sources and uses the approach of Ornstein and Weiss [117] (see also Shields [132]). The second step removes the requirement for ergodicity. This result will later be seen to provide an information theoretic interpretation of the ergodic decomposition. The third step extends the result to AMS processes by showing that such processes inherit limiting sample entropies from their stationary mean. The later extension of these results to more general relative entropy and information densities will closely parallel the proofs of the second and third steps for the finite case.

3.2 Stationary Ergodic Sources

This section is devoted to proving the entropy ergodic theorem for the special case of stationary ergodic sources. The result was originally proved by Breiman [19]. The original proof first used the martingale convergence theorem to infer the convergence of conditional probabilities of the form $m(X_0|X_{-1}, X_{-2}, \cdots, X_{-k})$ to $m(X_0|X_{-1}, X_{-2}, \cdots)$. This result was combined with an an extended form of the ergodic theorem stating that if $g_k \to g$ as $k \to \infty$ and if g_k is L^1-dominated ($\sup_k |g_k|$ is in L^1), then $1/n \sum_{k=0}^{n-1} g_k T^k$ has the same limit as $1/n \sum_{k=0}^{n-1} gT^k$. Combining these facts yields that that

$$\frac{1}{n}\ln m(X^n) = \frac{1}{n}\sum_{k=0}^{n-1}\ln m(X_k|X^k)$$

$$= \frac{1}{n}\sum_{k=0}^{n-1}\ln m(X_0|X_{-k}^k)T^k$$

has the same limit as

$$\frac{1}{n}\sum_{k=0}^{n-1}\ln m(X_0|X_{-1}, X_{-2}, \cdots)T^k$$

which, from the usual ergodic theorem, is the expectation

$$E(\ln m(X_0|\mathbf{X}^-) \equiv E(\ln m(X_0|X_{-1}, X_{-2}, \cdots)).$$

As suggested at the end of the preceeding chapter, this should be minus the conditional entropy $H(X_0|X_{-1}, X_{-2}, \cdots)$ which in turn should be the entropy rate $\bar{H}_X$. This approach has three shortcomings: it requires a result from martingale theory which has not been proved here or in the companion volume [50], it requires an extended ergodic theorem which has similarly not been proved here, and it requires a more advanced definition of entropy which has not yet been introduced. Another approach is the sandwich proof of Algoet and Cover [7]. They show without using martingale theory or the extended ergodic theorem that $1/n \sum_{i=0}^{n-1} \ln m(X_0|X_{-i}^{i})T^i$ is asymptotically sandwiched between the entropy rate of a kth order Markov approximation:

$$\frac{1}{n}\sum_{i=k}^{n-1} \ln m(X_0|X_{-k}^{k})T^i \to E_m[\ln m(X_0|X_{-k}^{k})] = -H(X_0|X_{-k}^{k})$$

and

$$\frac{1}{n}\sum_{i=k}^{n-1} \ln m(X_0|X_{-1}, X_{-2}, \cdots)T^i \to E_m[\ln m(X_0|X_1, \cdots)]$$

$$= -H(X_0|X_{-1}, X_{-2}, \cdots).$$

By showing that these two limits are arbitrarily close as $k \to \infty$, the result is proved. The drawback of this approach for present purposes is that again the more advanced notion of conditional entropy given the infinite past is required. Algoet and Cover's proof that the above two entropies are asymptotically close involves martingale theory, but this can be avoided by using Corollary 5.2.4 as will be seen.

The result can, however, be proved without martingale theory, the extended ergodic theorem, or advanced notions of entropy using the approach of Ornstein and Weiss [117], which is the approach we shall take in this chapter. In a later chapter when the entropy ergodic theorem is generalized to nonfinite alphabets and the convergence of entropy and information densities is proved, the sandwich approach will be used since the appropriate general definitions of entropy will have been developed and the necessary side results will have been proved.

Lemma 3.2.1: Given a finite alphabet source $\{X_n\}$ with a stationary ergodic distribution m, we have that

$$\lim_{n\to\infty} \frac{-\ln m(X^n)}{n} = h;\ m-\text{a.e.},$$

where $h(x)$ is the invariant function defined by

$$h(x) = \bar{H}_m(X).$$

Proof: Define

$$h_n(x) = -\ln m(X^n)(x) = -\ln m(x^n)$$

and

$$\underline{h}(x) = \liminf_{n\to\infty} \frac{1}{n} h_n(x) = \liminf_{n\to\infty} \frac{-\ln m(x^n)}{n}.$$

Since $m((x_0, \cdots, x_{n-1})) \leq m((x_1, \cdots, x_{n-1}))$, we have that

$$h_n(x) \geq h_{n-1}(Tx).$$

Dividing by n and taking the limit infimum of both sides shows that $\underline{h}(x) \geq \underline{h}(Tx)$. Since the $n^{-1}h_n$ are nonnegative and uniformly integrable (Lemma 2.3.6), we can use Fatou's lemma to deduce that $\underline{h}$ and hence also $\underline{h}T$ are integrable with respect to m. Integrating with respect to the stationary measure m yields

$$\int dm(x)\underline{h}(x) = \int dm(x)\underline{h}(Tx)$$

which can only be true if

$$\underline{h}(x) = \underline{h}(Tx); m-\text{a.e.},$$

that is, if $\underline{h}$ is an invariant function with m-probability one. If $\underline{h}$ is invariant almost everywhere, however, it must be a constant with probability one since m is ergodic (Lemma 6.7.1 of [50]). Since it has a finite integral (bounded by $\bar{H}_m(X)$), $\underline{h}$ must also be finite. Henceforth we consider $\underline{h}$ to be a finite constant.

We now proceed with steps that resemble those of the proof of the ergodic theorem in Section 7.2 of [50]. Fix $\epsilon > 0$. We also choose for later use a $\delta > 0$ small enough to have the following properties: If A is the alphabet of X_0 and $||A||$ is the finite cardinality of the alphabet, then

$$\delta \ln ||A|| < \epsilon, \tag{3.2.1}$$

and

$$-\delta\ln\delta - (1-\delta)\ln(1-\delta) \equiv h_2(\delta) < \epsilon. \tag{3.2.2}$$

The latter property is possible since $h_2(\delta) \to 0$ as $\delta \to 0$.

Define the random variable $n(x)$ to be the smallest integer n for which $n^{-1}h_n(x) \leq \underline{h}+\epsilon$. By definition of the limit infimum there must be infinitely many n for which this is true and hence $n(x)$ is everywhere finite. Define the set of "bad" sequences by $B = \{x : n(x) > N\}$ where N is chosen so large that $m(B) < \delta/2$. Still mimicking the proof of the ergodic theorem, we define a bounded modification of $n(x)$ by

$$\tilde{n}(x) = \begin{cases} n(x) & x \notin B \\ 1 & x \in B \end{cases}$$

so that $\tilde{n}(x) \leq N$ for all $x \in B^c$. We now parse the sequence into variable-length blocks. Iteratively define $n_k(x)$ by

$$n_0(x) = 0$$

$$n_1(x) = \tilde{n}(x)$$

$$n_2(x) = n_1(x) + \tilde{n}(T^{n_1(x)}x) = n_1(x) + l_1(x)$$

$$\vdots$$

$$n_{k+1}(x) = n_k(x) + \tilde{n}(T^{n_k(x)}x) = n_k(x) + l_k(x),$$

where $l_k(x)$ is the length of the kth block:

$$l_k(x) = \tilde{n}(T^{n_k(x)}x).$$

We have parsed a long sequence $x^L = (x_0, \cdots, x_{L-1})$, where $L >> N$, into blocks $x_{n_k(x)}, \cdots, x_{n_{k+1}(x)-1} = x_{n_k(x)}^{l_k(x)}$ which begin at time $n_k(x)$ and have length $l_k(x)$ for $k = 0, 1, \cdots$. We refer to this parsing as the *block decomposition* of a sequence. The kth block, which begins at time $n_k(x)$, must either have sample entropy satisfying

$$\frac{-\ln m(x_{n_k(x)}^{l_k(x)})}{l_k(x)} \leq \underline{h} + \epsilon \tag{3.2.3}$$

or, equivalently, probability at least

$$m(x_{n_k(x)}^{l_k(x)}) \geq e^{-l_k(x)(\underline{h}+\epsilon)}, \tag{3.2.4}$$

or it must consist of only a single symbol. Blocks having length 1 ($l_k = 1$) could have the correct sample entropy, that is,

$$\frac{-\ln m(x_{n_k(x)}^1)}{1} \leq \bar{h} + \epsilon,$$

or they could be bad in the sense that they are the first symbol of a sequence with $n > N$; that is,

$$n(T^{n_k(x)}x) > N,$$

or, equivalently,

$$T^{n_k(x)}x \in B.$$

Except for these bad symbols, each of the blocks by construction will have a probability which satisfies the above bound.

Define for nonnegative integers n and positive integers l the sets

$$S(n,l) = \{x : m(X_n^l(x)) \geq e^{-l(\underline{h}+\epsilon)}\},$$

that is, the collection of infinite sequences for which (3.2.2) and (3.2.3) hold for a block starting at n and having length l. Observe that for such blocks there cannot be more than $e^{l(\underline{h}+\epsilon)}$ distinct l-tuples for which the bound holds (lest the probabilities sum to something greater than 1). In symbols this is

$$||S(n,l)|| \leq e^{l(\underline{h}+\epsilon)}. \tag{3.2.5}$$

The ergodic theorem will imply that there cannot be too many single symbol blocks with $n(T^{n_k(x)}x) > N$ because the event has small probability. These facts will be essential to the proof.

Even though we write $\tilde{n}(x)$ as a function of the entire infinite sequence, we can determine its value by observing only the prefix x^N of x since either there is an $n \leq N$ for which $n^{-1}\ln m(x^n) \leq \underline{h} + \epsilon$ or there is not. Hence there is a function $\hat{n}(x^N)$ such that $\tilde{n}(x) = \hat{n}(x^N)$. Define the finite length sequence event $C = \{x^N : \hat{n}(x^N) = 1 \text{ and } -\ln m(x^1) > \underline{h}+\epsilon\}$, that is, C is the collection of all N-tuples x^N that are prefixes of bad infinite sequences, sequences x for which $n(x) > N$. Thus in particular,

$$x \in B \text{ if and only if } x^N \in C. \tag{3.2.6}$$

Now recall that we parse sequences of length $L >> N$ and define the set G_L of "good" L-tuples by

$$G_L = \{x^L : \frac{1}{L-N}\sum_{i=0}^{L-N-1} 1_C(x_i^N) \leq \delta\},$$

that is, G_L is the collection of all L-tuples which have fewer than $\delta(L-N) \leq \delta L$ time slots i for which x_i^N is a prefix of a bad infinite sequence. From (3.2.6) and the ergodic theorem for stationary ergodic sources we know that m-a.e. we get an x for which

$$\lim_{n\to\infty}\frac{1}{n}\sum_{i=0}^{n-1} 1_C(x_i^N) = \lim_{n\to\infty}\frac{1}{n}\sum_{i=0}^{n-1} 1_B(T^i x) = m(B) \leq \frac{\delta}{2}. \tag{3.2.7}$$

From the definition of a limit, this means that with probability 1 we get an x for which there is an $L_0 = L_0(x)$ such that

$$\frac{1}{L-N}\sum_{i=0}^{L-N-1} 1_C(x_i^N) \leq \delta; \text{ for all } L > L_0. \tag{3.2.8}$$

This follows simply because if the limit is less than $\delta/2$, there must be an L_0 so large that for larger L the time average is at least no greater than $2\delta/2 = \delta$. We can restate (3.2.8) as follows: with probability 1 we get an x for which $x^L \in G_L$ for all but a finite number of L. Stating this in negative fashion, we have one of the key properties required by the proof: If $x^L \in G_L$ for all but a finite number of L, then x^L cannot be in the complement G_L^c infinitely often, that is,

$$m(x : x^L \in G_L^c \text{ i.o.}) = 0. \tag{3.2.9}$$

We now change tack to develop another key result for the proof. For each L we bounded above the cardinality $||G_L||$ of the set of good L-tuples. By construction there are no more than δL bad symbols in an L-tuple in G_L and these can occur in any of at most

$$\sum_{k \leq \delta L} \binom{L}{k} \leq e^{h_2(\delta)L} \tag{3.2.10}$$

places, where we have used Lemma 2.3.5. Eq. (3.2.10) provides an upper bound on the number of ways that a sequence in G_L can be parsed by the given rules. The bad symbols and the final N symbols in the L-tuple can take on any of the $||A||$ different values in the alphabet. Eq. (3.2.5) bounds the number of finite length sequences that can occur in each of the remaining blocks and hence for any given block decomposition, the number of ways that the remaining blocks blocks can be filled is bounded above by

$$\prod_{k:T^{n_k(x)}x \notin B} e^{l_k(x)(\underline{h}+\epsilon)} = e^{\sum_k l_k(x)(\underline{h}+\epsilon)} = e^{L(\underline{h}+\epsilon)}, \tag{3.2.11}$$

regardless of the details of the parsing. Combining these bounds we have that

$$||G_L|| \leq e^{h_2(\delta)L} \times ||A||^{\delta L} \times ||A||^N \times e^{L(\underline{h}+\epsilon)} = e^{h_2(\delta)L+(\delta L+N)\ln||A||+L(\underline{h}+\epsilon)}$$

or

$$||G_L|| \leq e^{L(\underline{h}+\epsilon+h_2(\delta)+(\delta+\frac{N}{L})\ln||A||)}.$$

Since δ satisfies (3.2.1)–(3.2.2), we can choose L_1 large enough so that $N \ln ||A||/L_1 \leq \epsilon$ and thereby obtain

$$||G_L|| \leq e^{L(\underline{h}+4\epsilon)};\ L \geq L_1. \tag{3.2.12}$$

This bound provides the second key result in the proof of the lemma. We now combine (3.2.12) and (3.2.9) to complete the proof.

Let B_L denote a collection of L-tuples that are bad in the sense of having too large a sample entropy or, equivalently, too small a probability; that is if $x^L \in B_L$, then

$$m(x^L) \leq e^{-L(\underline{h}+5\epsilon)}$$

or, equivalently, for any x with prefix x^L

$$h_L(x) \geq \underline{h} + 5\epsilon.$$

The upper bound on $||G_L||$ provides a bound on the probability of $B_L \bigcap G_L$:

$$m(B_L \bigcap G_L) = \sum_{x^L \in B_L \bigcap G_L} m(x^L) \leq \sum_{x^L \in G_L} e^{-L(\underline{h}+5\epsilon)}$$

$$\leq ||G_L|| e^{-L(\underline{h}+5\epsilon)} \leq e^{-\epsilon L}.$$

Recall now that the above bound is true for a fixed $\epsilon > 0$ and for all $L \geq L_1$. Thus

$$\sum_{L=1}^{\infty} m(B_L \bigcap G_L) = \sum_{L=1}^{L_1-1} m(B_L \bigcap G_L) + \sum_{L=L_1}^{\infty} m(B_L \bigcap G_L)$$

$$\leq L_1 + \sum_{L=L_1}^{\infty} e^{-\epsilon L} < \infty$$

and hence from the Borel-Cantelli lemma (Lemma 4.6.3 of [50]) $m(x : x^L \in B_L \bigcap G_L \text{ i.o.}) = 0$. We also have from (3.2.8), however, that $m(x : x^L \in G_L^c \text{ i.o. }) = 0$ and hence $x^L \in G_L$ for all but a finite number of L. Thus $x^L \in B_L$ i.o. if and only if $x^L \in B_L \bigcap G_L$ i.o. As this latter event has zero probability, we have shown that $m(x : x^L \in B_L \text{ i.o.}) = 0$ and hence

$$\limsup_{L\to\infty} h_L(x) \leq \underline{h} + 5\epsilon.$$

Since ϵ is arbitrary we have proved that the limit supremum of the sample entropy $-n^{-1} \ln m(X^n)$ is less than or equal to the limit infimum and therefore that the limit exists and hence with m-probability 1

$$\lim_{n\to\infty} \frac{-\ln m(X^n)}{n} = \underline{h}. \tag{3.2.13}$$

Since the terms on the left in (3.2.13) are uniformly integrable from Lemma 2.3.6, we can integrate to the limit and apply Lemma 2.4.1 to find that

$$\underline{h} = \lim_{n\to\infty} \int dm(x) \frac{-\ln m(X^n(x))}{n} = \bar{H}_m(X),$$

which completes the proof of the lemma and hence also proves Theorem 3.1.1 for the special case of stationary ergodic measures. □

3.3 Stationary Nonergodic Sources

Next suppose that a source is stationary with ergodic decomposition $\{m_\lambda;\ \lambda \in \Lambda\}$ and ergodic component function ψ as in Theorem 1.8.3. The source will produce with probability one under m an ergodic component m_λ and Lemma 3.2.2 will hold for this ergodic component. In other words, we should have that

$$\lim_{n\to\infty} -\frac{1}{n} \ln m_\psi(X^n) = \bar{H}_{m_\psi}(X);\ m-\text{a.e.},$$

that is,

$$m(\{x : -\lim_{n\to\infty} \ln m_{\psi(x)}(x^n) = \bar{H}_{m_{\psi(x)}}(X)\}) = 1.$$

This argument is made rigorous in the following lemma.

Lemma 3.3.1: Suppose that $\{X_n\}$ is a stationary not necessarily ergodic source with ergodic component function ψ. Then

$$m(\{x : -\lim_{n\to\infty} \ln m_{\psi(x)}(x^n) = \bar{H}_{m_{\psi(x)}}(X)\}) = 1;\ m-\text{a.e.}. \tag{3.3.2}$$

Proof: Let

$$G = \{x : -\lim_{n\to\infty} \ln m_{\psi(x)}(x^n) = \bar{H}_{m_{\psi(x)}}(X)\}$$

and let G_λ denote the section of G at λ, that is,

$$G_\lambda = \{x : -\lim_{n\to\infty} \ln m_\lambda(x^n) = \bar{H}_{m_\lambda}(X)\}.$$

From the ergodic decomposition (e.g., Theorem 1.8.3 or [50], Theorem 8.5.1) and (1.6.8)

$$m(G) = \int dP_\psi(\lambda) m_\lambda(G),$$

where

$$m_\lambda(G) = m(G|\psi = \lambda) = m(G \bigcap \{x : \psi(x) = \lambda\} | \psi = \lambda)$$

$$= m(G_\lambda | \psi = \lambda) = m_\lambda(G_\lambda)$$

which is 1 for all λ from the stationary ergodic result. Thus

$$m(G) = \int dP_\psi(\lambda) m_\lambda(G_\lambda) = 1.$$

It is straightforward to verify that all of the sets considered are in fact measurable. □

Unfortunately it is not the sample entropy using the distribution of the ergodic component that is of interest, rather it is the original sample entropy for which we wish to prove convergence. The following lemma shows that the two sample entropies converge to the same limit and hence Lemma 3.3.1 will also provide the limit of the sample entropy with respect to the stationary measure.

Lemma 3.3.2: Given a stationary source $\{X_n\}$, let $\{m_\lambda;\ \lambda \in \Lambda\}$ denote the ergodic decomposition and ψ the ergodic component function of Theorem 1.8.3. Then

$$\lim_{n\to\infty} \frac{1}{n} \ln \frac{m_\psi(X^n)}{m(X^n)} = 0;\ m - \text{a.e.}$$

Proof: First observe that if $m(a^n)$ is 0, then from the ergodic decomposition with probability 1 $m_\psi(a^n)$ will also be 0. One part is easy. For any $\epsilon > 0$ we have from the Markov inequality that

$$m(\frac{1}{n} \ln \frac{m(X^n)}{m_\psi(X^n)} > \epsilon) = m(\frac{m(X^n)}{m_\psi(X^n)} > e^{n\epsilon})$$

$$\leq E_m(\frac{m(X^n)}{m_\psi(X^n)})e^{-n\epsilon}.$$

The expectation, however, can be evaluated as follows: Let $A_n^{(\lambda)} = \{a^n : m_\lambda(a^n) > 0\}$. Then

$$E_m\left(\frac{m(X^n)}{m_\psi(X^n)}\right) = \int dP_\psi(\lambda) \sum_{a^n \in A_n} \frac{m(a^n)}{m_\lambda(a^n)} m_\lambda(a^n)$$

$$= \int dP_\psi(\lambda) m(A_n^{(\lambda)}) \leq 1,$$

where P_ψ is the distribution of ψ. Thus

$$m(\frac{1}{n} \ln \frac{m(X^n)}{m_\psi(X^n)} > \epsilon) \leq e^{-n\epsilon}.$$

and hence

$$\sum_{n=1}^{\infty} m(\frac{1}{n} \ln \frac{m(X^n)}{m_\psi(X^n)} > \epsilon) < \infty$$

and hence from the Borel-Cantelli lemma

$$m(\frac{1}{n} \ln \frac{m(X^n)}{m_\psi(X^n)} > \epsilon \text{ i.o.}) = 0$$

and hence with m probability 1

$$\limsup_{n\to\infty} \frac{1}{n} \ln \frac{m(X^n)}{m_\psi(X^n)} \le \epsilon.$$

Since ϵ is arbitrary,

$$\limsup_{n\to\infty} \frac{1}{n} \ln \frac{m(X^n)}{m_\psi(X^n)} \le 0; \; m-\text{a.e.}. \tag{3.3.3}$$

For later use we restate this as

$$\liminf_{n\to\infty} \frac{1}{n} \ln \frac{m_\psi(X^n)}{m(X^n)} \ge 0; \; m-\text{a.e.}. \tag{3.3.4}$$

We now turn to the converse inequality. For any positive integer k, we can construct a stationary k-step Markov approximation to m as in Section 2.6, that is, construct a process $m^{(k)}$ with the conditional probabilities

$$m^{(k)}(X_n \in F|X^n) = m^{(k)}(X_n \in F|X^k_{n-k})$$

$$= m(X_n \in F|X^k_{n-k})$$

and the same kth order distributions $m^{(k)}(X^k \in F) = m(X^k \in F)$. Consider the probability

$$m(\frac{1}{n} \ln \frac{m^{(k)}(X^n)}{m(X^n)} \ge \epsilon) = m(\frac{m^{(k)}(X^n)}{m(X^n)} \ge e^{n\epsilon})$$

$$\le E_m(\frac{m^{(k)}(X^n)}{m(X^n)})e^{-n\epsilon}.$$

The expectation is evaluated as

$$\sum_{x^n} \frac{m^{(k)}(x^n)}{m(x^n)} m(x^n) = 1$$

and hence we again have using Borel-Cantelli that

$$\limsup_{n\to\infty} \frac{1}{n} \ln \frac{m^{(k)}(X^n)}{m(X^n)} \leq 0.$$

We can apply the usual ergodic theorem to conclude that with probability 1 under m

$$\limsup_{n\to\infty} \frac{1}{n} \ln \frac{1}{m(X^n)} \leq \lim_{n\to\infty} \frac{1}{n} \ln \frac{1}{m^{(k)}(X^n)} = E_{m_\psi}[-\ln m(X_k|X^k)].$$

Combining this result with (3.3.1) we have using Lemma 2.4.3 that

$$\limsup_{n\to\infty} \frac{1}{n} \ln \frac{m_\psi(X^n)}{m(X^n)} \leq -\bar{H}_{m_\psi}(X) - E_{m_\psi}[\ln m(X_k|X^k)].$$

$$= \bar{H}_{m_\psi||m^{(k)}}(X). \qquad (3.3.5)$$

This bound holds for any integer k and hence it must also be true that m-a.e. the following holds:

$$\limsup_{n\to\infty} \frac{1}{n} \ln \frac{m_\psi(X^n)}{m(X^n)} \leq \inf_k \bar{H}_{m_\psi||m^{(k)}}(X) \equiv \zeta. \qquad (3.3.6)$$

In order to evaluate ζ we apply the ergodic decomposition of relative entropy rate (Corollary 2.4.2) and the ordinary ergodic decomposition to write

$$\int dP_\psi \zeta = \int dP_\psi \inf_k \bar{H}_{m_\psi||m^{(k)}}(X)$$

$$\leq \inf_k \int dP_\psi \bar{H}_{m_\psi||m^{(k)}}(X) = \inf_k \bar{H}_{m||m^{(k)}}(X).$$

From Theorem 2.6.2, the right hand term is 0. If the integral of a nonnegative function is 0, the integrand must itself be 0 with probability one. Thus (3.3.6) becomes

$$\limsup_{n\to\infty} \frac{1}{n} \ln \frac{m_\psi(X^n)}{m(X^n)} \leq 0,$$

which with (3.3.4) completes the proof of the lemma. □

We shall later see that the quantity

$$i_n(X^n; \psi) = \frac{1}{n} \ln \frac{m_\psi(X^n)}{m(X^n)}$$

is the sample mutual information (in a generalized sense so that it applies to the usually non-discrete ψ) and hence the lemma states that the normalized sample mutual information between the process outputs and the ergodic component function goes to 0 as the number of samples goes to infinity.

The two previous lemmas immediately yield the following result.

Corollary 3.3.1: The conclusions of Theorem 3.1.1 hold for sources that are stationary.

3.4 AMS Sources

The principal idea required to extend the entropy theorem from stationary sources to AMS sources is contained in Lemma 3.4.2. It shows that an AMS source inherits sample entropy properties from an asymptotically dominating stationary source (just as it inherits ordinary ergodic properties from such a source). The result is originally due to Gray and Kieffer [54], but the proof here is somewhat different. The tough part here is handling the fact that the sample average being considered depends on a specific measure. From Theorem 1.7.1, the stationary mean of an AMS source dominates the original source on tail events, that is, events in $\mathcal{F}_\infty$. We begin by showing that certain important events can be recast as tail events, that is, they can be determined by looking at only samples in the arbitrarily distant future. The following result is of this variety: It implies that sample entropy is unaffected by the starting time.

Lemma 3.4.1: Let $\{X_n\}$ be a finite alphabet source with distribution m. Recall that $X_k^n = (X_k, X_{k+1}, \cdots, X_{k+n-1})$ and define the *information density*

$$i(X^k; X_k^{n-k}) = \ln \frac{m(X^n)}{m(X^k)m(X_k^{n-k})}.$$

Then

$$\lim_{n\to\infty} \frac{1}{n} i(X^k; X_k^{n-k}) = 0;\ m-\text{a.e..}$$

Comment: The lemma states that with probability 1 the per-sample mutual information density between the first k samples and future samples goes to zero in the limit. Equivalently, limits of $n^{-1}\ln m(X^n)$ will be the same as limits of $n^{-1}\ln m(X_k^{n-k})$ for any finite k. Note that the result does not require even that the source be AMS. The lemma is a direct consequence of Lemma 2.7.1.

Proof: Define the distribution $p = m_{X^k} \times m_{X_k, X_{k+1}, \cdots}$, that is, a distribution for which all samples after the first k are independent of the first k samples. Thus, in particular, $p(X^n) = m(X^k)m(X_k^n)$. We will show that $p >> m$, in which case the lemma will follow from Lemma 2.7.1. Suppose that $p(F) = 0$. If we denote $X_k^+ = X_k, X_{k+1}, \cdots$, then

$$0 = p(F) = \sum_{x^k} m(x^k) m_{X_k^+}(F_{x^k}),$$

where F_{x^k} is the section $\{x_k^+ : (x^k, x_k^+) = x \in F\}$. For the above relation to hold, we must have $m_{X_k^+}(F_{x^k}) = 0$ for all x^k with $m(x^k) \neq 0$. We also have, however, that

$$m(F) = \sum_{a^k} m(X^k = a^k, X_k^+ \in F_{a^k})$$

$$= \sum_{a^k} m(X^k = a^k | X_k^+ \in F_{a^k}) m(X_k^+ \in F_{a^k}).$$

But this sum must be 0 since the rightmost terms are 0 for all a^k for which $m(X^k = a^k)$ is not 0. (Observe that we must have $m(X^k = a^k | X_k^+ \in F_{a^k})$ $= 0$ if $m(X_k^+ \in F_{a^k}) \neq 0$ since otherwise $m(X^k = a^k) \geq m(X^k = a^k, X_k^+ \in F_{a^k}) > 0$, yielding a contradiction.) Thus $p >> m$ and the lemma is proved. □

For later use we note that we have shown that a joint distribution is dominated by a product of its marginals if one of the marginal distributions is discrete.

Lemma 3.4.2: Suppose that $\{X_n\}$ is an AMS source with distribution m and suppose that $\bar{m}$ is a stationary source that asymptotically dominates m (e.g., $\bar{m}$ is the stationary mean). If there is an invariant function h such that

$$\lim_{n\to\infty} -\frac{1}{n} \ln \bar{m}(X^n) = h;\ \bar{m} - \text{a.e.},$$

then also,

$$\lim_{n\to\infty} -\frac{1}{n} \ln m(X^n) = h;\ m - \text{a.e.}$$

Proof: For any k we can write using the chain rule for densities

$$-\frac{1}{n} \ln m(X^n) + \frac{1}{n} \ln m(X_k^{n-k}) = -\frac{1}{n} \ln m(X^k | X_k^{n-k})$$

$$= -\frac{1}{n} i(X^k; X_k^{n-k}) - \frac{1}{n} \ln m(X^k).$$

From the previous lemma and from the fact that $H_m(X^k) = -E_m \ln m(X^k)$ is finite, the right hand terms converge to 0 as $n \to \infty$ and hence for any k

$$\lim_{n\to\infty} -\frac{1}{n} \ln m(X^k | X_k^{n-k})$$

$$= \lim_{n\to\infty} (-\frac{1}{n} \ln m(X^n) + \frac{1}{n} \ln m(X_k^{n-k})) = 0; \ m - \text{a.e.}. \tag{3.4.1}$$

This implies that there is a subsequence $k(n) \to \infty$ such that

$$-\frac{1}{n} \ln m(X^{k(n)} | X_{k(n)}^{n-k(n)})$$

$$= -\frac{1}{n} \ln m(X^n) - \frac{1}{n} \ln m(X_{k(n)}^{n-k}(n)) \to 0; \ m - \text{a.e.}. \tag{3.4.2}$$

To see this, observe that (3.4.1) ensures that for each k there is an $N(k)$ large enough so that $N(k) > N(k-1)$ and

$$m(| - \frac{1}{N(k)} \ln m(X^k | X_k^{N(k)-k})| > 2^{-k}) \leq 2^{-k}. \tag{3.4.3}$$

Applying the Borel-Cantelli lemma implies that for any ϵ,

$$m(| - 1/N(k) \ln m(X^k | X_k^{N(k)-k})| > \epsilon \text{ i.o.}) = 0.$$

Now let $k(n) = k$ for $N(k) \leq n < N(k+1)$. Then

$$m(| - 1/n \ln m(X^{k(n)} | X_{k(n)}^{n-k(n)})| > \epsilon \text{ i.o.}) = 0$$

and therefore

$$\lim_{n\to\infty} \left(-\frac{1}{n} \ln m(X^n) + \frac{1}{n} \ln m(X_{k(n)}^{n-k(n)}) \right) = 0; \ m - \text{a.e.}$$

as claimed in (3.4.2).

In a similar manner we can also choose the sequence so that

$$\lim_{n\to\infty} \left(-\frac{1}{n} \ln \bar{m}(X^n) + \frac{1}{n} \ln \bar{m}(X_{k(n)}^{n-k(n)}) \right) = 0; \ \bar{m} - \text{a.e.},$$

that is, we can choose $N(k)$ so that (3.4.3) simultaneously holds for both m and $\bar{m}$. Invoking the entropy ergodic theorem for the stationary $\bar{m}$ (Corollary 3.3.1) we have therefore that

$$\lim_{n\to\infty} -\frac{1}{n} \ln \bar{m}(X_{k(n)}^{n-k(n)}) = \bar{h}; \ \bar{m} - \text{a.e.}. \tag{3.4.4}$$

From Markov's inequality (Lemma 4.4.3 of [50])

$$\bar{m}(-\frac{1}{n} \ln m(X_k^n) \leq -\frac{1}{n} \ln \bar{m}(X_k^n) - \epsilon) = \bar{m}(\frac{m(X_k^n)}{\bar{m}(X_k^n)} \geq e^{n\epsilon})$$

$$\leq e^{-n\epsilon} E_{\bar{m}} \frac{m(X_k^{n-k})}{\bar{m}(X_k^{n-k})} = e^{-n\epsilon} \sum_{x_k^{n-k}: \bar{m}(x_k^{n-k}) \neq 0} \frac{m(x_k^{n-k})}{\bar{m}(x_k^{n-k})} \bar{m}(x_k^{n-k}) \leq e^{-n\epsilon}.$$

Hence taking $k = k(n)$ and again invoking the Borel-Cantelli lemma we have that

$$\bar{m}(-\frac{1}{n} \ln m(X_{k(n)}^{n-k(n)}) \leq -\frac{1}{n} \ln \bar{m}(X_{k(n)}^{n-k(n)}) - \epsilon \text{ i.o.}) = 0$$

or, equivalently, that

$$\liminf_{n\to\infty} -\frac{1}{n} \ln \frac{m(X_{k(n)}^{n-k(n)})}{\bar{m}(X_{k(n)}^{n-k(n)})} \geq 0; \ \bar{m} - \text{a.e..} \tag{3.4.5}$$

Therefore from (3.4.4)

$$\liminf_{n\to\infty} -\frac{1}{n} \ln m(X_{k(n)}^{n-k(n)}) \geq h; \ \bar{m} - \text{a.e..} \tag{3.4.6}$$

The above event is in the tail σ-field $\mathcal{F}_\infty = \bigcap_n \sigma(X_n, X_{n+1}, \cdots)$ since it can be determined from $X_{k(n)}, \cdots$ for arbitrarily large n and since h is invariant. Since $\bar{m}$ dominates m on the tail σ-field (Theorem 1.7.2), we have also

$$\liminf_{n\to\infty} -\frac{1}{n} \ln m(X_{k(n)}^{n-k(n)}) \geq h; \ m - \text{a.e.}$$

and hence by (3.4.2)

$$\liminf_{n\to\infty} -\frac{1}{n} \ln m(X^n) \geq h; \ m - \text{a.e.}$$

which proves half of the lemma.

Since

$$\bar{m}(\lim_{n\to\infty} -\frac{1}{n} \ln \bar{m}(X^n) \neq h) = 0$$

and since $\bar{m}$ asymptotically dominates m (Theorem 1.7.1), given $\epsilon > 0$ there is a k such that

$$m(\lim_{n\to\infty} -\frac{1}{n} \ln \bar{m}(X_k^n) = h) \geq 1 - \epsilon.$$

Again applying Markov's inequality and the Borel-Cantelli lemma as in the development of (3.4.4) we have that

$$\liminf_{n\to\infty} -\frac{1}{n} \ln \frac{\bar{m}(X_k^n)}{m(X_k^n)} \geq 0; \ m - \text{a.e,}$$

which implies that

$$m(\limsup_{n\to\infty} -\frac{1}{n}\ln m(X_k^n) \leq h) \geq 1-\epsilon$$

and hence also that

$$m(\limsup_{n\to\infty} -\frac{1}{n}\ln m(X^n) \leq h) \geq 1-\epsilon.$$

Since ϵ can be made arbitrarily small, this proves that m-a.e.

$$\limsup -n^{-1}\ln m(X^n) \leq h,$$

which completes the proof of the lemma. □

The lemma combined with Corollary 3.3.1 completes the proof of Theorem 3.1.1. □

3.5 The Asymptotic Equipartition Property

Since convergence almost everywhere implies convergence in probability, Theorem 3.1.2 has the following implication: Suppose that $\{X_n\}$ is an AMS ergodic source with entropy rate $\bar{H}$. Given $\epsilon > 0$ there is an N such that for all $n > N$ the set

$$\begin{aligned} G_n &= \{x^n : |n^{-1}h_n(x) - \bar{H}| \geq \epsilon\} \\ &= \{x^n : e^{-n(\bar{H}+\epsilon)} \leq m(x^n) \leq e^{-n(\bar{H}-\epsilon)}\} \end{aligned}$$

has probability greater then $1-\epsilon$. Furthermore, as in the proof of the theorem, there can be no more than $e^{n(\bar{H}+\epsilon)}$ n-tuples in G_n. Thus there are two sets of n-tuples: a "good" set of approximately $e^{n\bar{H}}$ n-tuples having approximately equal probability of $e^{-n\bar{H}}$ and the complement of this set which has small total probability. The set of good sequences are often referred to as "typical sequences" in the information theory literature and in this form the theorem is called the asymptotic equipartition property or the AEP.

As a first information theoretic application of an ergodic theorem, we consider a simple coding scheme called an "almost noiseless source code." As we often do, we consider logarithms to the base 2 when considering specific coding applications. Suppose that a random process $\{X_n\}$ has a finite alphabet A with cardinality $||A||$ and entropy rate $\bar{H}$. Suppose that $\bar{H} < \log ||A||$, e.g., A might have 16 symbols, but the entropy rate is slightly less than 2 bits per symbol rather than $\log 16 = 4$. Larger alphabets cost money in either storage or communication applications. For example, to communicate a source with a 16 letter alphabet sending one letter per second without using any coding and using a binary communication system we would need to send 4 binary symbols (or four *bits*) for each source letter and hence 4 bits per second would be required. If the alphabet only had 4 letters, we would need to send only 2 bits per second. The question is the following: Since our source has an alphabet of size 16 but an entropy rate of less than 2, can we code the original source into a new source with an alphabet of only 4 letters so as to communicate the source at the smaller rate and yet have the receiver be able to recover the original source? The AEP suggests a technique for accomplishing this provided we are willing to tolerate occasional errors.

We construct a code of the original source by first picking a small ϵ and a δ small enough so that $\bar{H}+\delta < 2$. Choose a large enough n so that the AEP holds giving a set G_n of good sequences as above with probability greater than $1-\epsilon$. Index this collection of fewer than $2^{n(\bar{H}+\delta)} < 2^{2n}$ sequences using binary $2n$-tuples. The source X_k is parsed into blocks of length n as $X_{kn}^n = (X_{kn}, X_{kn+1}, \cdots, X_{(k+1)n})$ and each block is encoded into a binary $2n$-tuple as follows: If the source n-tuple is in G_n, the codeword is its binary $2n$-tuple index. Select one of the unused binary $2n$-tuples as the error index and whenever an n-tuple is not in G_n, the error index is the codeword. The receiver or decoder than uses the received index and decodes it as the appropriate n-tuple in G_n. If the error index is received, the decoder can declare an arbitrary source sequence or just declare an error. With probability at least $1-\epsilon$ a source n-tuple at a particular time will be in G_n and hence it will be correctly decoded. We can make this probability as small as desired by taking n large enough, but we cannot in general make it 0.

The above simple scheme is an example of a block coding scheme. If considered as a mapping from sequences into sequences, the map is not stationary, but it is block stationary in the sense that shifting an input block by n results in a corresponding block shift of the encoded sequence by $2n$ binary symbols.

Chapter 4

Information Rates I

4.1 Introduction

Before proceeding to generalizations of the various measures of information, entropy, and divergence to nondiscrete alphabets, we consider several properties of information and entropy rates of finite alphabet processes. We show that codes that produce similar outputs with high probability yield similar rates and that entropy and information rate, like ordinary entropy and information, are reduced by coding. The discussion introduces a basic tool of ergodic theory–the partition distance–and develops several versions of an early and fundamental result from information theory–Fano's inequality. We obtain an ergodic theorem for information densities of finite alphabet processes as a simple application of the general Shannon-McMillan-Breiman theorem coupled with some definitions. In Chapter 6 these results easily provide L^1 ergodic theorems for information densities for more general processes.

4.2 Stationary Codes and Approximation

We consider the behavior of entropy when codes or measurements are taken on the underlying random variables. We have seen that entropy is a continuous function with respect to the underlying measure. We now wish to fix the measure and show that entropy is a continuous function with respect to the underlying measurement.

Say we have two finite alphabet measurements f and g on a common probability space having a common alphabet A. Suppose that $\mathcal{Q}$ and $\mathcal{R}$ are the corresponding partitions. A common metric or distance measure on partitions in ergodic theory is

$$|\mathcal{Q} - \mathcal{R}| = \frac{1}{2} \sum_i P(Q_i \Delta R_i), \tag{4.2.1}$$

which in terms of the measurements (assuming they have distinct values on distinct atoms) is just $\Pr(f \neq g)$. If we consider f and g as two codes on a common space, random variable, or random process (that is, finite alphabet mappings), then the partition distance can also be considered as a form of distance between the codes. The following lemma shows that entropy of partitions or measurements is continuous with respect to this distance. The result is originally due to Fano and is called Fano's inequality [37].

Lemma 4.2.1: Given two finite alphabet measurements f and g on a common probability space $(\Omega, \mathcal{B}, P)$ having a common alphabet A or, equivalently, the given corresponding partitions $\mathcal{Q} = \{f^{-1}(a); a \in A\}$ and $\mathcal{R} = \{g^{-1}(a); a \in A\}$, define the error probability $P_e = |\mathcal{Q} - \mathcal{R}| = \Pr(f \neq g)$. Then

$$H(f|g) \leq h_2(P_e) + P_e \ln(||A|| - 1)$$

and

$$|H(f) - H(g)| \leq h_2(P_e) + P_e \ln(M - 1)$$

and hence entropy is continuous with respect to partition distance for a fixed measure.

Proof: Let $M = ||A||$ and define a measurement

$$r : A \times A \to \{0, 1, \cdots, M - 1\}$$

by $r(a, b) = 0$ if $a = b$ and $r(a, b) = i$ if $a \neq b$ and a is the ith letter in the alphabet $A_b = A - b$. If we know g and we know $r(f, g)$, then clearly we know f since either $f = g$ (if $r(f, g)$ is 0) or, if not, it is equal to the $r(f, g)$th letter in the alphabet A with g removed. Since f can be considered a function of g and $r(f, g)$,

$$H(f|g, r(f, g)) = 0$$

and hence

$$H(f, g, r(f, g)) = H(f|g, r(f, g)) + H(g, r(f, g)) = H(g, r(f, g)).$$

Similarly

$$H(f, g, r(f, g)) = H(f, g).$$

From Lemma 2.3.2

$$H(f, g) = H(g, r(f, g)) \leq H(g) + H(r(f, g))$$

or

$$H(f,g) - H(g) = H(f|g) \leq H(r(f,g))$$
$$= -P(r=0)\ln P(r=0) - \sum_{i=1}^{M-1} P(r=i)\ln P(r=i).$$

Since $P(r=0) = 1 - P_e$ and since $\sum_{i\neq 0} P(r=i) = P_e$, this becomes

$$H(f|g) \leq -(1-P_e)\ln(1-P_e) - P_e \sum_{i=1}^{M-1} \frac{P(r=i)}{P_e} \ln \frac{P(r=i)}{P_e} - P_e \ln P_e$$
$$\leq h_2(P_e) + P_e \ln(M-1)$$

since the entropy of a random variable with an alphabet of size $M-1$ is no greater than $\ln(M-1)$. This proves the first inequality. Since $H(f) \leq H(f,g) = H(f|g) + H(g)$, this implies

$$H(f) - H(g) \leq h_2(P_e) + P_e \ln(M-1).$$

Interchanging the roles of f and g completes the proof. □

The lemma can be used to show that related information measures such as mutual information and conditional mutual information are also continuous with respect to the partition metric. The following corollary provides useful extensions. Similar extensions may be found in Csiszár and Körner [26].

Corollary 4.2.1: Given two sequences of measurements $\{f_n\}$ and $\{g_n\}$ with finite alphabet A on a common probability space, define

$$P_e^{(n)} = \frac{1}{n}\sum_{i=0}^{n-1} \Pr(f_i \neq g_i).$$

Then

$$\frac{1}{n}H(f^n|g^n) \leq P_e^{(n)} \ln(||A||-1) + h_2(P_e^{(n)})$$

and

$$|\frac{1}{n}H(f^n) - \frac{1}{n}H(g^n)| \leq P_e^{(n)} \ln(||A||-1) + h_2(P_e^{(n)}).$$

If $\{f_n, g_n\}$ are also AMS and hence the limit

$$\bar{P}_e = \lim_{n\to\infty} P_e^{(n)}$$

exists, then if we define

$$\bar{H}(f|g) = \lim_{n\to\infty} \frac{1}{n}H(f^n|g^n) = \lim_{n\to\infty}\frac{1}{n}(H(f^n,g^n) - H(g^n)),$$

where the limits exist since the processes are AMS, then

$$\bar{H}(f|g) \leq \bar{P}_e \ln(||A|| - 1) + h_2(\bar{P}_e)$$

$$|\bar{H}(f) - \bar{H}(g)| \leq \bar{P}_e \ln(||A|| - 1) + h_2(\bar{P}_e).$$

Proof: From the chain rule for entropy (Corollary 2.5.1), Lemma 2.5.2, and Lemma 4.2.1

$$H(f^n|g^n) = \sum_{i=0}^{n-1} H(f_i|f^i, g^n) \leq \sum_{i=0}^{n-1} H(f_i|g^i)$$

$$\leq \sum_{i=0}^{n-1} H(f_i|g_i) \leq \sum_{i=0}^{n-1} (\Pr(f_i \neq g_i) \ln(||A|| - 1) + h_2(\Pr(f_i \neq g_i)))$$

from the previous lemma. Dividing by n yields the first inequality which implies the second as in the proof of the previous lemma. If the processes are jointly AMS, then the limits exist and the entropy rate results follows from the continuity of h_2 by taking the limit. □

The per-symbol probability of error $P_e^{(n)}$ has an alternative form. Recall that the (average) Hamming distance between two vectors is the number of positions in which they differ, i.e.,

$$d_H^{(1)}(x_0, y_0) = 1 - \delta_{x_0,y_0},$$

where $\delta_{a,b}$ is the Kronecker delta function (0 if $a = b$ and 1 otherwise), and

$$d_H^{(n)}(x^n, y^n) = \sum_{i=0}^{n-1} d_H^{(1)}(x_i, y_i).$$

We have then that

$$P_e^{(n)} = E\left(\frac{1}{n} d_H^{(n)}(f^n, g^n)\right),$$

the normalized average Hamming distance.

The next lemma and corollary provide a useful tool for approximating complicated codes by simpler ones.

Lemma 4.2.2: Given a probability space $(\Omega, \mathcal{B}, P)$ suppose that $\mathcal{F}$ is a generating field: $\mathcal{B} = \sigma(\mathcal{F})$. Suppose that $\mathcal{B}$-measurable $\mathcal{Q}$ is a partition of Ω and $\epsilon > 0$. Then there is a partition $\mathcal{Q}'$ with atoms in $\mathcal{F}$ such that $|\mathcal{Q} - \mathcal{Q}'| \leq \epsilon$.

Proof: Let $||A|| = K$. From Theorem 1.2.1 given $\gamma > 0$ we can find sets $R_i \in \mathcal{F}$ such that $P(Q_i \Delta R_i) \leq \gamma$ for $i = 1, 2, \cdots, K - 1$. The remainder of the proof consists of set theoretic manipulations showing that we can construct the desired partition from the R_i by removing overlapping pieces. The algebra is given for completeness, but it can be skipped. Form a partition from the sets as

$$Q'_i = R_i - \bigcup_{j=1}^{i-1} R_j, i = 1, 2, \cdots, K-1$$

$$Q'_K = (\bigcup_{i=1}^{K-1} Q'_i)^c.$$

For $i < K$

$$P(Q_i \Delta Q'_i) = P(Q_i \bigcup Q'_i) - P(Q_i \bigcap Q'_i)$$

$$\leq P(Q_i \bigcup R_i) - P(Q_i \bigcap (R_i - \bigcup_{j<i} R_j)). \tag{4.2.2}$$

The rightmost term can be written as

$$P(Q_i \bigcap (R_i - \bigcup_{j<i} R_j)) = P((Q_i \bigcap R_i) - (\bigcup_{j<i} Q_i \bigcap R_i \bigcap R_j))$$

$$= P(Q_i \bigcap R_i) - P(\bigcup_{j<i} Q_i \bigcap R_i \bigcap R_j), \tag{4.2.3}$$

where we have used the fact that a set difference is unchanged if the portion being removed is intersected with the set it is being removed from and we have used the fact that $P(F - G) = P(F) - P(G)$ if $G \subset F$. Combining (4.2.2) and (4.2.3) we have that

$$P(Q_i \Delta Q'_i) \leq P(Q_i \bigcup R_i) - P(Q_i \bigcap R_i) + P(\bigcup_{j<i} Q_i \bigcap R_i \bigcap R_j)$$

$$= P(Q_i \Delta R_i) + P(\bigcup_{j<i} Q_i \bigcap R_i \bigcap R_j) \leq \gamma + \sum_{j<i} P(Q_i \bigcap R_j).$$

For $j \neq i$, however, we have that

$$P(Q_i \bigcap R_j) = P(Q_i \bigcap R_j \bigcap Q_j^c) \leq P(R_j \bigcap Q_j^c)$$

$$\leq P(R_j \Delta Q_j) \leq \gamma,$$

which with the previous equation implies that

$$P(Q_i \Delta Q'_i) \leq K\gamma; i = 1, 2, \cdots, K-1.$$

For the remaining atom:

$$P(Q_K \Delta Q'_K) = P(Q_K \bigcap Q'^c_K \bigcup Q^c_K \bigcap Q'_K). \tag{4.2.4}$$

We have

$$Q_K \bigcap Q'^c_K = Q_K \bigcap (\bigcup_{j<K} Q'_j) = Q_K \bigcap (\bigcup_{j<K} Q'_j \bigcap Q^c_j),$$

where the last equality follows since points in Q'_j that are also in Q_j cannot contribute to the intersection with Q_K since the Q_j are disjoint. Since $Q'_j \bigcap Q^c_j \subset Q'_j \Delta Q_j$ we have

$$Q_K \bigcap Q'^c_K \subset Q_K \bigcap (\bigcup_{j<K} Q'_j \Delta Q_j) \subset \bigcup_{j<K} Q'_j \Delta Q_j.$$

A similar argument shows that

$$Q^c_K \bigcap Q'_K \subset \bigcup_{j<k} Q'_j \Delta Q_j$$

and hence with (4.2.4)

$$P(Q_K \Delta Q'_K) \leq P(\bigcup_{j<K} Q_j \Delta Q'_j) \leq \sum_{j<K} P(Q_j \Delta Q'_j) \leq K^2 \gamma.$$

To summarize, we have shown that

$$P(Q_i \Delta Q'_i) \leq K^2 \gamma; \; i = 1, 2, \cdots, K$$

If we now choose γ so small that $K^2 \gamma \leq \epsilon / K$, the lemma is proved. □

Corollary 4.2.2: Let $(\Omega, \mathcal{B}, P)$ be a probability space and $\mathcal{F}$ a generating field. Let $f : \Omega \to A$ be a finite alphabet measurement. Given $\epsilon > 0$ there is a measurement $g : \Omega \to A$ that is measurable with respect to $\mathcal{F}$ (that is, $g^{-1}(a) \in \mathcal{F}$ for all $a \in A$) for which $P(f \neq g) \leq \epsilon$.

Proof: Follows from the previous lemma by setting $\mathcal{Q} = \{f^{-1}(a);\, a \in A\}$, choosing $\mathcal{Q}'$ from the lemma, and then assigning g for atom Q'_i in $\mathcal{Q}'$ the same value that f takes on in atom Q_i in $\mathcal{Q}$. Then

$$P(f \neq g) = \frac{1}{2} \sum_i P(Q_i \Delta Q'_i) \leq \epsilon. \quad \square$$

We now develop applications of the previous results which relate the idea of the entropy of a dynamical system with the entropy rate of a random process. The result is not required for later coding theorems, but it provides insight into the connections between entropy as considered in ergodic theory and entropy as used in information theory. In addition, the development involves some ideas of coding and approximation which are useful in proving the ergodic theorems of information theory used to prove coding theorems.

Let $\{X_n\}$ be a random process with alphabet A_X. Let A_X^∞ denote the one or two-sided sequence space. Consider the dynamical system $(\Omega, \mathcal{B}, P, T)$ defined by $(A_X^\infty, \mathcal{B}(A_X)^\infty, P, T)$, where P is the process distribution and T the shift. Recall from Section 2.2 that a stationary coding or infinite length sliding block coding of $\{X_n\}$ is a measurable mapping $f: A_X^\infty \to A_f$ into a finite alphabet which produces an encoded process $\{f_n\}$ defined by

$$f_n(x) = f(T^n x);\ x \in A_X^\infty.$$

The entropy $H(P,T)$ of the dynamical system was defined by

$$H(P,T) = \sup_f \bar{H}_P(f),$$

the supremum of the entropy rates of finite alphabet stationary codings of the original process. We shall soon show that if the original alphabet is finite, then the entropy of the dynamical system is exactly the entropy rate of the process. First, however, we require several preliminary results, some of independent interest.

Lemma 4.2.3: If f is a stationary coding of an AMS process, then the process $\{f_n\}$ is also AMS. If the input process is ergodic, then so is $\{f_n\}$.

Proof: Suppose that the input process has alphabet A_X and distribution P and that the measurement f has alphabet A_f. Define the sequence mapping $\bar{f}: A_X^\infty \to A_f^\infty$ by $\bar{f}(x) = \{f_n(x);\ n \in \mathcal{T}\}$, where $f_n(x) = f(T^n x)$ and T is the shift on the input sequence space A_X^∞. If T also denotes the shift on the output space, then by construction $\bar{f}(Tx) = T\bar{f}(x)$ and hence for any output event F, $\bar{f}^{-1}(T^{-1}F) = T^{-1}\bar{f}^{-1}(F)$. Let m denote the process distribution for the encoded process. Since $m(F) = P(\bar{f}^{-1}(F))$ for any event $F \in \mathcal{B}(A_f)^\infty$, we have using the stationarity of the mapping f that

$$\lim_{n\to\infty} \frac{1}{n}\sum_{i=0}^{n-1} m(T^{-i}F) = \lim_{n\to\infty} \frac{1}{n}\sum_{i=0}^{n-1} P(\bar{f}^{-1}(T^{-i}F))$$

$$= \lim_{n\to\infty} \frac{1}{n}\sum_{i=0}^{n-1} P(T^{-i}\bar{f}^{-1}(F)) = \bar{P}(\bar{f}^{-1}(F)),$$

where $\bar{P}$ is the stationary mean of P. Thus m is AMS. If G is an invariant output event, then $f^{-1}(G)$ is also invariant since $T^{-1}\bar{f}^{-1}(G) = \bar{f}^{-1}(T^{-1}G)$.

Hence if input invariant sets can only have probability 1 or 0, the same is true for output invariant sets. □

The lemma and Theorem 3.1.1 immediately yields the following:

Corollary 4.2.3: If f is a stationary coding of an AMS process, then

$$\bar{H}(f) = \lim_{n\to\infty} \frac{1}{n} H(f^n),$$

that is, the limit exists.

For later use the next result considers general standard alphabets. A stationary code f is a *scalar quantizer* if there is a map $q : A_X \to A_f$ such that $f(x) = q(x_0)$. Intuitively, f depends on the input sequence only through the current symbol. Mathematically, f is measurable with respect to $\sigma(X_0)$. Such codes are effectively the simplest possible and have no memory or dependence on the future.

Lemma 4.2.4: Let $\{X_n\}$ be an AMS process with standard alphabet A_X and distribution m. Let f be a stationary coding of the process with finite alphabet A_f. Fix $\epsilon > 0$. If the process is two-sided, then there is a scalar quantizer $q : A_X \to A_q$, an integer N, and a mapping $g : A_q^N \to A_f$ such that

$$\lim_{n\to\infty} \frac{1}{n} \sum_{i=0}^{n-1} \Pr(f_i \neq g(q(X_{i-N}), q(X_{i-N+1}), \cdots, q(X_{i+N}))) \leq \epsilon.$$

If the process is one-sided, then there is a scalar quantizer $q : A_X \to A_q$, an integer N, and a mapping $g : A_q^N \to A_f$ such that

$$\lim_{n\to\infty} \frac{1}{n} \sum_{i=0}^{n-1} \Pr(f_i \neq g(q(X_i), q(X_{i+1}), \cdots, q(X_{i+N-1}))) \leq \epsilon.$$

Comment: The lemma states that any stationary coding of an AMS process can be approximated by a code that depends only on a finite number of quantized inputs, that is, by a coding of a finite window of a scalar quantized version of the original process. In the special case of a finite alphabet input process, the lemma states that an arbitrary stationary coding can be well approximated by a coding depending only on a finite number of the input symbols.

Proof: Suppose that $\bar{m}$ is the stationary mean and hence for any measurements f and g

$$\bar{m}(f_0 \neq g_0) = \lim_{n\to\infty} \frac{1}{n} \sum_{n=0}^{n-1} \Pr(f_i \neq g_i).$$

Let q_n be an asymptotically accurate scalar quantizer in the sense that $\sigma(q_n(X_0))$ asymptotically generates $\mathcal{B}(A_X)$. (Since A_X is standard this exists. If A_X is finite, then take $q(a) = a$.) Then $\mathcal{F}_n = \sigma(q_n(X_i);\ i = 0, 1, 2, \cdots, n-1)$ asymptotically generates $\mathcal{B}(A_X)^\infty$ for one-sided processes and $\mathcal{F}_n = \sigma(q_n(X_i);\ i = -n, \cdots, n)$ does the same for two-sided processes. Hence from Corollary 4.2.2 given ϵ we can find a sufficiently large n and a mapping g that is measurable with respect to $\mathcal{F}_n$ such that $\bar{m}(f \neq g) \leq \epsilon$. Since g is measurable with respect to $\mathcal{F}_n$, it must depend on only the finite number of quantized samples that generate $\mathcal{F}_n$. (See, e.g., Lemma 5.2.1 of [50].) This proves the lemma. □

Combining the lemma and Corollary 4.2.1 immediately yields the following corollary, which permits us to study the entropy rate of general stationary codes by considering codes which depend on only a finite number of inputs (and hence for which the ordinary entropy results for random vectors can be applied).

Corollary 4.2.4: Given a stationary coding f of an AMS process let $\mathcal{F}_n$ be defined as above. Then given $\epsilon > 0$ there exists for sufficiently large n a code g measurable with respect to $\mathcal{F}_n$ such that

$$|\bar{H}(f) - \bar{H}(g)| \leq \epsilon.$$

The above corollary can be used to show that entropy rate, like entropy, is reduced by coding. The general stationary code is approximated by a code depending on only a finite number of inputs and then the result that entropy is reduced by mapping (Lemma 2.3.3) is applied.

Corollary 4.2.5: Given an AMS process $\{X_n\}$ with finite alphabet A_X and a stationary coding f of the process, then

$$\bar{H}(X) \geq \bar{H}(f),$$

that is, stationary coding reduces entropy rate.

Proof: For integer n define $\mathcal{F}_n = \sigma(X_0, X_1, \cdots, X_n)$ in the one-sided case and $\sigma(X_{-n}, \cdots, X_n)$ in the two-sided case. Then $\mathcal{F}_n$ asymptotically generates $\mathcal{B}(A_X)^\infty$. Hence given a code f and an $\epsilon > 0$ we can choose using the finite alphabet special case of the previous lemma a large k and a $\mathcal{F}_k$-measurable code g such that $|\bar{H}(f) - \bar{H}(g)| \leq \epsilon$. We shall show that $\bar{H}(g) \leq \bar{H}(X)$, which will prove the lemma. To see this in the one-sided case observe that g is a function of X^k and hence g^n depends only on X^{n+k} and hence

$$H(g^n) \leq H(X^{n+k})$$

and hence

$$\bar{H}(g) = \lim_{n\to\infty} \frac{1}{n} H(g^n) \leq \lim_{n\to\infty} \frac{1}{n} \frac{n}{n+k} H(X^{n+k}) = \bar{H}(X).$$

In the two-sided case g depends on $\{X_{-k},\cdots,X_k\}$ and hence g_n depends on $\{X_{-k},\cdots,X_{n+k}\}$ and hence

$$H(g^n) \leq H(X_{-k},\cdots,X_{-1},X_0,\cdots,X_{n+k}) \leq H(X_{-k},\cdots,X_{-1})+H(X^{n+k}).$$

Dividing by n and taking the limit completes the proof as before. □

Theorem 4.2.1: Let $\{X_n\}$ be a random process with alphabet A_X. Let A_X^∞ denote the one or two-sided sequence space. Consider the dynamical system $(\Omega,\mathcal{B},P,T)$ defined by $(A_X^\infty,\mathcal{B}(A_X)^\infty,P,T)$, where P is the process distribution and T is the shift. Then

$$H(P,T)=\bar{H}(X).$$

Proof: From (2.2.4), $H(P,T) \geq \bar{H}(X)$. Conversely suppose that f is a code which yields $\bar{H}(f) \geq H(P,T)-\epsilon$. Since f is a stationary coding of the process $\{X_n\}$, the previous corollary implies that $\bar{H}(f) \leq \bar{H}(X)$, which completes the proof. □

4.3 Information Rate of Finite Alphabet Processes

Let $\{(X_n,Y_n)\}$ be a one-sided random process with finite alphabet $A \times B$ and let $((A\times B)^{\mathcal{Z}_+},\mathcal{B}(A\times B)^{\mathcal{Z}_+})$ be the corresponding one-sided sequence space of outputs of the pair process. We consider X_n and Y_n to be the sampling functions on the sequence spaces A^∞ and B^∞ and (X_n,Y_n) to be the pair sampling function on the product space, that is, for $(x,y)\in A^\infty\times B^\infty$, $(X_n,Y_n)(x,y)=(X_n(x),Y_n(y))=(x_n,y_n)$. Let p denote the process distribution induced by the original space on the process $\{(X_n,Y_n)\}$. Analogous to entropy rate we can define the mutual information rate (or simply information rate) of a finite alphabet pair process by

$$\bar{I}(X,Y)=\limsup_{n\to\infty}\frac{1}{n}I(X^n,Y^n).$$

The following lemma follows immediately from the properties of entropy rates of Theorems 2.4.1 and 3.1.1 since for AMS finite alphabet processes

$$\bar{I}(X;Y)=\bar{H}(X)+\bar{H}(Y)-\bar{H}(X,Y)$$

and since from (3.1.4) the entropy rate of an AMS process is the same as that of its stationary mean. Analogous to Theorem 3.1.1 we define the random variables $p(X^n,Y^n)$ by $p(X^n,Y^n)(x,y)=p(X^n=x^n,Y^n=y^n)$, $p(X^n)$ by $p(X^n)(x,y)=p(X^n=x^n)$, and similarly for $p(Y^n)$.

Lemma 4.3.1: Suppose that $\{X_n, Y_n\}$ is an AMS finite alphabet random process with distribution p and stationary mean $\bar{p}$. Then the limits supremum defining information rates are limits and

$$\bar{I}_p(X,Y) = \bar{I}_{\bar{p}}(X,Y).$$

$\bar{I}_p$ is an affine function of the distribution p. If $\bar{p}$ has ergodic decomposition $\bar{p}_{xy}$, then

$$\bar{I}_p(X,Y) = \int d\bar{p}(x,y)\bar{I}_{\bar{p}_{xy}}(X,Y).$$

If we define the information density

$$i_n(X^n,Y^n) = \ln \frac{p(X^n,Y^n)}{p(X^n)p(Y^n)},$$

then

$$\lim_{n\to\infty} \frac{1}{n} i_n(X^n,Y^n) = \bar{I}_{\bar{p}_{xy}}(X,Y)$$

almost everywhere with respect to $\bar{p}$ and p and in $L^1(p)$.

The following lemmas follow either directly from or similarly to the corresponding results for entropy rate of the previous section.

Lemma 4.3.2: Suppose that $\{X_n, Y_n, X'_n, Y'_n\}$ is an AMS process and

$$\bar{P} = \lim_{n\to\infty} \frac{1}{n} \sum_{i=0}^{n-1} \Pr((X_i,Y_i) \neq (X'_i, Y'_i)) \leq \epsilon$$

(the limit exists since the process is AMS). Then

$$|\bar{I}(X;Y) - \bar{I}(X';Y')| \leq 3(\epsilon \ln(||A||-1) + h_2(\epsilon)).$$

Proof: The inequality follows from Corollary 4.2.1 since

$$\bar{|}(X;Y) - \bar{I}(X';Y')| \leq$$

$$|\bar{H}(X) - \bar{H}(X')| + |\bar{H}(Y) - \bar{H}(Y')| + |\bar{H}(X,Y) - \bar{H}(X',Y')|$$

and since $\Pr((X_i,Y_i) \neq (X_i', Y_i')) = \Pr(X_i \neq X_i' \text{ or } Y_i \neq Y_i')$ is no smaller than $\Pr(X_i \neq X_i')$ or $\Pr(Y_i \neq Y_i')$. □

Corollary 4.3.1: Let $\{X_n, Y_n\}$ be an AMS process and let f and g be stationary measurements on X and Y, respectively. Given $\epsilon > 0$ there is an N sufficiently large, scalar quantizers q and r, and mappings f' and g' which depend only on $\{q(X_0), \cdots, q(X_{N-1})\}$ and $\{r(Y_0), \cdots, r(Y_{N-1})\}$ in the one-sided case and $\{q(X_{-N}), \cdots, q(X_N)\}$ and $\{r(Y_{-N}), \cdots, r(Y_N)\}$ in the two-sided case such that

$$|\bar{I}(f;g) - \bar{I}(f';g')| \leq \epsilon.$$

Proof: Choose the codes f' and g' from Lemma 4.2.4 and apply the previous lemma. □

Lemma 4.3.3: If $\{X_n, Y_n\}$ is an AMS process and f and g are stationary codings of X and Y, respectively, then

$$\bar{I}(X;Y) \geq \bar{I}(f;g).$$

Proof: This is proved as Corollary 4.2.5 by first approximating f and g by finite-window stationary codes, applying the result for mutual information (Lemma 2.5.2), and then taking the limit. □

Chapter 5

Relative Entropy

5.1 Introduction

A variety of information measures have been introduced for finite alphabet random variables, vectors, and processes:entropy, mutual information, relative entropy, conditional entropy, and conditional mutual information. All of these can be expressed in terms of divergence and hence the generalization of these definitions to infinite alphabets will follow from a general definition of divergence. Many of the properties of generalized information measures will then follow from those of generalized divergence.

In this chapter we extend the definition and develop the basic properties of divergence, including the formulas for evaluating divergence as expectations of information densities and as limits of divergences of finite codings. We also develop several inequalities for and asymptotic properties of divergence. These results provide the groundwork needed for generalizing the ergodic theorems of information theory from finite to standard alphabets. The general definitions of entropy and information measures originated in the pioneering work of Kolmogorov and his colleagues Gelfand, Yaglom, Dobrushin, and Pinsker [45] [90] [32] [125].

5.2 Divergence

Given a probability space $(\Omega, \mathcal{B}, P)$ (not necessarily with finite alphabet) and another probability measure M on the same space, define the *divergence of P with respect to M* by

$$D(P||M) = \sup_{\mathcal{Q}} H_{P||M}(\mathcal{Q}) = \sup_{f} D(P_f||M_f), \tag{5.2.1}$$

where the first supremum is over all finite measurable partitions $\mathcal{Q}$ of Ω and the second is over all finite alphabet measurements on Ω. The two forms have the same interpretation: the divergence is the supremum of the relative entropies or divergences obtainable by finite alphabet codings of the sample space. The partition form is perhaps more common when considering divergence *per se*, but the measurement or code form is usually more intuitive when considering entropy and information. This section is devoted to developing the basic properties of divergence, all of which will yield immediate corollaries for the measures of information.

The first result is a generalization of the divergence inequality that is a trivial consequence of the definition and the finite alphabet special case.

Lemma 5.2.1: *The Divergence Inequality:*

For any two probability measures P and M

$$D(P||M) \geq 0$$

with equality if and only if $P = M$.

Proof: Given any partition $\mathcal{Q}$, Theorem 2.3.1 implies that

$$\sum_{Q \in \mathcal{Q}} P(Q) \ln \frac{P(Q)}{M(Q)} \geq 0$$

with equality if and only if $P(Q) = M(Q)$ for all atoms Q of the partition. Since $D(P||Q)$ is the supremum over all such partitions, it is also nonnegative. It can be 0 only if P and M assign the same probabilities to all atoms in all partitions (the supremum is 0 only if the above sum is 0 for all partitions) and hence the divergence is 0 only if the measures are identical. □

As in the finite alphabet case, Lemma 5.2.1 justifies interpreting divergence as a form of distance or dissimilarity between two probability measures. It is not a true distance or metric in the mathematical sense since it is not symmetric and it does not satisfy the triangle inequality. Since it is nonnegative and equals zero only if two measures are identical, the divergence is a ***distortion measure*** as considered in information theory [51], which is a generalization of the notion of distance. This view often provides interpretations of the basic properties of divergence. We shall develop several relations between the divergence and other distance measures. The reader is referred to Csiszár [25] for a development of the distance-like properties of divergence.

The following two lemmas provide means for computing divergences and studying their behavior. The first result shows that the supremum can be confined to partitions with atoms in a generating field. This will provide a means for computing divergences by approximation or limits. The result is due to Dobrushin and is referred to as Dobrushin's theorem. The second result shows that the divergence can be evaluated as the expectation of an entropy density defined as the logarithm of the Radon-Nikodym derivative of one measure relative to the other. This result is due to Gelfand, Yaglom, and Perez. The proofs largely follow the translator's remarks in Chapter 2 of Pinsker [125] (which in turn follows Dobrushin [32]).

Lemma 5.2.2: Suppose that $(\Omega, \mathcal{B})$ is a measurable space where $\mathcal{B}$ is generated by a field $\mathcal{F}$, $\mathcal{B} = \sigma(\mathcal{F})$. Then if P and M are two probability measures on this space,

$$D(P||M) = \sup_{\mathcal{Q} \subset \mathcal{F}} H_{P||M}(\mathcal{Q}).$$

Proof: From the definition of divergence, the right-hand term above is clearly less than or equal to the divergence. If P is not absolutely continuous with respect to M, then we can find a set F such that $M(F) = 0$ but $P(F) \neq 0$ and hence the divergence is infinite. Approximating this event by a field element F_0 by applying Theorem 1.2.1 simultaneously to M and G will yield a partition $\{F_0, F_0^c\}$ for which the right hand side of the previous equation is arbitrarily large. Hence the lemma holds for this case. Henceforth assume that $M >> P$.

Fix $\epsilon > 0$ and suppose that a partition $\mathcal{Q} = \{Q_1, \cdots, Q_K\}$ yields a relative entropy close to the divergence, that is,

$$H_{P||M}(\mathcal{Q}) = \sum_{i=1}^{K} P(Q_i) \ln \frac{P(Q_i)}{M(Q_i)} \geq D(P||M) - \epsilon/2.$$

We will show that there is a partition, say $\mathcal{Q}'$ with atoms in $\mathcal{F}$ which has almost the same relative entropy, which will prove the lemma. First observe that $P(Q) \ln[P(Q)/M(Q)]$ is a continuous function of $P(Q)$ and $M(Q)$ in the sense that given $\epsilon/(2K)$ there is a sufficiently small $\delta > 0$ such that if $|P(Q) - P(Q')| \leq \delta$ and $|M(Q) - M(Q')| \leq \delta$, then provided $M(Q) \neq 0$

$$|P(Q) \ln \frac{P(Q)}{M(Q)} - P(Q') \ln \frac{P(Q')}{M(Q')}| \leq \frac{\epsilon}{2K}.$$

If we can find a partition $\mathcal{Q}'$ with atoms in $\mathcal{F}$ such that

$$|P(Q_i') - P(Q_i)| \leq \delta, \ |M(Q_i') - M(Q_i)| \leq \delta, \ i = 1, \cdots, K, \tag{5.2.2}$$

then

$$|H_{P||M}(\mathcal{Q}') - H_{P||M}(\mathcal{Q})| \leq \sum_i |P(Q_i) \ln \frac{P(Q_i)}{M(Q_i)} - P(Q_i') \ln \frac{P(Q_i')}{M(Q_i')}|$$

$$\leq K\frac{\epsilon}{2K} = \frac{\epsilon}{2}$$

and hence

$$H_{P||M}(\mathcal{Q}') \geq D(P||M) - \epsilon$$

which will prove the lemma. To find the partition $\mathcal{Q}'$ satisfying (5.2.2), let m be the mixture measure $P/2 + M/2$. As in the proof of Lemma 4.2.2, we can find a partition $\mathcal{Q}' \subset \mathcal{F}$ such that $m(Q_i \Delta Q_i') \leq K^2\gamma$ for $i = 1, 2, \cdots, K$, which implies that

$$P(Q_i \Delta Q_i') \leq 2K^2\gamma;\ i = 1, 2, \cdots, K,$$

and

$$M(Q_i \Delta Q_i') \leq 2K^2\gamma;\ i = 1, 2, \cdots, K.$$

If we now choose γ so small that $2K^2\gamma \leq \delta$, then (5.2.2) and hence the lemma follow from the above and the fact that

$$|P(F) - P(G)| \leq P(F\Delta G).\square \tag{5.2.3}$$

Lemma 5.2.3: Given two measures P and M on a common probability space $(\Omega, \mathcal{B})$, if P is not absolutely continuous with respect to M, then

$$D(P||M) = \infty.$$

If $P << M$ (e.g., if $D(P||M) < \infty$), then the Radon-Nikodym derivative $f = dP/dM$ exists and

$$D(P||M) = \int \ln f(\omega)dP(\omega) = \int f(\omega)\ln f(\omega)dM(\omega).$$

The quantity $\ln f$ (if it exists) is called the *entropy density* or *relative entropy density* of P with respect to M.

Proof: The first statement was shown in the proof of the previous lemma. If P is not absolutely continuous with respect to M, then there is a set Q such that $M(Q) = 0$ and $P(Q) > 0$. The relative entropy for the partition $\mathcal{Q} = \{Q, Q^c\}$ is then infinite, and hence so is the divergence.

Assume that $P << M$ and let $f = dP/dM$. Suppose that Q is an event for which $M(Q) > 0$ and consider the conditional cumulative distribution function for the real random variable f given that $\omega \in Q$:

$$F_Q(u) = \frac{M(\{f < u\} \bigcap Q)}{M(Q)};\ u \in (-\infty, \infty).$$

Observe that the expectation with respect to this distribution is

$$E_M(f|Q) = \int_0^\infty u\, dF_Q(u) = \frac{1}{M(Q)}\int_Q f(\omega)\, dM(\omega) = \frac{P(Q)}{M(Q)}.$$

We also have that

$$\int_0^\infty u \ln u \, dF_Q(u) = \frac{1}{M(Q)} \int_Q f(\omega) \ln f(\omega) \, dM(\omega),$$

where the existence of the integral is ensured by the fact that $u \ln u \geq -e^{-1}$.

Applying Jensen's inequality to the convex $\bigcup$ function $u \ln u$ yields the inequality

$$\frac{1}{M(Q)} \int_Q \ln f(\omega) \, dP(\omega) = \frac{1}{M(Q)} \int_Q f(\omega) \ln f(\omega) \, dM(\omega) = \int_0^\infty u \ln u \, dF_Q(u)$$

$$\geq [\int_0^\infty u \, dF_Q(u)] \ln[\int_0^\infty u \, dF_Q(u)] = \frac{P(Q)}{M(Q)} \ln \frac{P(Q)}{M(Q)}.$$

We therefore have that for any event Q with $M(Q) > 0$ that

$$\int_Q \ln f(\omega) \, dP(\omega) \geq P(Q) \ln \frac{P(Q)}{M(Q)}. \tag{5.2.4}$$

Now let $\mathcal{Q} = \{Q_i\}$ be a finite partition and we have

$$\int \ln f(\omega) dP(\omega) = \sum_i \int_{Q_i} \ln f(\omega) \, dP(\omega)$$

$$\geq \sum_{i:P(Q_i) \neq 0} \int_{Q_i} \ln f(\omega) \, dP(\omega) = \sum_i P(Q_i) \ln \frac{P(Q_i)}{M(Q_i)},$$

where the inequality follows from (5.2.4) since $P(Q_i) \neq 0$ implies that $M(Q_i) \neq 0$ since $M >> P$. This proves that

$$D(P||M) \leq \int \ln f(\omega) \, dP(\omega).$$

To obtain the converse inequality, let q_n denote the asymptotically accurate quantizers of Section 1.6. From (1.6.3)

$$\int \ln f(\omega) \, dP(\omega) = \lim_{n \to \infty} \int q_n(\ln f(\omega)) \, dP(\omega).$$

For fixed n the quantizer q_n induces a partition of Ω into $2n2^n + 1$ atoms $\mathcal{Q}$. In particular, there are $2n2^n - 1$ "good" atoms such that for ω, ω' inside the atoms we have that $|\ln f(\omega) - \ln f(\omega')| \leq 2^{-(n-1)}$. The remaining two atoms group ω for which $\ln f(\omega) \geq n$ or $\ln f(\omega) < -n$. Defining the shorthand $P(\ln f < -n) = P(\{\omega : \ln f(\omega) < -n\})$, we have then that

$$\sum_{Q \in \mathcal{Q}} P(Q) \ln \frac{P(Q)}{M(Q)} = \sum_{\text{good } Q} P(Q) \ln \frac{P(Q)}{M(Q)}$$

$$+P(\ln f \geq n) \ln \frac{P(\ln f \geq n)}{M(\ln f \geq n)} + P(\ln f < -n) \ln \frac{P(\ln f < -n)}{M(\ln f < -n)}.$$

The rightmost two terms above are bounded below as

$$P(\ln f \geq n) \ln \frac{P(\ln f \geq n)}{M(\ln f \geq n)} + P(\ln f < -n) \ln \frac{P(\ln f < -n)}{M(\ln f < -n)}$$

$$\geq P(\ln f \geq n) \ln P(\ln f \geq n) + P(\ln f < -n) \ln P(\ln f < -n).$$

Since $P(\ln f \geq n)$ and $P(\ln f < -n) \to 0$ as $n \to \infty$ and since $x \ln x \to 0$ as $x \to 0$, given ϵ we can choose n large enough to ensure that the above term is greater than $-\epsilon$. This yields the lower bound

$$\sum_{Q \in \mathcal{Q}} P(Q) \ln \frac{P(Q)}{M(Q)} \geq \sum_{\text{good } Q} P(Q) \ln \frac{P(Q)}{M(Q)} - \epsilon.$$

Fix a good atom Q and define $\bar{h} = \sup_{\omega \in Q} \ln f(\omega)$ and $\underline{h} = \inf_{\omega \in Q} \ln f(\omega)$ and note that by definition of the good atoms

$$\bar{h} - \underline{h} \leq 2^{-(n-1)}.$$

We now have that

$$P(Q)\bar{h} \geq \int_Q \ln f(\omega)\, dP(\omega)$$

and

$$M(Q)e^{\underline{h}} \leq \int_Q f(\omega) dM(\omega) = P(Q).$$

Combining these we have that

$$P(Q) \ln \frac{P(Q)}{M(Q)} \geq P(Q) \ln \frac{P(Q)}{P(Q)e^{-\underline{h}}} = P(Q)\underline{h}$$

$$\geq P(Q)(\bar{h} - 2^{-(n-1)}) \geq \int_Q \ln f(\omega) dP(\omega) - P(Q)2^{-(n-1)}.$$

Therefore

$$\sum_{Q \in \mathcal{Q}} P(Q) \ln \frac{P(Q)}{M(Q)} \geq \sum_{\text{good } Q} P(Q) \ln \frac{P(Q)}{M(Q)} - \epsilon$$

$$\geq \sum_{\text{good } Q} \int_Q \ln f(\omega)\, dP - 2^{-(n-1)} - \epsilon$$

$$= \int_{\omega:|\ln f(\omega)|\leq n} \ln f(\omega)\, dP(\omega) - 2^{-(n-1)} - \epsilon.$$

Since this is true for arbitrarily large n and arbitrarily small ϵ,

$$D(P||Q) \geq \int \ln f(\omega) dP(\omega),$$

completing the proof of the lemma. □

It is worthwhile to point out two examples for the previous lemma. If P and M are discrete measures with corresponding pmf's p and q, than the Radon-Nikodym derivative is simply $dP/dM(\omega) = p(\omega)/m(\omega)$ and the lemma gives the known formula for the discrete case. If P and M are both probability measures on Euclidean space $\mathcal{R}^n$ and if both measures are absolutely continuous with respect to Lebesgue measure, then there exists a density f called a *probability density function* or *pdf* such that

$$P(F) = \int_F f(x) dx,$$

where dx means $dm(x)$ with m Lebesgue measure. (Lebesgue measure assigns each set its volume.) Similarly, there is a pdf g for M. In this case,

$$D(P||M) = \int_{\mathcal{R}^n} f(x) \ln \frac{f(x)}{g(x)} dx. \tag{5.2.5}$$

The following immediate corollary to the previous lemma provides a formula that is occasionally useful for computing divergences.

Corollary 5.2.1: Given three probability distributions $M >> Q >> P$, then

$$D(P||M) = D(P||Q) + E_P(\ln \frac{dQ}{dM}).$$

Proof: From the chain rule for Radon-Nikodym derivatives (e.g., Lemma 5.7.3 of [50])

$$\frac{dP}{dM} = \frac{dP}{dQ}\frac{dQ}{dM}$$

and taking expectations using the previous lemma yields the corollary. □

The next result is a technical result that shows that given a mapping on a space, the divergence between the induced distributions can be computed from the restrictions of the original measures to the sub-σ-field induced by the mapping. As part of the result, the relation between the induced Radon-Nikodym derivative and the original derivative is made explicit.

Recall that if P is a probability measure on a measurable space $(\Omega, \mathcal{B})$ and if $\mathcal{F}$ is a sub-σ-field of $\mathcal{B}$, then the restriction $P_{\mathcal{F}}$ of P to $\mathcal{F}$ is the probability measure on the measurable space $(\Omega, \mathcal{F})$ defined by $P_{\mathcal{F}}(G) = P(G)$, for all $G \in \mathcal{F}$. In other words, we can use either the probability measures on the new space or the restrictions of the probability measures on the old space to compute the divergence. This motivates considering the properties of divergences of restrictions of measures, a useful generality in that it simplifies proofs. The following lemma can be viewed as a bookkeeping result relating the divergence and the Radon-Nikodym derivatives in the two spaces.

Lemma 5.2.4: (a) Suppose that M, P are two probability measures on a space $(\Omega, \mathcal{B})$ and that X is a measurement mapping this space into $(A, \mathcal{A})$. Let P_X and M_X denote the induced distributions (measures on $(A, \mathcal{A})$) and let $P_{\sigma(X)}$ and $M_{\sigma(X)}$ denote the restrictions of P and M to $\sigma(X)$, the sub-σ-field of $\mathcal{B}$ generated by X. Then

$$D(P_X || M_X) = D(P_{\sigma(X)} || M_{\sigma(X)}).$$

If the Radon-Nikodym derivative $f = dP_X/dM_X$ exists (e.g., the above divergence is finite), then define the function $f(X) : \Omega \to [0, \infty)$ by

$$f(X)(\omega) = f(X(\omega)) = \frac{dP_X}{dM_X}(X(\omega));$$

then with probability 1 under both M and P

$$f(X) = \frac{dP_{\sigma(X)}}{dM_{\sigma(X)}}.$$

(b) Suppose that $P << M$. Then for any sub-σ-field $\mathcal{F}$ of $\mathcal{B}$, we have that

$$\frac{dP_{\mathcal{F}}}{dM_{\mathcal{F}}} = E_M(\frac{dP}{dM} | \mathcal{F}).$$

Thus the Radon-Nikodym derivative for the restrictions is just the conditional expectation of the original Radon-Nikodym derivative.

Proof: The proof is mostly algebra: $D(P_{\sigma(X)} || M_{\sigma(X)})$ is the supremum over all finite partitions $\mathcal{Q}$ with elements in $\sigma(X)$ of the relative entropy $H_{P_{\sigma(X)} || M_{\sigma(X)}}(\mathcal{Q})$. Each element $Q \in \mathcal{Q} \subset \sigma(X)$ corresponds to a unique set $Q' \in \mathcal{A}$ via $Q = X^{-1}(Q')$ and hence to each $\mathcal{Q} \subset \sigma(X)$ there is a corresponding partition $\mathcal{Q}' \subset \mathcal{A}$. The corresponding relative entropies are equal, however, since

$$H_{P_X||M_X}(\mathcal{Q}') = \sum_{Q' \in \mathcal{Q}'} P_f(Q') \ln \frac{P_X(Q')}{M_X(Q')}$$

$$= \sum_{Q' \in \mathcal{Q}'} P(X^{-1}(Q')) \ln \frac{P(X^{-1}(Q'))}{M(X^{-1}(Q'))} = \sum_{Q \in \mathcal{Q}} P_X(Q) \ln \frac{P_X(Q)}{M_X(Q)}$$

$$= H_{P_{\sigma(X)}||M_{\sigma(X)}}(\mathcal{Q}).$$

Taking the supremum over the partitions proves that the divergences are equal. If the derivative is $f = dP_X/dM_X$, then $f(X)$ is measurable since it is a measurable function of a measurable function. In addition, it is measurable with respect to $\sigma(X)$ since it depends on ω only through $X(\omega)$. For any $F \in \sigma(X)$ there is a $G \in \mathcal{A}$ such that $F = X^{-1}(G)$ and

$$\int_F f(X) dM_{\sigma(X)} = \int_F f(X) dM = \int_G f dM_X$$

from the change of variables formula (see, e.g., Lemma 4.4.7 of [50]). Thus

$$\int_F f(X) dM_{\sigma(X)} = P_X(G) = P_{\sigma(X)}(X^{-1}(G)) = P_{\sigma(X)}(F),$$

which proves that $f(X)$ is indeed the claimed derivative with probability 1 under M and hence also under P.

The variation quoted in part (b) is proved by direct verification using iterated expectation. If $G \in \mathcal{F}$, then using iterated expectation we have that

$$\int_G E_M(\frac{dP}{dM}|\mathcal{F})\, dM_{\mathcal{F}} = \int E_M(1_G \frac{dP}{dM}|\mathcal{F})\, dM_{\mathcal{F}}$$

Since the argument of the integrand is $\mathcal{F}$-measurable (see, e.g., Lemma 5.3.1 of [50]), invoking iterated expectation (e.g., Corollary 5.9.3 of [50]) yields

$$\int_G E_M(\frac{dP}{dM}|\mathcal{F})\, dM_{\mathcal{F}} = \int E_M(1_G \frac{dP}{dM}|\mathcal{F})\, dM$$

$$= E(1_G \frac{dP}{dM}) = P(G) = P_{\mathcal{F}}(G),$$

proving that the conditional expectation is the claimed derivative. □

Part (b) of the Lemma was pointed out to the author by Paul Algoet.

Having argued above that restrictions of measures are useful when finding divergences of random variables, we provide a key trick for treating such restrictions.

Lemma 5.2.5: Let $M >> P$ be two measures on a space $(\Omega, \mathcal{B})$. Suppose that $\mathcal{F}$ is a sub-σ-field and that $P_{\mathcal{F}}$ and $M_{\mathcal{F}}$ are the restrictions of P and M to $\mathcal{F}$ Then there is a measure S such that $M >> S >> P$ and

$$\frac{dP}{dS} = \frac{dP/dM}{dP_{\mathcal{F}}/dM_{\mathcal{F}}},$$

$$\frac{dS}{dM} = \frac{dP_{\mathcal{F}}}{dM_{\mathcal{F}}},$$

and

$$D(P||S) + D(P_{\mathcal{F}}||M_{\mathcal{F}}) = D(P||M). \tag{5.2.6}$$

Proof: If $M >> P$, then clearly $M_{\mathcal{F}} >> P_{\mathcal{F}}$ and hence the appropriate Radon-Nikodym derivatives exist. Define the set function S by

$$S(F) = \int_F \frac{dP_{\mathcal{F}}}{dM_{\mathcal{F}}}\, dM = \int_F E_M(\frac{dP}{dM}|\mathcal{F})\, dM,$$

using part (b) of the previous lemma. Thus $M >> S$ and $dS/dM = dP_{\mathcal{F}}/dM_{\mathcal{F}}$. Observe that for $F \in \mathcal{F}$, iterated expectation implies that

$$S(F) = E_M(E_M(1_F\frac{dP}{dM}|\mathcal{F})) = E_M(1_F\frac{dP}{dM})$$
$$= P(F) = P_{\mathcal{F}}(F);\; F \in \mathcal{F}$$

and hence in particular that $S(\Omega)$ is 1 so that $dP_{\mathcal{F}}/dM_{\mathcal{F}}$ is integrable and S is indeed a probability measure on $(\Omega, \mathcal{B})$. (In addition, the restriction of S to $\mathcal{F}$ is just $P_{\mathcal{F}}$.) Define

$$g = \frac{dP/dM}{dP_{\mathcal{F}}/dM_{\mathcal{F}}}.$$

This is well defined since with M probability 1, if the denominator is 0, then so is the numerator. Given $F \in \mathcal{B}$ the Radon-Nikodym theorem (e.g., Theorem 5.6.1 of [50]) implies that

$$\int_F g dS = \int 1_F g \frac{dS}{dM} dM = \int 1_F \frac{dP/dM}{dP_{\mathcal{F}}/dM_{\mathcal{F}}} dP_{\mathcal{F}}/dM_{\mathcal{F}} dM = P(F),$$

that is, $P << S$ and

$$\frac{dP}{dS} = \frac{dP/dM}{dP_{\mathcal{F}}/dM_{\mathcal{F}}},$$

proving the first part of the lemma. The second part follows by direct verification:

$$D(P||M) = \int \ln \frac{dP}{dM}\, dP = \int \ln \frac{dP_{\mathcal{F}}}{dM_{\mathcal{F}}}\, dP + \int \ln \frac{dP/dM}{dP_{\mathcal{F}}/dM_{\mathcal{F}}}\, dP$$

$$= \int \ln \frac{dP_{\mathcal{F}}}{dM_{\mathcal{F}}} dP_{\mathcal{F}} + \int \ln \frac{dP}{dS} dP = D(P_{\mathcal{F}}||M_{\mathcal{F}}) + D(P||S). \quad \Box$$

The two previous lemmas and the divergence inequality immediately yield the following result for $M >> P$. If M does not dominate P, then the result is trivial.

Corollary 5.2.2: Given two measures M, P on a space $(\Omega, \mathcal{B})$ and a sub-σ-field $\mathcal{F}$ of $\mathcal{B}$, then

$$D(P||M) \geq D(P_{\mathcal{F}}||M_{\mathcal{F}}).$$

If f is a measurement on the given space, then

$$D(P||M) \geq D(P_f||M_f).$$

The result is obvious for finite fields $\mathcal{F}$ or finite alphabet measurements f from the definition of divergence. The general result for arbitrary measurable functions could also have been proved by combining the corresponding finite alphabet result of Corollary 2.3.1 and an approximation technique. As above, however, we will occasionally get results comparing the divergences of measures and their restrictions by combining the trick of Lemma 5.2.5 with a result for a single divergence.

The following corollary follows immediately from Lemma 5.2.2 since the union of a sequence of asymptotically generating sub-σ-fields is a generating field.

Corollary 5.2.3: Suppose that M, P are probability measures on a measurable space $(\Omega, \mathcal{B})$ and that $\mathcal{F}_n$ is an asymptotically generating sequence of sub-σ-fields and let P_n and M_n denote the restrictions of P and M to $\mathcal{F}_n$ (e.g., $P_n = P_{\mathcal{F}_n}$). Then

$$D(P_n||M_n) \uparrow D(P||M).$$

There are two useful special cases of the above corollary which follow immediately by specifying a particular sequence of increasing sub-σ-fields. The following two corollaries give these results.

Corollary 5.2.4: Let M, P be two probability measures on a measurable space $(\Omega, \mathcal{B})$. Suppose that f is an A-valued measurement on the space. Assume that $q_n : A \to A_n$ is a sequence of measurable mappings into finite sets A_n with the property that the sequence of fields $\mathcal{F}_n = \mathcal{F}(q_n(f))$ generated by the sets $\{q_n^{-1}(a);\ a \in A_n\}$ asymptotically generate $\sigma(f)$. (For example, if the original space is standard let $\mathcal{F}_n$ be a basis and let q_n map the points in the ith atom of $\mathcal{F}_n$ into i.) Then

$$D(P_f||M_f) = \lim_{n\to\infty} D(P_{q_n(f)}||M_{q_n(f)}).$$

The corollary states that the divergence between two distributions of a random variable can be found as a limit of quantized versions of the random variable. Note that the limit could also be written as

$$\lim_{n\to\infty} H_{P_f\|M_f}(q_n).$$

In the next corollary we consider increasing sequences of random variables instead of increasing sequences of quantizers, that is, more random variables (which need not be finite alphabet) instead of ever finer quantizers. The corollary follows immediately from Corollary 5.2.3 and Lemma 5.2.4.

Corollary 5.2.5: Suppose that M and P are measures on the sequence space corresponding to outcomes of a sequence of random variables $X_0, X_1, \cdots$ with alphabet A. Let $\mathcal{F}_n = \sigma(X_0, \cdots, X_{n-1})$, which asymptotically generates the σ-field $\sigma(X_0, X_1, \cdots)$. Then

$$\lim_{n\to\infty} D(P_{X^n}\|M_{X^n}) = D(P\|M).$$

We now develop two fundamental inequalities involving entropy densities and divergence. The first inequality is from Pinsker [125]. The second is an improvement of an inequality of Pinsker [125] by Csiszár [24] and Kullback [91]. The second inequality is more useful when the divergence is small. Coupling these inequalities with the trick of Lemma 5.2.5 provides a simple generalization of an inequality of [48] and will provide easy proofs of L^1 convergence results for entropy and information densities. A key step in the proof involves a notion of distance between probability measures and is of interest in its own right. Given two probability measures M, P on a common measurable space $(\Omega, \mathcal{B})$, the *variational distance* between them is defined by

$$d(P, M) \equiv \sup_{\mathcal{Q}} \sum_{Q\in\mathcal{Q}} |P(Q) - M(Q)|,$$

where the supremum is over all finite measurable partitions. We will proceed by stating first the end goal, the two inequalities involving divergence, as a lemma, and then state two lemmas giving the basic required properties of the variational distance. The lemmas will be proved in a different order.

Lemma 5.2.6: Let P and M be two measures on a common probability space $(\Omega, \mathcal{B})$ with $P << M$. Let $f = dP/dM$ be the Radon-Nikodym derivative and let $h = \ln f$ be the entropy density. Then

$$D(P\|M) \leq \int |h| dP \leq D(P\|M) + \frac{2}{e}, \tag{5.2.7}$$

$$\int |h| dP \leq D(P||M) + \sqrt{2D(P||M)}. \tag{5.2.8}$$

Lemma 5.2.7: Given two probability measures M, P on a common measurable space $(\Omega, \mathcal{B})$, the variational distance is given by

$$d(P, M) = 2 \sup_{F \in \mathcal{B}} |P(F) - M(F)|. \tag{5.2.9}$$

Furthermore, if S is a measure for which $P << S$ and $M << S$ ($S = (P + M)/2$, for example), then also

$$d(P, M) = \int |\frac{dP}{dS} - \frac{dM}{dS}| dS \tag{5.2.10}$$

and the supremum in (5.2.9) is achieved by the set

$$F = \{\omega : \frac{dP}{dS}(\omega) > \frac{dM}{dS}(\omega)\}.$$

Lemma 5.2.8

$$d(P, M) \leq \sqrt{2D(P||M)}.$$

Proof of Lemma 5.2.7: First observe that for any set F we have for the partition $\mathcal{Q} = \{F, F^c\}$ that

$$d(P, M) \geq \sum_{Q \in \mathcal{Q}} |P(Q) - M(Q)| = 2|P(F) - M(F)|$$

and hence

$$d(P, M) \geq 2 \sup_{F \in \mathcal{B}} |P(F) - M(F)|.$$

Conversely, suppose that $\mathcal{Q}$ is a partition which approximately yields the variational distance, e.g.,

$$\sum_{Q \in \mathcal{Q}} |P(Q) - M(Q)| \geq d(P, M) - \epsilon$$

for $\epsilon > 0$. Define a set F as the union of all of the Q in $\mathcal{Q}$ for which $P(Q) \geq M(Q)$ and we have that

$$\sum_{Q \in \mathcal{Q}} |P(Q) - M(Q)| = P(F) - M(F) + M(F^c) - P(F^c) = 2(P(F) - M(F))$$

and hence

$$d(P,M) - \epsilon \leq \sup_{F \in \mathcal{B}} 2|P(F) - M(F)|.$$

Since ϵ is arbitrary, this proves the first statement of the lemma.

Next suppose that a measure S dominating both P and M exists and define the set

$$F = \{\omega : \frac{dP}{dS}(\omega) > \frac{dM}{dS}(\omega)\}$$

and observe that

$$\int |\frac{dP}{dS} - \frac{dM}{dS}|\, dS = \int_F (\frac{dP}{dS} - \frac{dM}{dS})\, dS - \int_{F^c} (\frac{dP}{dS} - \frac{dM}{dS})\, dS$$

$$= P(F) - M(F) - (P(F^c) - M(F^c)) = 2(P(F) - M(F)).$$

From the definition of F, however,

$$P(F) = \int_F \frac{dP}{dS}\, dS \geq \int_F \frac{dM}{dS}\, dS = M(F)$$

so that $P(F) - M(F) = |P(F) - M(F)|$. Thus we have that

$$\int |\frac{dP}{dS} - \frac{dM}{dS}|\, dS = 2|P(F) - M(F)| \leq 2 \sup_{G \in \mathcal{B}} |P(G) - M(G)| = d(P,M).$$

To prove the reverse inequality, assume that $\mathcal{Q}$ approximately yields the variational distance, that is, for $\epsilon > 0$ we have

$$\sum_{Q \in \mathcal{Q}} |P(Q) - M(Q)| \geq d(P,M) - \epsilon.$$

Then

$$\sum_{Q \in \mathcal{Q}} |P(Q) - M(Q)| = \sum_{Q \in \mathcal{Q}} |\int_Q (\frac{dP}{dS} - \frac{dM}{dS})\, dS|$$

$$\leq \sum_{Q \in \mathcal{Q}} \int_Q |\frac{dP}{dS} - \frac{dM}{dS}|\, dS = \int |\frac{dP}{dS} - \frac{dM}{dS}|\, dS$$

which, since ϵ is arbitrary, proves that

$$d(P,M) \leq \int |\frac{dP}{dS} - \frac{dM}{dS}|\, dS,$$

Combining this with the earlier inequality proves (5.2.10). We have already seen that this upper bound is actually achieved with the given choice of F, which completes the proof of the lemma. □

Proof of Lemma 5.2.8: Assume that $M >> P$ since the result is trivial otherwise because the right-hand side is infinite. The inequality will follow from the first statement of Lemma 5.2.7 and the following inequality: Given $1 \geq p, m \geq 0$,

$$p \ln \frac{p}{m} + (1-p) \ln \frac{1-p}{1-m} - 2(p-m)^2 \geq 0. \tag{5.2.11}$$

To see this, suppose the truth of (5.2.11). Since F can be chosen so that $2(P(F) - M(F))$ is arbitrarily close to $d(P, M)$, given $\epsilon > 0$ choose a set F such that $[2(P(F) - M(F))]^2 \geq d(P, M)^2 - 2\epsilon$. Since $\{F, F^c\}$ is a partition,

$$D(P||M) - \frac{d(P, M)^2}{2}$$

$$\geq P(F) \ln \frac{P(F)}{M(F)} + (1 - P(F)) \ln \frac{1 - P(F)}{1 - M(F)} - 2(P(F) - M(F))^2 - \epsilon.$$

If (5.2.11) holds, then the right-hand side is bounded below by $-\epsilon$, which proves the lemma since ϵ is arbitrarily small. To prove (5.2.11) observe that the left-hand side equals zero for $p = m$, has a negative derivative with respect to m for $m < p$, and has a positive derivative with respect to m for $m > p$. (The derivative with respect to m is $(m-p)[1-4m(1-m)]/[m(1-m)$.) Thus the left hand side of (5.2.11) decreases to its minimum value of 0 as m tends to p from above or below. □

Proof of Lemma 5.2.6: The magnitude entropy density can be written as

$$|h(\omega)| = h(\omega) + 2h(\omega)^- \tag{5.2.12}$$

where $a^- = -\min(a, 0)$. This inequality immediately gives the trivial left-hand inequality of (5.2.7). The right-hand inequality follows from the fact that

$$\int h^- dP = \int f[\ln f]^- dM$$

and the elementary inequality $a \ln a \geq -1/e$.

The second inequality will follow from (5.2.12) if we can show that

$$2 \int h^- dP \leq \sqrt{2D(P||M)}.$$

Let F denote the set $\{h \leq 0\}$ and we have from (5.2.4) that

$$2 \int h^- dP = -2 \int_F h dP \leq -2P(F) \ln \frac{P(F)}{M(F)}$$

and hence using the inequality $\ln x \leq x - 1$ and Lemma 5.2.7

$$2\int h^- dP \leq 2P(F)\ln\frac{M(F)}{P(F)} \leq 2(M(F) - P(F))$$

$$\leq d(P, M) \leq \sqrt{2D(P||M)},$$

completing the proof. □

Combining Lemmas 5.2.6 and 5.2.5 yields the following corollary, which generalizes Lemma 2 of [54]:

Corollary 5.2.6: Let P and M be two measures on a space $(\Omega, \mathcal{B})$. Suppose that $\mathcal{F}$ is a sub-σ-field and that $P_{\mathcal{F}}$ and $M_{\mathcal{F}}$ are the restrictions of P and M to $\mathcal{F}$. Assume that $M >> P$. Define the entropy densities $h = \ln dP/dM$ and $h' = \ln dP_{\mathcal{F}}/dM_{\mathcal{F}}$. Then

$$\int |h - h'|\, dP \leq D(P||M) - D(P_{\mathcal{F}}||M_{\mathcal{F}}) + \frac{2}{e}, \tag{5.2.13}$$

and

$$\int |h - h'|\, dP \leq D(P||M) -$$

$$D(P_{\mathcal{F}}||M_{\mathcal{F}}) + \sqrt{2D(P||M) - 2D(P_{\mathcal{F}}||M_{\mathcal{F}})}. \tag{5.2.14}$$

Proof: Choose the measure S as in Lemma 5.2.5 and then apply Lemma 5.2.6 with S replacing M. □

Variational Description of Divergence

As in the discrete case, divergence has a variational characterization that is a fundamental property for its applications to large deviations theory [143] [31]. We again take a detour to state and prove the property without delving into its applications.

Suppose now that P and M are two probability measures on a common probability space, say $(\Omega, \mathcal{B})$, such that $M >> P$ and hence the density

$$f = \frac{dP}{dM}$$

is well defined. Suppose that Φ is a real-valued random variable defined on the same space, which we explicitly require to be finite-valued (it cannot assume ∞ as a value) and to have finite cumulant generating function:

$$E_M(e^{\Phi}) < \infty.$$

Then we can define a probability measure M^{ϕ} by

$$M^{\Phi}(F) = \int_F \frac{e^{\Phi}}{E_M(e^{\Phi})} dM \tag{5.2.15}$$

and observe immediately that by construction $M >> M^{\Phi}$ and

$$\frac{dM^{\Phi}}{dM} = \frac{e^{\Phi}}{E_M(e^{\Phi})}.$$

The measure M^{Φ} is called a "tilted" distribution. Furthermore, by construction $dM^{\Phi}/dM \neq 0$ and hence we can write

$$\int_F \frac{f}{e^{\phi}/E_M(e^{\Phi})} dQ = \int_F \frac{f}{e^{\phi}/E_M(e^{\Phi})} \frac{dM^{\Phi}}{dM} dM = \int_F f dM = P(F)$$

and hence $P << M^{\Phi}$ and

$$\frac{dP}{dM^{\Phi}} = \frac{f}{e^{\phi}/E_M(e^{\Phi})}.$$

We are now ready to state and prove the principal result of this section, a variational characterization of divergence.

Theorem 5.2.1: Suppose that $M >> P$. Then

$$D(P||M) = \sup_{\Phi} \left(E_P\Phi - \ln(E_M(e^{\Phi})) \right), \tag{5.2.16}$$

where the supremum is over all random variables Φ for which Φ is finite-valued and e^{Φ} is M-integrable.

Proof: First consider the random variable Φ defined by $\Phi = \ln f$ and observe that

$$E_P\Phi - \ln(E_M(e^{\Phi})) = \int dP \ln f - \ln(\int dM f)$$

$$= D(P||M) - \ln \int dP = D(P||M).$$

This proves that the supremum over all Φ is no smaller than the divergence. To prove the other half observe that for any Φ,

$$H(P||M) - \left(E_P\Phi - \ln E_M(e^{\Phi})\right) = E_P \left(\ln \frac{dP/dM}{dP/dM^{\Phi}} \right),$$

where M^{Φ} is the tilted distribution constructed above. Since $M >> M^{\Phi} >> P$, we have from the chain rule for Radon-Nikodym derivatives that

$$H(P||M) - \left(E_P\Phi - \ln E_M(e^{\Phi})\right) = E_P \ln \frac{dP}{dM^{\Phi}} = D(P||M^{\Phi}) \geq 0$$

from the divergence inequality, which completes the proof. Note that equality holds and the supremum is achieved if and only if $M^{\Phi} = P$. □

5.3 Conditional Relative Entropy

Lemmas 5.2.4 and 5.2.5 combine with basic properties of conditional probability in standard spaces to provide an alternative form of Lemma 5.2.5 in terms of random variables that gives an interesting connection between the densities for combinations of random variables and those for individual random variables. The results are collected in Theorem 5.3.1. First, however, several definitions are required. Let X and Y be random variables with standard alphabets A_X and A_Y and σ-fields $\mathcal{B}_{A_X}$ and $\mathcal{B}_{A_Y}$, respectively. Let P_{XY} and M_{XY} be two distributions on $(A_X \times A_Y, \mathcal{B}_{A_X \times A_Y})$ and assume that $M_{XY} >> P_{XY}$. Let M_Y and P_Y denote the induced marginal distributions, e.g., $M_Y(F) = M_{XY}(A_X \times F)$. Define the (nonnegative) densities (Radon-Nikodym derivatives):

$$f_{XY} = \frac{dP_{XY}}{dM_{XY}}, f_Y = \frac{dP_Y}{dM_Y}$$

so that

$$P_{XY}(F) = \int_F f_{XY} dM_{XY};\ F \in \mathcal{B}_{A_X \times A_Y}$$

$$P_Y(F) = \int_F f_Y dM_Y;\ F \in \mathcal{B}_{A_Y}.$$

Note that $M_{XY} >> P_{XY}$ implies that $M_Y >> P_Y$ and hence f_Y is well defined if f_{XY} is. Define also the *conditional density*

$$f_{X|Y}(x|y) = \begin{cases} \frac{f_{XY}(x,y)}{f_Y(y)}; & \text{if } f_Y(y) > 0 \\ 1; & \text{otherwise}. \end{cases}$$

Suppose now that the entropy density

$$h_Y = \ln f_Y$$

exists and define the *conditional entropy density* or *conditional relative entropy density* by

$$h_{X|Y} = \ln f_{X|Y}.$$

Again suppose that these densities exist, we (tentatively) define the *conditional relative entropy*

$$H_{P||M}(X|Y) = E \ln f_{X|Y} = \int dP_{XY}(x,y) \ln f_{X|Y}(x|y)$$

$$= \int dM_{XY}(x,y) f_{XY}(x,y) \ln f_{X|Y}(x|y). \qquad (5.3.1)$$

if the expectation exists. Note that unlike unconditional relative entropies, the above definition of conditional relative entropy requires the existence of densities. Although this is sufficient in many of the applications and is convenient for the moment, it is not sufficiently general to handle all the cases we will encounter. In particular, there will be situations where we wish to define a conditional relative entropy $H_{P||M}(X|Y)$ even though it is not true that $M_{XY} >> P_{XY}$. Hence at the end of this section we will return to this question and provide a general definition that agrees with the current one when the appropriate densities exist and that shares those properties not requiring the existence of densities, e.g., the chain rule for relative entropy. An alternative approach to a general definition for conditional relative entropy can be found in Algoet [6].

The previous construction immediately yields the following lemma providing chain rules for densities and relative entropies.

Lemma 5.3.1:

$$f_{XY} = f_{X|Y} f_Y,$$

$$h_{XY} = h_{X|Y} + h_Y,$$

and hence

$$D(P_{XY}||M_{XY}) = H_{P||M}(X|Y) + D(P_Y||M_Y), \qquad (5.3.2)$$

or, equivalently,

$$H_{P||M}(X,Y) = H_{P||M}(Y) + H_{P||M}(X|Y), \qquad (5.3.3)$$

a chain rule for relative entropy analogous to that for ordinary entropy. Thus if $H_{P||M}(Y) < \infty$ so that the indeterminate form $\infty - \infty$ is avoided, then

$$H_{P||M}(X|Y) = H_{P||M}(X,Y) - H_{P||M}(Y).$$

Since the alphabets are standard, there is a regular version of the conditional probabilities of X given Y under the distribution M_{XY}; that is, for each $y \in B$ there is a probability measure $M_{X|Y}(F|y)$; $F \in \mathcal{B}_A$ for fixed $F \in \mathcal{B}_{A_X}$ $M_{X|Y}(F|y)$ is a measurable function of y and such that for all $G \in \mathcal{B}_{A_Y}$

$$M_{XY}(F \times G) = E(1_G(Y) M_{X|Y}(F|Y)) = \int_G M_{X|Y}(F|y) dM_Y(y).$$

Lemma 5.3.2: Given the previous definitions, define the set $\bar{B} \in \mathcal{B}_B$ to be the set of y for which

$$\int_A f_{X|Y}(x|y) dM_{X|Y}(x|y) = 1.$$

Define $P_{X|Y}$ for $y \in \bar{B}$ by

$$P_{X|Y}(F|y) = \int_F f_{X|Y}(x|y) dM_{X|Y}(x|y); \; F \in \mathcal{B}_A$$

and let $P_{X|Y}(.|y)$ be an arbitrary fixed probability measure on $(A, \mathcal{B}_A)$ for all $y \notin \bar{B}$. Then $M_Y(\bar{B}) = 1$, $P_{X|Y}$ is a regular conditional probability for X given Y under the distribution P_{XY}, and

$$P_{X|Y} << M_{X|Y}; \; M_Y - \text{a.e.},$$

that is, $M_Y(\{y : P_{X|Y}(\cdot|y) << M_{X|Y}(\cdot|y)\}) = 1$. Thus if $P_{XY} << M_{XY}$, we can choose regular conditional probabilities under both distributions so that with probability one under M_Y the conditional probabilities under P are dominated by those under M and

$$\frac{dP_{X|Y}}{dM_{X|Y}}(x|y) \equiv \frac{dP_{X|Y}(\cdot|y)}{dM_{X|Y}(\cdot|y)}(x) = f_{X|Y}(x|y); \; x \in A.$$

Proof: Define for each $y \in B$ the set function

$$G_y(F) = \int_F f_{X|Y}(x|y) dM_{X|Y}(x|y); \; F \in \mathcal{B}_A.$$

We shall show that $G_y(F)$, $y \in B$, $F \in \mathcal{B}_A$ is a version of a regular conditional probability of X given Y under P_{XY}. First observe using iterated expectation and the fact that conditional expectations are expectations with respect to conditional probability measures ([50], Section 5.9) that for any $F \in \mathcal{B}_B$

$$\int_F [\int_A f_{X|Y}(x|y) dM_{X|Y}(x|y)] \, dM_Y(y) = E(1_F(Y) E[1_A(X) f_{X|Y}|Y])$$

$$= E(1_F(Y) 1_A(X) \frac{f_{XY}}{f_Y} 1_{f_Y>0}) = \int 1_{A\times F} \frac{1}{f_Y} 1_{\{f_Y>0\}} f_{XY} \, dM_{XY}$$

$$= \int_{A\times F} \frac{1}{f_Y} 1_{\{f_Y>0\}} \, dP_{XY} = \int_F \frac{1}{f_Y} 1_{\{f_Y>0\}} \, dP_Y \int_F \frac{1}{f_Y} \, dP_Y,$$

where the last step follows since since the function being integrated depends only on Y and hence is measurable with respect to $\sigma(Y)$ and therefore its expectation can be computed from the restriction of P_{XY} to $\sigma(Y)$ (see, for example, Lemma 5.3.1 of [50]) and since $P_Y(f_Y > 0) = 1$. We can compute this last expectation, however, using M_Y as

$$\int_F \frac{1}{f_Y} dP_Y = \int_F \frac{1}{f_Y} f_Y dM_Y = \int_F dM_Y = M_Y(F)$$

which yields finally that

$$\int_F [\int_A f_{X|Y}(x|y)\, dM_{X|Y}(x|y)]\, dM_Y(y) = M_Y(F); \quad \text{all } F \in \mathcal{B}_B.$$

If

$$\int_F g(y) dM_Y(y) = \int_F 1 dM_Y(y), \quad \text{all } F \in \mathcal{B}_B,$$

however, it must also be true that $g = 1$ M_Y-a.e. (See, for example, Corollary 5.3.1 of [50].) Thus we have M_Y-a.e. and hence also P_Y-a.e. that

$$\int_A f_{X|Y}(x|y) dM_{X|Y}(x|y)] dM_Y(y) = 1;$$

that is, $M_Y(\bar{B}) = 1$. For $y \in \bar{B}$, it follows from the basic properties of integration that G_y is a probability measure on $(A, \mathcal{B}_A)$ (see Corollary 4.4.3 of [50]).

By construction, $P_{X|Y}(\cdot|y) << M_{X|Y}(\cdot|y)$ for all $y \in \bar{B}$ and hence this is true with probability 1 under M_Y and P_Y. Furthermore, by construction

$$\frac{dP_{X|Y}(\cdot|y)}{dM_{X|Y}(\cdot|y)}(x) = f_{X|Y}(x|y).$$

To complete the proof we need only show that $P_{X|Y}$ is indeed a version of the conditional probability of X given Y under P_{XY}. To do this, fix $G \in \mathcal{B}_A$ and observe for any $F \in \mathcal{B}_B$ that

$$\int_F P_{X|Y}(G|y)\, dP_Y(y) = \int_F [\int_G f_{X|Y}(x|y) dM_{X|Y}(x|y)]\, dP_Y(y)$$

$$= \int_F [\int_G f_{X|Y}(x|y)\, dM_{X|Y}(x|y)] f_Y(y)\, dM_Y(y)$$

$$= E[1_F(Y) f_Y E[1_G(X) f_{X|Y}|Y] = E_M[1_{G\times F} f_{XY}],$$

again using iterated expectation. This immediately yields

$$\int_F P_{X|Y}(G|y)\, dP_Y(y) = \int_{G\times F} f_{XY} dM_{XY} = \int_{G\times F} dP_{XY} = P_{XY}(G \times F),$$

which proves that $P_{X|Y}(G|y)$ is a version of the conditional probability of X given Y under P_{XY}, thereby completing the proof. □

Theorem 5.3.1: Given the previous definitions with $M_{XY} >> P_{XY}$, define the distribution S_{XY} by

$$S_{XY}(F \times G) = \int_G M_{X|Y}(F|y)dP_Y(y), \tag{5.3.4}$$

that is, S_{XY} has P_Y as marginal distribution for Y and $M_{X|Y}$ as the conditional distribution of X given Y. Then the following statements are true:

1. $M_{XY} >> S_{XY} >> P_{XY}$.
2. $dS_{XY}/dM_{XY} = f_Y$ and $dP_{XY}/dS_{XY} = f_{X|Y}$.
3. $D(P_{XY}||M_{XY}) = D(P_Y||M_Y)+D(P_{XY}||S_{XY})$, and hence $D(P_{XY}||M_{XY})$ exceeds $D(P_Y||M_Y)$ by an amount $D(P_{XY}||S_{XY}) = H_{P||M}(X||Y)$.

Proof: To apply Lemma 5.2.5 define $P = P_{XY}$, $M = M_{XY}$, $\mathcal{F} = \sigma(Y)$, $P' = P_{\sigma(Y)}$, and $M' = M_{\sigma(Y)}$. Define S by

$$S(F \times G) = \int_{F\times G} \frac{dP_{\sigma(Y)}}{dM_{\sigma(Y)}} dM_{XY},$$

for $F \in \mathcal{B}_A$ and $G \in \mathcal{B}_B$. We begin by showing that $S = S_{XY}$. All of the properties will then follow from Lemma 5.2.5.

For $F \in \mathcal{B}_{A_X}$ and $G \in \mathcal{B}_{A_Y}$

$$S(F \times G) = \int_{F\times G} \frac{dP_{\sigma(Y)}}{dM_{\sigma(Y)}} dM_{XY} = E\left(1_{F\times G}\frac{dP_{\sigma(Y)}}{dM_{\sigma(Y)}}\right),$$

where the expectation is with respect to M_{XY}. Using Lemma 5.2.4 and iterated conditional expectation (c.f. Corollary 5.9.3 of [50]) yields

$$E\left(1_{F\times G}\frac{dP_{\sigma(Y)}}{dM_{\sigma(Y)}}\right) = E\left(1_F(X)1_G(Y)\frac{dP_Y}{dM_Y}(Y)\right)$$

$$= E\left(1_G(Y)\frac{dP_Y}{dM_Y}(Y)E[1_F(X)|Y]\right) = E\left(1_G(Y)\frac{dP_Y}{dM_Y}(Y)M_{X|Y}(F|Y)\right)$$

$$\int M_{X|Y}(F|y)\frac{dP_Y}{dM_Y}(Y)dM_Y(y) = \int_G M_{X|Y}(F|y)\, dP_Y(y),$$

proving that $S = S_{XY}$. Thus Lemma 5.5.2 implies that $M_{XY} >> S_{XY} >> P_{XY}$, proving the first property.

From Lemma 5.2.4, $dP'/dM' = dP_{\sigma(Y)}/dM_{\sigma(Y)} = dP_Y/dM_Y = f_Y$, proving the first equality of property 2. This fact and the first property imply the second equality of property 2 from the chain rule of Radon-Nikodym derivatives. (See, e.g., Lemma 5.7.3 of [50].) Alternatively, the second equality of the second property follows from Lemma 5.2.5 since

$$\frac{dP_{XY}}{dS_{XY}} = \frac{dP_{XY}/dM_{XY}}{dM_{XY}/dS_{XY}} = \frac{f_{XY}}{f_Y}.$$

Corollary 5.2.1 therefore implies that $D(P_{XY}||M_{XY}) = D(P_{XY}||S_{XY}) + D(S_{XY}||M_{XY})$, which with Property 2, Lemma 5.2.3, and the definition of relative entropy rate imply Property 3. □

It should be observed that it is not necessarily true that $D(P_{XY}||S_{XY}) \geq D(P_X||M_X)$ and hence that $D(P_{XY}||M_{XY}) \geq D(P_X||M_X) + D(P_Y||M_Y)$ as one might expect since in general $S_X \neq M_X$. These formulas will, however, be true in the special case where $M_{XY} = M_X \times M_Y$.

We next turn to an extension and elaboration of the theorem when there are three random variables instead of two. This will be a crucial generalization for our later considerations of processes, when the three random variables will be replaced by the current output, a finite number of previous outputs, and the infinite past.

Suppose that $M_{XYZ} >> P_{XYZ}$ are two distributions for three standard alphabet random variables X, Y, and Z taking values in measurable spaces $(A_X, \mathcal{B}_{A_X})$, $(A_Y, \mathcal{B}_{A_Y})$, $(A_Z, \mathcal{B}_{A_Z})$, respectively. Observe that the absolute continuity implies absolute continuity for the restrictions, e.g., $M_{XY} >> P_{XY}$ and $M_Y >> P_Y$. Define the Radon-Nikodym derivatives f_{XYZ}, f_{YZ}, f_Y, etc. in the obvious way; for example,

$$f_{XYZ} = \frac{dP_{XYZ}}{dM_{XYZ}}.$$

Let h_{XYZ}, h_{YZ}, h_Y, etc., denote the corresponding relative entropy densities, e.g.,

$$h_{XYZ} = \ln f_{XYZ}.$$

Define as previously the conditional densities

$$f_{X|YZ} = \frac{f_{XYZ}}{f_{YZ}};\ f_{X|Y} = \frac{f_{XY}}{f_Y},$$

the conditional entropy densities

$$h_{X|YZ} = \ln f_{X|YZ};\ h_{X|Y} = \ln f_{X|Y},$$

and the conditional relative entropies

$$H_{P||M}(X|Y) = E(\ln f_{X|Y})$$

and

$$H_{P||M}(X|Y,Z) = E(\ln f_{X|YZ}).$$

By construction (or by double use of Lemma 5.3.1) we have the following chain rules for conditional relative entropy and its densities.

Lemma 5.3.3:

$$f_{XYZ} = f_{X|YZ} f_{Y|Z} f_Z,$$

$$h_{XYZ} = h_{X|YZ} + h_{Y|Z} + h_Z,$$

and hence

$$H_{P||M}(X,Y,Z) = H_{P||M}(X|YZ) + H_{P||M}(Y|Z) + H_{P||M}(Z).$$

Corollary 5.3.1: Given a distribution P_{XY}, suppose that there is a product distribution $M_{XY} = M_X \times M_Y >> P_{XY}$. Then

$$M_{XY} >> P_X \times P_Y >> P_{XY},$$

$$\frac{dP_{XY}}{d(P_X \times P_Y)} = \frac{f_{XY}}{f_X f_Y} = \frac{f_{X|Y}}{f_X},$$

$$\frac{d(P_X \times P_Y)}{dM_{XY}} = f_X f_Y,$$

$$D(P_{XY}||P_X \times P_Y) + H_{P||M}(X) = H_{P||M}(X|Y),$$

and

$$D(P_X \times P_Y||M_{XY}) = H_{P||M}(X) + H_{P||M}(Y).$$

Proof: First apply Theorem 5.3.1 with $M_{XY} = M_X \times M_Y$. Since M_{XY} is a product measure, $M_{X|Y} = M_X$ and $M_{XY} >> S_{XY} = M_X \times P_Y >> P_{XY}$ from the theorem. Next we again apply Theorem 5.3.1, but this time the roles of X and Y in the theorem are reversed and we replace M_{XY} in the theorem statement by the current $S_{XY} = M_X \times P_Y$ and we replace S_{XY} in the theorem statement by

$$S'_{XY}(F \times G) = \int_F S_{Y|X}(G|x)\, dP_X(x) = P_X(F)P_Y(G);$$

that is, $S'_{XY} = P_X \times P_Y$. We then conclude from the theorem that $S'_{XY} = P_X \times P_Y >> P_{XY}$, proving the first statement. We now have that

$$M_{XY} = M_X \times M_Y >> P_X \times P_Y >> P_{XY}$$

and hence the chain rule for Radon-Nikodym derivatives (e.g., Lemma 5.7.3 of [50]) implies that

$$f_{XY} = \frac{dP_{XY}}{dM_{XY}} = \frac{dP_{XY}}{d(P_X \times P_Y)} \frac{d(P_X \times P_Y)}{d(M_X \times M_Y)}.$$

It is straightforward to verify directly that

$$\frac{d(P_X \times P_Y)}{d(M_X \times M_Y)} = \frac{dP_X}{dM_X}\frac{dP_Y}{dM_Y} = f_X f_Y$$

and hence

$$f_{XY} = \frac{dP_{XY}}{d(P_X \times P_Y)})f_X f_Y,$$

as claimed. Taking expectations using Lemma 5.2.3 then completes the proof (as in the proof of Corollary 5.2.1.) □

The lemma provides an interpretation of the product measure $P_X \times P_Y$. This measure yields independent random variables with the same marginal distributions as P_{XY}, which motivates calling $P_X \times P_Y$ the *independent approximation* or *memoryless approximation* to P_{XY}. The next corollary further enhances this name by showing that $P_X \times P_Y$ is the best such approximation in the sense of yielding the minimum divergence with respect to the original distribution.

Corollary 5.3.2: Given a distribution P_{XY} let $\mathcal{M}$ denote the class of all product distributions for XY; that is, if $M_{XY} \in \mathcal{M}$, then $M_{XY} = M_X \times M_Y$. Then

$$\inf_{M_{XY} \in \mathcal{M}} D(P_{XY}||M_{XY}) = D(P_{XY}||P_X \times P_Y).$$

Proof: We need only consider those M yielding finite divergence (since if there are none, both sides of the formula are infinite and the corollary is trivially true). Then

$$D(P_{XY}||M_{XY}) = D(P_{XY}||P_X \times P_Y) + D(P_X \times P_Y||M_{XY})$$

$$\geq D(P_{XY}||P_X \times P_Y)$$

with equality if and only if $D(P_X \times P_Y||M_{XY}) = 0$, which it will be if $M_{XY} = P_X \times P_Y$. □

Recall that given random variables (X, Y, Z) with distribution M_{XYZ}, then $X \to Y \to Z$ is a Markov chain (with respect to M_{XYZ}) if for any event $F \in \mathcal{B}_{A_Z}$ with probability one

$$M_{Z|YX}(F|y, x) = M_{Z|Y}(F|y).$$

If this holds, we also say that X and Z are conditionally independent given Y. Equivalently, if we define the distribution $M_{X\times Z|Y}$ by

$$M_{X\times Z|Y}(F_X \times F_Z \times F_Y) = \int_{F_y} M_{X|Y}(F_X|y)M_{Z|Y}(F_Z|y)dM_Y(y);$$

$$F_X \in \mathcal{B}_{A_X};\ F_Z \in \mathcal{B}_{A_Z};\ F_Y \in \mathcal{B}_{A_Y};$$

then $Z \to Y \to X$ is a Markov chain if $M_{X\times Z|Y} = M_{XYZ}$. (See Section 5.10 of [50].) This construction shows that a Markov chain is symmetric in the sense that $X \to Y \to Z$ if and only if $Z \to Y \to X$.

Note that for any measure M_{XYZ}, $X \to Y \to Z$ is a Markov chain under $M_{X\times Z|Y}$ by construction.

The following corollary highlights special properties of the various densities and relative entropies when the dominating measure is a Markov chain. It will lead to the idea of a Markov approximation to an arbitrary distribution on triples extending the independent approximation of the previous corollary.

Corollary 5.3.3: Given a probability space, suppose that $M_{XYZ} >> P_{XYZ}$ are two distributions for a random vector (X,Y,Z) with the property that $Z \to Y \to X$ forms a Markov chain under M. Then

$$M_{XYZ} >> P_{X\times Z|Y} >> P_{XYZ}$$

and

$$\frac{dP_{XYZ}}{dP_{X\times Z|Y}} = \frac{f_{X|YZ}}{f_{X|Y}} \tag{5.3.5}$$

$$\frac{dP_{X\times Z|Y}}{dM_{XYZ}} = f_{YZ}f_{X|Y}. \tag{5.3.6}$$

Thus

$$\ln\frac{dP_{XYZ}}{dP_{X\times Z|Y}} + h_{X|Y} = h_{X|YZ}$$

$$\ln\frac{dP_{X\times Z|Y}}{dM_{XYZ}} = h_{YZ} + h_{X|Y}$$

and taking expectations yields

$$D(P_{XYZ}||P_{X\times Z|Y}) + H_{P||M}(X|Y) = H_{P||M}(X|YZ) \tag{5.3.7}$$

$$D(P_{X\times Z|Y}||M_{XYZ}) = D(P_{YZ}||M_{YZ}) + H_{P||M}(X|Y). \tag{5.3.8}$$

Furthermore,

$$P_{X\times Z|Y} = \overline{P_{X|Y}P_{YZ}}, \tag{5.3.9}$$

that is,

$$P_{X\times Z|Y}(F_X \times F_Z \times F_Y) = \int_{F_Y\times F_Z} P_{X|Y}(F_X|y)dP_{ZY}(z,y). \tag{5.3.10}$$

Lastly, if $Z \to Y \to X$ is a Markov chain under M, then it is also a Markov chain under P if and only if

$$h_{X|Y} = h_{X|YZ} \tag{5.3.11}$$

in which case

$$H_{P||M}(X||Y) = H_{P||M}(X||YZ). \tag{5.3.12}$$

Proof: Dcfinc

$$g(x,y,z) = \frac{f_{X|YZ}(x|y,z)}{f_{X|Y}(x|y)} = \frac{f_{XYZ}(x,y,z)}{f_{YZ}(y,z)}\frac{f_Y(y)}{f_{XY}(x,y)}$$

and simplify notation by defining the measure $Q = P_{X\times Z|Y}$. Note that $Z \to Y \to X$ is a Markov chain with respect to Q. To prove the first statement of the Corollary requires proving the following relation:

$$P_{XYZ}(F_X \times F_Y \times F_Z) = \int_{F_X\times F_Y\times F_Z} gdQ;$$

$$\text{all } F_X \in \mathcal{B}_{A_X}, F_Z \in \mathcal{B}_{A_Z}, F_Y \in \mathcal{B}_{A_Y}.$$

From iterated expectation with respect to Q (e.g., Section 5.9 of [50])

$$E(g1_{F_X}(X)1_{F_Z}(Z)1_{F_Y}(Y)) = E(1_{F_Y}(Y)1_{F_Z}(Z)E(g1_{F_X}(X)|YZ))$$

$$= \int 1_{F_Y}(y)1_{F_Z}(z)(\int_{F_X} g(x,y,z)\, dQ_{X|YZ}(x|y,z))\, dQ_{YZ}(y,z).$$

Since $Q_{YZ} = P_{YZ}$ and $Q_{X|YZ} = P_{X|Y}$ Q-a.e. by construction, the previous formula implies that

$$\int_{F_X\times F_Y\times F_Z} g\, dQ = \int_{F_Y\times F_Z} dP_{YZ} \int_{F_X} gdP_{X|Y}.$$

This proves (5.3.9. Since $M_{XYZ} >> P_{XYZ}$, we also have that $M_{XY} >> P_{XY}$ and hence application of Theorem 5.3.1 yields

$$\int_{F_X\times F_Y\times F_Z} gdQ = \int_{F_Y\times F_Z} dP_{YZ} \int_{F_X} gf_{X|Y}dM_{X|Y}$$

$$= \int_{F_Y \times F_Z} dP_{YZ} \int_{F_X} f_{X|YZ} dM_{X|Y}.$$

By assumption, however, $M_{X|Y} = M_{X|YZ}$ a.e. and therefore

$$\int_{F_X \times F_Y \times F_Z} g\, dQ = \int_{F_Y \times F_Z} dP_{YZ} \int_{F_X} f_{X|YZ}\, dM_{X|YZ}$$

$$= \int_{F_Y \times F_Z} dP_{YZ} \int_{F_X} dP_{X|YZ} = P_{XYZ}(F_X \times F_Y \times F_Z),$$

where the final step follows from iterated expectation. This proves (5.3.5 and that $Q >> P_{XYZ}$.

To prove (5.3.6) we proceed in a similar manner and replace g by $f_{X|Y} f_{ZY}$ and replace Q by $M_{XYZ} = M_{X \times Y|Z}$. Also abbreviate $P_{X \times Y|Z}$ to $\hat{P}$. As in the proof of (5.3.5) we have since $Z \rightarrow Y \rightarrow X$ is a Markov chain under M that

$$\int_{F_X \times F_Y \times F_Z} g\, dQ = \int_{F_Y \times F_Z} dM_{YZ} \int_{F_X} g\, dM_{X|Y}$$

$$= \int_{F_Y \times F_Z} f_{ZY}\, dM_{YZ} \left(\int_{F_X} f_{X|Y}\, dM_{X|Y} \right) = \int_{F_Y \times F_Z} dP_{YZ} \left(\int_{F_X} f_{X|Y}\, dM_{X|Y} \right).$$

From Theorem 5.3.1 this is

$$\int_{F_Y \times F_Z} P_{X|Y}(F_X|y)\, dP_{YZ}.$$

But $P_{YZ} = \hat{P}_{YZ}$ and

$$P_{X|Y}(F_X|y) = \hat{P}_{X|Y}(F_X|y) = \hat{P}_{X|YZ}(F_X|yz)$$

since $\hat{P}$ yields a Markov chain. Thus the previous formula is $\hat{P}(F_X \times F_Y \times F_Z)$, proving (5.3.6) and the corresponding absolute continuity.

If $Z \rightarrow Y \rightarrow X$ is a Markov chain under both M and P, then $P_{X \times Z|Y} = P_{XYZ}$ and hence

$$\frac{dP_{XYZ}}{dP_{X \times Z|Y}} = 1 = \frac{f_{X|YZ}}{f_{X|Y}},$$

which implies (5.3.11). Conversely, if (5.3.11) holds, then $f_{X|YZ} = f_{X|Y}$ which with (5.3.5) implies that $P_{XYZ} = P_{X \times Z|Y}$, proving that $Z \rightarrow Y \rightarrow X$ is a Markov chain under P. □

The previous corollary and one of the constructions used will prove important later and hence it is emphasized now with a definition and another corollary giving an interesting interpretation.

Given a distribution P_{XYZ}, define the distribution $P_{X\times Z|Y}$ as the *Markov approximation* to P_{XYZ}. Abbreviate $P_{X\times Z|Y}$ to $\hat{P}$. The definition has two motivations. First, the distribution $\hat{P}$ makes $Z \to Y \to X$ a Markov chain which has the same initial distribution $\hat{P}_{ZY} = P_{ZY}$ and the same conditional distribution $\hat{P}_{X|Y} = P_{X|Y}$, the only difference is that $\hat{P}$ yields a Markov chain, that is, $\hat{P}_{X|ZY} = \hat{P}_{X|Y}$. The second motivation is the following corollary which shows that of all Markov distributions, $\hat{P}$ is the closest to P in the sense of minimizing the divergence.

Corollary 5.3.4: Given a distribution $P = P_{XYZ}$, let $\mathcal{M}$ denote the class of all distributions for XYZ for which $Z \to Y \to X$ is a Markov chain under M_{XYZ} ($M_{XYZ} = M_{X\times Z|Y}$). Then

$$\inf_{M_{XYZ}\in\mathcal{M}} D(P_{XYZ}||M_{XYZ}) = D(P_{XYZ}||P_{X\times Z|Y});$$

that is, the infimum is a minimum and it is achieved by the Markov approximation.

Proof: If no M_{XYZ} in the constraint set satisfies $M_{XYZ} >> P_{XYZ}$, then both sides of the above equation are infinite. Hence confine interest to the case $M_{XYZ} >> P_{XYZ}$. Similarly, if all such M_{XYZ} yield an infinite divergence, we are done. Hence we also consider only M_{XYZ} yielding finite divergence. Then the previous corollary implies that $M_{XYZ} >> P_{X\times Z|Y} >> P_{XYZ}$ and hence

$$\begin{aligned} D(P_{XYZ}||M_{XYZ}) &= D(P_{XYZ}||P_{X\times Z|Y}) + D(P_{X\times Z|Y}||M_{XYZ}) \\ &\geq D(P_{XYZ}||P_{X\times Z|Y}) \end{aligned}$$

with equality if and only if

$$D(P_{X\times Z|Y}||M_{XYZ}) = D(P_{YZ}||M_{YZ}) + H_{P||M}(X|Y) = 0.$$

But this will be zero if M is the Markov approximation to P since then $M_{YZ} = P_{YZ}$ and $M_{X|Y} = P_{X|Y}$ by construction. □

Generalized Conditional Relative Entropy

We now return to the issue of providing a general definition of conditional relative entropy, that is, one which does not require the existence of the densities or, equivalently, the absolute continuity of the underlying measures. We require, however, that the general definition reduce to that considered thus far when the densities exist so that all of the results of this section will remain valid when applicable. The general definition takes advantage of the of the basic construction of the early part of this section. Once again let M_{XY} and P_{XY} be two measures, where we no longer assume that $M_{XY} >> P_{XY}$. Define as in Theorem 5.3.1 the modified measure S_{XY} by

$$S_{XY}(F \times G) = \int_G M_{X|Y}(F|y)dP_Y(y); \tag{5.3.13}$$

that is, S_{XY} has the same Y marginal as P_{XY} and the same conditional distribution of X given Y as M_{XY}. We now replace the previous definition by the following: The *conditional relative entropy* is defined by

$$H_{P||M}(X|Y) = D(P_{XY}||S_{XY}). \tag{5.3.14}$$

If $M_{XY} >> P_{XY}$ as before, then from Theorem 5.3.1 this is the same quantity as the original definition and there is no change. The divergence of (5.3.14), however, is well-defined even if it is not true that $M_{XY} >> P_{XY}$ and hence the densities used in the original definition do not work. The key question is whether or not the chain rule

$$H_{P||M}(Y) + H_{P||M}(X|Y) = H_{P||M}(XY) \tag{5.3.15}$$

remains valid in the more general setting. It has already been proven in the case that $M_{XY} >> P_{XY}$, hence suppose this does not hold. In this case, if it is also true that $M_Y >> P_Y$ does not hold, then both the marginal and joint relative entropies will be infinite and (5.3.15) again must hold since the conditional relative entropy is nonnegative. Thus we need only show that the formula holds for the case where $M_Y >> P_Y$ but it is not true that $M_{XY} >> P_{XY}$. By assumption there must be an event F for which

$$M_{XY}(F) = \int M_{X|Y}(F_y)\, dM_Y(y) = 0$$

but

$$P_{XY} = \int P_{X|Y}(F_y)\, dP_Y(y) \neq 0,$$

where $F_y = \{(x,y) : (x,y) \in F\}$ is the section of F at F_y. Thus $M_{X|Y}(F_y) = 0$ M_Y-a.e. and hence also P_Y-a.e. since $M_Y >> P_Y$. Thus

$$S_{XY}(F) = \int M_{X|Y}(F_y)\, dP_Y(y) = 0$$

and hence it is not true that $S_{XY} >> P_{XY}$ and therefore

$$D(P_{XY}||S_{XY}) = \infty,$$

which proves that the chain rule holds in the general case.

It can happen that P_{XY} is not absolutely continuous with respect to M_{XY}, and yet $D(P_{XY}||S_{XY}) < \infty$ and hence $P_{XY} << S_{XY}$ and hence

$$H_{P||M}(X|Y) = \int dP_{XY} \ln \frac{dP_{XY}}{dS_{XY}},$$

in which case it makes sense to define the conditional density

$$f_{X|Y} \equiv \frac{dP_{XY}}{dS_{XY}}$$

so that exactly as in the original tentative definition in terms of densities (5.3.1) we have that

$$H_{P||M}(X|Y) = \int dP_{XY} \ln f_{X|Y}.$$

Note that this allows us to define a meaningful conditional density even though the joint density f_{XY} does not exist! If the joint density does exist, then the conditional density reduces to the previous definition from Theorem 5.3.1.

We summarize the generalization in the following theorem.

Theorem 5.3.2 The conditional relative entropy defined by (5.3.14) and (5.3.13) agrees with the definition (5.3.1) in terms of densities and satisfies the chain rule (5.3.15). If the conditional relative entropy is finite, then

$$H_{P||M}(X|Y) = \int dP_{XY} \ln f_{X|Y},$$

where the conditional density is defined by

$$f_{X|Y} \equiv \frac{dP_{XY}}{dS_{XY}}.$$

If $M_{XY} >> P_{XY}$, then this reduces to the usual definition

$$f_{X|Y} = \frac{f_{XY}}{f_Y}.$$

The generalizations can be extended to three or more random variables in the obvious manner.

5.4 Limiting Entropy Densities

We now combine several of the results of the previous section to obtain results characterizing the limits of certain relative entropy densities.

Lemma 5.4.1: Given a probability space $(\Omega, \mathcal{B})$ and an asymptotically generating sequence of sub-σ-fields $\mathcal{F}_n$ and two measures $M >> P$, let $P_n = P_{\mathcal{F}_n}$, $M_n = M_{\mathcal{F}_n}$ and let $h_n = \ln dP_n/dM_n$ and $h = \ln dP/dM$ denote the entropy densities. If $D(P||M) < \infty$, then

$$\lim_{n\to\infty} \int |h_n - h|\, dP = 0,$$

that is, $h_n \to h$ in L^1. Thus the entropy densities h_n are uniformly integrable.

Proof: Follows from the Corollaries 5.2.3 and 5.2.6. □

The following lemma is Lemma 1 of Algoet and Cover [7].

Lemma 5.4.2: Given a sequence of nonnegative random variables $\{f_n\}$ defined on a probability space $(\Omega, \mathcal{B}, P)$, suppose that

$$E(f_n) \le 1; \quad \text{all } n.$$

Then

$$\limsup_{n\to\infty} \frac{1}{n} \ln f_n \le 0.$$

Proof: Given any $\epsilon > 0$ the Markov inequality and the given assumption imply that

$$P(f_n > e^{n\epsilon}) \le \frac{E(f_n)}{e^{n\epsilon}} \le e^{-n\epsilon}.$$

We therefore have that

$$P(\frac{1}{n} \ln f_n \ge \epsilon) \le e^{-n\epsilon}$$

and therefore

$$\sum_{n=1}^{\infty} P(\frac{1}{n} \ln f_n \ge \epsilon) \le \sum_{n=1}^{\infty} e^{-n\epsilon} = \frac{1}{e^{\epsilon-1}} < \infty,$$

Thus from the Borel-Cantelli lemma (Lemma 4.6.3 of [50]), $P(n^{-1}h_n \ge \epsilon \text{ i.o.}) = 0$. Since ϵ is arbitrary, the lemma is proved. □

The lemma easily gives the first half of the following result, which is also due to Algoet and Cover [7], but the proof is different here and does not use martingale theory. The result is the generalization of Lemma 2.7.1.

Theorem 5.4.1: Given a probability space $(\Omega, \mathcal{B})$ and an asymptotically generating sequence of sub-σ-fields $\mathcal{F}_n$, let M and P be two probability measures with their restrictions $M_n = M_{\mathcal{F}_n}$ and $P_n = P_{\mathcal{F}_n}$. Suppose that $M_n >> P_n$ for all n and define $f_n = dP_n/dM_n$ and $h_n = \ln f_n$. Then

$$\limsup_{n\to\infty} \frac{1}{n} h_n \le 0, M - \text{a.e.}$$

and

$$\liminf_{n\to\infty} \frac{1}{n} h_n \geq 0, P-\text{a.e.}.$$

If it is also true that $M >> P$ (e.g., $D(P||M) < \infty$), then

$$\lim_{n\to\infty} \frac{1}{n} h_n = 0, P-\text{a.e.}.$$

Proof: Since

$$E_M f_n = E_{M_n} f_n = 1,$$

the first statement follows from the previous lemma. To prove the second statement consider the probability

$$P(-\frac{1}{n}\ln\frac{dP_n}{M_n} > \epsilon) = P_n(-\frac{1}{n}\ln f_n > \epsilon) = P_n(f_n < e^{-n\epsilon})$$

$$= \int_{f_n < e^{-n\epsilon}} dP_n = \int_{f_n < e^{-n\epsilon}} f_n\, dM_n$$

$$< e^{-n\epsilon} \int_{f_n < e^{-n\epsilon}} dM_n = e^{-n\epsilon} M_n(f_n < e^{-n\epsilon}) \leq e^{-n\epsilon}.$$

Thus it has been shown that

$$P(\frac{1}{n} h_n < -\epsilon) \leq e^{-n\epsilon}$$

and hence again applying the Borel-Cantelli lemma we have that

$$P(n^{-1} h_n \leq -\epsilon \text{ i.o.}) = 0$$

which proves the second claim of the theorem.

If $M >> P$, then the first result also holds P-a.e., which with the second result proves the final claim. □

Barron [9] provides an additional property of the sequence h_n/n. If $M >> P$, then the sequence h_n/n is dominated by an integrable function.

5.5 Information for General Alphabets

We can now use the divergence results of the previous sections to generalize the definitions of information and to develop their basic properties. We assume now that all random variables and processes are defined on a common underlying probability space $(\Omega, \mathcal{B}, P)$. As we have seen how all of the various information quantities–entropy, mutual information, conditional mutual information–can be expressed in terms of divergence in the finite case, we immediately have definitions for the general case. Given two random variables X and Y, define the average mutual information between them by

$$I(X;Y) = D(P_{XY}||P_X \times P_Y), \tag{5.5.1}$$

where P_{XY} is the joint distribution of the random variables X and Y and $P_X \times P_Y$ is the product distribution.

Define the entropy of a single random variable X by

$$H(X) = I(X;X). \tag{5.5.2}$$

From the definition of divergence this implies that

$$I(X;Y) = \sup_{\mathcal{Q}} H_{P_{XY}||P_X \times P_Y}(\mathcal{Q}).$$

from Dobrushin's theorem (Lemma 5.2.2), the supremum can be taken over partitions whose elements are contained in generating field. Letting the generating field be the field of all rectangles of the form $F \times G$, $F \in \mathcal{B}_{A_X}$ and $G \in \mathcal{B}_{A_Y}$, we have the following lemma which is often used as a definition for mutual information.

Lemma 5.5.1:

$$I(X;Y) = \sup_{q,r} I(q(X);\, r(Y)),$$

where the supremum is over all quantizers q and r of A_X and A_Y. Hence there exist sequences of increasingly fine quantizers $q_n : A_X \to A_n$ and $r_n : A_Y \to B_n$ such that

$$I(X;Y) = \lim_{n \to \infty} I(q_n(X);\, r_n(Y)).$$

Applying this result to entropy we have that

$$H(X) = \sup_{q} H(q(X)),$$

where the supremum is over all quantizers.

By "increasingly fine" quantizers is meant that the corresponding partitions $\mathcal{Q}_n = \{q_n^{-1}(a);\ a \in A_n\}$ are successive refinements, e.g., atoms in $\mathcal{Q}_n$ are unions of atoms in $\mathcal{Q}_{n+1}$. (If this were not so, a new quantizer could be defined for which it was true.) There is an important drawback to the lemma (which will shortly be removed in Lemma 5.5.5 for the special case where the alphabets are standard): the quantizers which approach the suprema may depend on the underlying measure P_{XY}. In particular, a sequence of quantizers which work for one measure need not work for another.

Given a third random variable Z, let A_X, A_Y, and A_Z denote the alphabets of X, Y, and Z and define the conditional average mutual information

$$I(X;Y|Z) = D(P_{XYZ}||P_{X\times Y|Z}). \tag{5.5.3}$$

This is the extension of the discrete alphabet definition of (2.5.4) and it makes sense only if the distribution $P_{X\times Y|Z}$ exists, which is the case if the alphabets are standard but may not be the case otherwise. We shall later provide an alternative definition due to Wyner [152] that is valid more generally and equal to the above when the spaces are standard.

Note that $I(X;Y|Z)$ can be interpreted using Corollary 5.3.4 as the divergence between P_{XYZ} and its Markov approximation.

Combining these definitions with Lemma 5.2.1 yields the following generalizations of the discrete alphabet results.

Lemma 5.5.2: Given two random variables X and Y, then

$$I(X;Y) \geq 0$$

with equality if and only if X and Y are independent. Given three random variables X, Y, and Z, then

$$I(X;Y|Z) \geq 0$$

with equality if and only if $Y \to Z \to X$ form a Markov chain.

Proof: The first statement follow from Lemma 5.2.1 since X and Y are independent if and only if $P_{XY} = P_X \times P_Y$. The second statement follows from (5.5.3) and the fact that $Y \to Z \to X$ is a Markov chain if and only if $P_{XYZ} = P_{X\times Y|Z}$ (see, e.g., Corollary 5.10.2 of [50]). □

The properties of divergence provide means of computing and approximating these information measures. From Lemma 5.2.3, if $I(X;Y)$ is finite, then

$$I(X;Y) = \int \ln \frac{dP_{XY}}{d(P_X \times P_Y)}\, dP_{XY} \tag{5.5.4}$$

and if $I(X;Y|Z)$ is finite, then

$$I(X;Y|Z) = \int \ln \frac{dP_{XYZ}}{dP_{X\times Y|Z}}\, dP_{XYZ}. \tag{5.5.5}$$

For example, if X, Y are two random variables whose distribution is absolutely continuous with respect to Lebesgue measure $dxdy$ and hence which have a pdf $f_{XY}(x,y) = dP_{XY}(xy)/dxdy$, then

$$I(X;Y) = \int dxdy f_{XY}(xy) \ln \frac{f_{XY}(x,y)}{f_X(x) f_Y(y)},$$

where f_X and f_Y are the marginal pdf's, e.g.,

$$f_X(x) = \int f_{XY}(x,y)\,dy = \frac{dP_X(x)}{dx}.$$

In the cases where these densities exist, we define the information densities

$$i_{X;Y} = \ln \frac{dP_{XY}}{d(P_X \times P_Y)}$$

$$i_{X;Y|Z} = \ln \frac{dP_{XYZ}}{dP_{X\times Y|Z}}. \tag{5.5.6}$$

The results of Section 5.3 can be used to provide conditions under which the various information densities exist and to relate them to each other. Corollaries 5.3.1 and 5.3.2 combined with the definition of mutual information immediately yield the following two results.

Lemma 5.5.3: Let X and Y be standard alphabet random variables with distribution P_{XY}. Suppose that there exists a product distribution $M_{XY} = M_X \times M_Y$ such that $M_{XY} >> P_{XY}$. Then

$$M_{XY} >> P_X \times P_Y >> P_{XY},$$

$$i_{X;Y} = \ln(f_{XY}/f_X f_Y) = \ln(f_{X|Y}/f_X)$$

and

$$I(X;Y) + H_{P||M}(X) = H_{P||M}(X|Y). \tag{5.5.7}$$

Comment: This generalizes the fact that $I(X;Y) = H(X) - H(X|Y)$ for the finite alphabet case. The sign reversal results from the difference in definitions of relative entropy and entropy. Note that this implies that unlike ordinary entropy, relative entropy is *increased* by conditioning, at least when the reference measure is a product measure.

The previous lemma provides an apparently more general test for the existence of a mutual information density than the requirement that $P_X \times P_Y >> P_{XY}$, it states that if P_{XY} is dominated by *any* product measure, then it is also dominated by the product of its own marginals and hence the densities exist. The generality is only apparent, however, as the given condition implies from Corollary 5.3.1 that the distribution is dominated by its independent approximation. Restating Corollary 5.3.1 in terms of mutual information yields the following.

Corollary 5.5.1: Given a distribution P_{XY} let $\mathcal{M}$ denote the collection of all product distributions $M_{XY} = M_X \times M_Y$. Then

$$I(X;Y) = \inf_{M_{XY}\in\mathcal{M}} H_{P||M}(X|Y) = \inf_{M_{XY}\in\mathcal{M}} D(P_{XY}||M_{XY}).$$

The next result is an extension of Lemma 5.5.3 to conditional information densities and relative entropy densities when three random variables are considered. It follows immediately from Corollary 5.3.3 and the definition of conditional information density.

Lemma 5.5.4: (The chain rule for relative entropy densities) Suppose that $M_{XYZ} >> P_{XYZ}$ are two distributions for three standard alphabet random variables and that $Z \to Y \to X$ is a Markov chain under M_{XYZ}. Let $f_{X|YZ}$, $f_{X|Y}$, $h_{X|YZ}$, and $h_{X|Y}$ be as in Section 5.3. Then $P_{X\times Z|Y} >> P_{XYZ}$,

$$h_{X|YZ} = i_{X;Z|Y} + h_{X|Y} \tag{5.5.8}$$

and

$$H_{P||M}(X|Y,Z) = I(X;Z|Y) + H_{P||M}(X|Y). \tag{5.5.9}$$

Thus, for example,

$$H_{P||M}(X|Y,Z) \geq H_{P||M}(X|Y).$$

As with Corollary 5.5.1, the lemma implies a variational description of conditional mutual information. The result is just a restatement of Corollary 5.3.4.

Corollary 5.5.2: Given a distribution P_{XYZ} let $\mathcal{M}$ denote the class of all distributions for XYZ under which $Z \to Y \to X$ is a Markov chain, then

$$I(X;Z|Y) = \inf_{M_{XYZ}\in\mathcal{M}} H_{P||M}(X|Y,Z) = \inf_{M_{XYZ}\in\mathcal{M}} D(P_{XYZ}||M_{XYZ}),$$

and the minimum is achieved by $M_{XYZ} = P_{X\times Z|Y}$.

The following Corollary relates the information densities of the various information measures and extends Kolmogorov's equality to standard alphabets.

Corollary 5.5.3: (The chain rule for information densities and Kolmogorov's formula.) Suppose that X,Y, and Z are random variables with standard alphabets and distribution P_{XYZ}. Suppose also that there exists a distribution $M_{XYZ} = M_X \times M_{YZ}$ such that $M_{XYZ} >> P_{XYZ}$. (This is true, for example, if $P_X \times P_{YZ} >> P_{XYZ}$.) Then the information densities $i_{X;Z|Y}$, $i_{X;Y}$, and $i_{X;(YZ)}$ exist and are related by

$$i_{X;Z|Y} + i_{X;Y} = i_{X;(Y,Z)} \tag{5.5.10}$$

and

$$I(X;Z|Y) + I(X;Y) = I(X;(Y,Z)). \tag{5.5.11}$$

Proof: If $M_{XYZ} = M_X \times M_{YZ}$, then $Z \to Y \to X$ is trivially a Markov chain since $M_{X|YZ} = M_{X|Y} = M_X$. Thus the previous lemma can be applied to this M_{XYZ} to conclude that $P_{X\times Z|Y} >> P_{XYZ}$ and that (5.5.8) holds. We also have that $M_{XY} = M_X \times M_Y >> P_{XY}$. Thus all of the densities exist. Applying Lemma 5.5.3 to the product measures $M_{XY} = M_X \times M_Y$ and $M_{X(YZ)} = M_X \times M_{YZ}$ in (5.5.8) yields

$$i_{X;Y|Z} = h_{X|YZ} - h_{X|Y} = \ln f_{X|YZ} - \ln f_{X|Y}$$

$$= \ln \frac{f_{X|YZ}}{f_X} - \ln \frac{f_{X|Y}}{f_X} = i_{X;YZ} - i_{X;Y}.$$

Taking expectations completes the proof. □

The previous Corollary implies that if $P_X \times P_{YZ} >> P_{XYZ}$, then also $P_{X\times Z|Y} >> P_{XYZ}$ and $P_X \times P_Y >> P_{XY}$ and hence that the existence of $i_{X;YZ}$ implies that of $i_{X;Z|Y}$ and $i_{X;Y}$. The following result provides a converse to this fact: the existence of the latter two densities implies that of the first. The result is due to Dobrushin [32]. (See also Theorem 3.6.1 of Pinsker [125] and the translator's comments.)

Corollary 5.5.4: If $P_{X\times Z|Y} >> P_{XYZ}$ and $P_X \times P_Y >> P_{XY}$, then also $P_X \times P_{YZ} >> P_{XYZ}$ and

$$\frac{dP_{XYZ}}{d(P_X \times P_{YZ})} = \frac{dP_{XY}}{d(P_X \times P_Y)}.$$

Thus the conclusions of Corollary 5.5.3 hold.

Proof: The key to the proof is the demonstration that

$$\frac{dP_{XY}}{d(P_X \times P_Y)} = \frac{dP_{X\times Z|Y}}{d(P_X \times P_{YZ})}, \tag{5.5.12}$$

which implies that $P_X \times P_{YZ} >> P_{X\times Z|Y}$. Since it is assumed that $P_{X\times Z|Y} >> P_{XYZ}$, the result then follows from the chain rule for Radon-Nikodym derivatives.

Eq. (5.5.12) will be proved if it is shown that for all $F_X \in \mathcal{B}_{A_X}$, $F_Y \in \mathcal{B}_{A_Y}$, and $F_Z \in \mathcal{B}_{A_Z}$,

$$P_{X\times Z|Y}(F_X \times F_Z \times F_Y) = \int_{F_X\times F_Z\times F_Y} \frac{dP_{XY}}{d(P_X \times P_Y)} d(P_X \times P_{YZ}). \tag{5.5.13}$$

The thrust of the proof is the demonstration that for any measurable non-negative function $f(x,z)$

$$\int_{z\in F_Z} f(x,y)\, d(P_X \times P_{YZ})(x,y,z)$$

$$= \int f(x,y)P_{Z|Y}(F_Z|y)d(P_X \times P_Y)(x,y). \tag{5.5.14}$$

The lemma will then follow by substituting

$$f(x,y) = \frac{dP_{XY}}{d(P_X \times P_Y)}(x,y)1_{F_X}(x)1_{F_Y}(y)$$

into (5.5.14) to obtain (5.5.13).

To prove (5.5.14) first consider indicator functions of rectangles: $f(x,y) = 1_{F_X}(x)1_{F_Y}(y)$. Then both sides of (5.5.14) equal $P_X(F_X)P_{YZ}(F_Y \times F_Y)$ from the definitions of conditional probability and product measures. In particular, from Lemma 5.10.1 of [50] the left-hand side is

$$\int_{z \in F_Z} 1_{F_X}(x)1_{F_Y}(y)\, d(P_X \times P_{YZ})(x,y,z)$$

$$= (\int 1_{F_X} dP_X)(\int 1_{F_Y \times F_Z}\, dP_{YZ}) = P_X(F)P_{YZ}(F_Y \times F_Z)$$

and the right-hand side is

$$\int 1_{F_X}(x)1_{F_Y}(y)P_{Z|Y}(F_Z|y)\, d(P_X \times P_Y)(x,y)$$

$$= (\int 1_{F_X}(x)\, dP_X(x))(\int 1_{F_Y}(y)P_{Z|Y}(F_Z|y)\, dP_Y(y))$$

$$= P_X(F)P_{YZ}(F_Y \times F_Z),$$

as claimed. This implies (5.5.14) holds also for simple functions and hence also for positive functions by the usual approximation arguments. □

Note that Kolmogorov's formula (5.5.9) gives a formula for computing conditional mutual information as

$$I(X;Z|Y) = I(X;(Y,Z)) - I(X;Y).$$

The formula is only useful if it is not indeterminate, that is, not of the form $\infty - \infty$. This will be the case if $I(Y;Z)$ (the smaller of the two mutual informations) is finite.

Corollary 5.2.5 provides a means of approximating mutual information by that of finite alphabet random variables. Assume now that the random variables X, Y have standard alphabets. For, say, random variable X with alphabet A_X there must then be an asymptotically generating sequence of finite fields $\mathcal{F}_X(n)$ with atoms $\mathcal{A}_X(n)$, that is, all of the members of $\mathcal{F}_X(n)$ can be written as unions of disjoint sets in $\mathcal{A}_X(n)$ and $\mathcal{F}_X(n) \uparrow \mathcal{B}_{A_X}$; that is, $\mathcal{B}_{A_X} = \sigma(\bigcup_n \mathcal{F}_X(n))$. The atoms $\mathcal{A}_X(n)$ form a partition of the alphabet of X.

Consider the divergence result of Corollary 5.2.5. with $P = P_{XY}$, $M = P_X \times P_Y$ and quantizer $q^{(n)}(x,y) = (q_X^{(n)}(x), q_Y^{(n)}(y))$. Consider the limit $n \to \infty$. Since $\mathcal{F}_X(n)$ asymptotically generates $\mathcal{B}_{A_X}$ and $\mathcal{F}_Y(n)$ asymptotically generates $\mathcal{B}_{A_Y}$ and since the pair σ-field $\mathcal{B}_{A_X \times A_Y}$ is generated by rectangles, the field generated by all sets of the form $F_X \times F_Y$ with $F_X \in \mathcal{F}_X(n)$, some n, and $F_Y \in \mathcal{F}_Y(m)$, some m, generates $\mathcal{B}_{A_X \times A_Y}$. Hence Corollary 5.2.5 yields the first result of the following lemma. The second is a special case of the first. The result shows that the quantizers of Lemma 5.5.1 can be chosen in a manner not depending on the underlying measure if the alphabets are standard.

Lemma 5.5.5: Suppose that X and Y are random variables with standard alphabets defined on a common probability space. Suppose that $q_X^{(n)}$, $n = 1, 2, \cdots$ is a sequence of quantizers for A_X such that the corresponding partitions asymptotically generate $\mathcal{B}_{A_X}$. Define quantizers for Y similarly. Then for any distribution P_{XY}

$$I(X;Y) = \lim_{n\to\infty} I(q_X^{(n)}(X); q_Y^{(n)}(Y))$$

and

$$H(X) = \lim_{n\to\infty} H(q_X^{(n)}(X));$$

that is, the same quantizer sequence works for all distributions.

An immediate application of the lemma is the extension of the convexity properties of Lemma 2.5.4 to standard alphabets.

Corollary 5.5.5: Let μ denote a distribution on a space $(A_X, \mathcal{B}_{A_X})$, and let ν be a regular conditional distribution $\nu(F|x) = \Pr(Y \in F|X = x)$, $x \in A_X$, $F \in \mathcal{B}_{A_Y}$. Let $\mu\nu$ denote the resulting joint distribution. Let $I_{\mu\nu} = I_{\mu\nu}(X;Y)$ be the average mutual information. Then $I_{\mu\nu}$ is a convex $\bigcup$ function of ν and a convex $\bigcap$ function of μ.

Proof: Follows immediately from Lemma 5.5.5 and the finite alphabet result Lemma 2.5.4. □

Next consider the mutual information $I(f(X), g(Y))$ for arbitrary measurable mappings f and g of X and Y. From Lemma 5.5.2 applied to the random variables $f(X)$ and $g(Y)$, this mutual information can be approximated arbitrarily closely by $I(q_1(f(X)); q_2(g(Y)))$ by an appropriate choice of quantizers q_1 and q_2. Since the composition of q_1 and f constitutes a finite quantization of X and similarly $q_2 g$ is a quantizer for Y, we must have that

$$I(f(X); g(Y)) \approx I(q_1(f(X)); q_2(g(Y)) \leq I(X;Y).$$

Making this precise yields the following corollary.

Corollary 5.5.6: If f is a measurable function of X and g is a measurable function of Y, then

$$I(f(X), g(Y)) \leq I(X;Y).$$

The corollary states that mutual information is reduced by any measurable mapping, whether finite or not. For practice we point out another proof of this basic result that directly applies a property of divergence. Let $P = P_{XY}$, $M = P_X \times P_Y$, and define the mapping $r(x,y) = (f(x), g(y))$. Then from Corollary 5.2.2 we have

$$I(X;Y) = D(P||M) \geq D(P_r||M_r) \geq D(P_{f(X),g(Y)}||M_{f(X),g(Y)}).$$

But $M_{f(X),g(Y)} = P_{f(X)} \times P_{g(Y)}$ since

$$M_{f(X),g(Y)}(F_X \times F_Z) = M(f^{-1}(F_X)\bigcap g^{-1}(F_Y)$$

$$= P_X(f^{-1}(F_X)) \times P_Y(g^{-1}(F_Y)) = P_{f(X)}(F_X) \times P_{g(Y)}(F_Y).$$

Thus the previous inequality yields the corollary. □

For the remainder of this section we focus on conditional entropy and information.

Although we cannot express mutual information as a difference of ordinary entropies in the general case (since the entropies of nondiscrete random variables are generally infinite), we can obtain such a representation in the case where one of the two variables is discrete. Suppose we are given a joint distribution P_{XY} and that X is discrete. We can choose a version of the conditional probability given Y so that $p_{X|Y}(x|y) = P(X = x|Y = y)$ is a valid pmf (considered as a function of x for fixed y) with P_Y probability 1. (This follows from Corollary 5.8.1 of [50] since the alphabet of X is discrete; the alphabet of Y need not be even standard.) Define

$$H(X|Y = y) = \sum_x p_{X|Y}(x|y) \ln \frac{1}{p_{X|Y}(x|y)}$$

and

$$H(X|Y) = \int H(X|Y = y)\, dP_Y(y).$$

Note that this agrees with the formula of Section 2.5 in the case that both alphabets are finite. The following result is due to Wyner [152].

Lemma 5.5.6: If X, Y are random variables and X has a finite alphabet, then

$$I(X;Y) = H(X) - H(X|Y).$$

Proof: We first claim that $p_{X|Y}(x|y)/p_X(x)$ is a version of $dP_{XY}/d(P_X \times P_Y)$. To see this observe that for $F \in \mathcal{B}(A_X \times A_Y)$, letting F_y denote the section $\{x : (x,y) \in F\}$ we have that

$$\int_F \frac{p_{X|Y}(x|y)}{p_X(x)} d(P_X \times P_Y) = \int \sum_{x \in F_y} \frac{p_{X|Y}(x|y)}{p_X(x)} p_X(x) dP_Y(y)$$

$$= \int dP_Y(y) \sum_{x \in F_y} p_{X|Y}(x|y) = \int dP_Y(y) P_X(F_y|y) = P_{XY}(F).$$

Thus

$$I(X;Y) = \int \ln(\frac{p_{X|Y}(x|y)}{p_X(x)})\, dP_{XY}$$

$$= H(X) + \int dP_Y(y) \sum_x p_{X|Y}(x|y) \ln p_{X|Y}(x|y). \quad \Box$$

We now wish to study the effects of quantizing on conditional information. As discussed in Section 2.5, it is not true that $I(X;Y|Z)$ is always greater than $I(f(X);q(Y)|r(Z))$ and hence that $I(X;Y|Z)$ can be written as a supremum over all quantizers and hence the definition of (5.5.3) and the formula (5.5.5) do not have the intuitive counterpart of a limit of informations of quantized values. We now consider an alternative (and more general) definition of conditional mutual information due to Wyner [152]. The definition has the form of a supremum over quantizers and does not require the existence of the probability measure $P_{X \times Y|Z}$ and hence makes sense for alphabets that are not standard. Given P_{XYZ} and any finite measurements f and g on X and Y, we can choose a version of the conditional probability given $Z = z$ so that

$$p_z(a,b) = \Pr(f(X) = a, g(Y) = b|Z = z)$$

is a valid pmf with probability 1 (since the alphabets of f and g are finite and hence standard a regular conditional probability exists from Corollary 5.8.1 of [50]). For such finite measurements we can define

$$I(f(X);g(Y)|Z = z) = \sum_{a \in A_f} \sum_{b \in A_g} p_z(a,b) \ln \frac{p_z(a,b)}{\sum_{a'} p_z(a',b) \sum_{b'} p_z(a,b')},$$

that is, the ordinary discrete average mutual information with respect to the distribution p_z.

Lemma 5.5.7: Define

$$I'(X;Y|Z) = \sup_{f,g} \int dP_Z(z) I(f(X);g(Y)|Z = z),$$

where the supremum is over all quantizers. Then there exist sequences of quantizers (as in Lemma 5.5.5) such that

$$I'(X;Y|Z) = \lim_{n\to\infty} I'(q_m(X);r_m(Y)|Z).$$

I' satisfies Kolmogorov's formula, that is,

$$I'(X;Y|Z) = I((X,Z);Y) - I(Y;Z).$$

If the alphabets are standard, then

$$I(X;Y|Z) = I'(X;Y|Z).$$

Comment: The main point here is that conditional mutual information can be expressed as a supremum or limit of quantizers. The other results simply point out that the two conditional mutual informations have the same relation to ordinary mutual information and are (therefore) equal when both are defined. The proof follows Wyner [152].

Proof: First observe that for any quantizers q and r of A_f and A_g we have from the usual properties of mutual information that

$$I(q(f(X));r(g(Y))|Z=z) \leq I(f(X);g(Y)|Z=z)$$

and hence integrating we have that

$$I'(q(f(X));r(g(Y))|Z) = \int I(q(f(X));r(g(Y))|Z=z)\,dP_Z(z) \quad (5.5.15)$$

$$\leq \int I(f(X);g(Y)|Z=z)\,dP_Z(z)$$

and hence taking the supremum over all q and r to get $I'(f(X);g(Y)|Z)$ yields

$$I'(f(X);g(Y)|Z) = \int I(f(X);g(Y)|Z=z)\,dP_Z(z). \quad (5.5.16)$$

so that (5.5.15) becomes

$$I'(q(f(X));r(g(Y))|Z) \leq I'(f(X);g(Y)|Z) \quad (5.5.17)$$

for any quantizers q and r and the definition of I' can be expressed as

$$I'(X;Y|Z) = \sup_{f,g} I'(f(X);g(Y)|Z), \quad (5.5.18)$$

where the supremum is over all quantizers f and g. This proves the first part of the lemma since the supremum can be approached by a sequence of quantizers. Next observe that

$$I'(f(X);g(Y)|Z) = \int I(f(X);g(Y)|Z=z)\,dP_Z(z)$$

$$= H(g(Y)|Z) - H(g(Y)|f(X),Z).$$

Since we have from Lemma 5.5.6 that

$$I(g(Y);Z) = H(g(Y)) - H(g(Y)|Z),$$

we have by adding these equations and again using Lemma 5.5.6 that

$$I(g(Y);Z) + I'(f(X);g(Y)|Z) = H(g(Y)) - H(g(Y)|f(X),Z)$$

$$= I((f(X),Z);g(Y)).$$

Taking suprema over both sides over all quantizers f and g yields the relation

$$I(X;Z) + I'(X;Y|Z) = I((X,Z);Y),$$

proving Kolmogorov's formula. Lastly, if the spaces are standard, then from Kolmogorov's inequality for the original definition (which is valid for the standard space alphabets) combined with the above formula implies that

$$I'(X;Y|Z) = I((X,Z);Y) - I(X;Z) = I(X;Y|Z).\square$$

5.6 Some Convergence Results

We now combine the convergence results for divergence with the definitions and properties of information densities to obtain some convergence results for information densities. Unlike the results to come for relative entropy rate and information rate, these are results involving the information between a sequence of random variables and a fixed random variable.

Lemma 5.6.1: Given random variables X and $Y_1, Y_2, \cdots$ defined on a common probability space,

$$\lim_{n\to\infty} I(X;(Y_1,Y_2,\cdots,Y_n)) = I(X;(Y_1,Y_2,\cdots)).$$

If in addition $I(X;(Y_1,Y_2,\cdots)) < \infty$ and hence $P_X \times P_{Y_1,Y_2,\cdots} >> P_{X,Y_1,Y_2,\cdots}$, then

$$i_{X;Y_1,Y_2,\cdots,Y_n} \underset{n\to\infty}{\to} i_{X;Y_1,Y_2,\cdots}$$

in L^1.

Proof: The first result follows from Corollary 5.2.5 with $X, Y_1, Y_2, \cdots, Y_{n-1}$ replacing X^n, P being the distribution $P_{X,Y_1,\cdots}$, and M being the product distribution $P_X \times P_{Y_1,Y_2,\cdots}$. The density result follows from Lemma 5.4.1. □

Corollary 5.6.1: Given random variables X, Y, and $Z_1, Z_2, \cdots$ defined on a common probability space, then

$$\lim_{n\to\infty} I(X;Y|Z_1,Z_2,\cdots,Z_n) = I(X;Y|Z_1,Z_2,\cdots).$$

If

$$I((X,Z_1,\cdots);Y) < \infty,$$

(e.g., if Y has a finite alphabet and hence $I((X,Z_1,\cdots);Y) \leq H(Y) < \infty$), then also

$$i_{X;Y|Z_1,\cdots,Z_n} \underset{n\to\infty}{\to} i_{X;Y|Z_1,\cdots} \tag{5.6.1}$$

in L^1.

Proof: From Kolmogorov's formula

$$I(X;Y|Z_1,Z_2,\cdots,Z_n) =$$

$$I(X;(Y,Z_1,Z_2,\cdots,Z_n)) - I(X;Z_1,\cdots,Z_n) \geq 0. \tag{5.6.2}$$

From the previous lemma, the first term on the left converges as $n \to \infty$ to $I(X;(Y,Z_1,\cdots))$ and the second term on the right is the negative of a term converging to $I(X;(Z_1,\cdots))$. If the first of these limits is finite, then the difference in (5.6.2) converges to the difference of these terms, which gives (5.6.1). From the chain rule for information densities, the conditional information density is the difference of the information densities:

$$i_{X;Y|Z_1,\cdots,Z_n} = i_{X;(Y,Z_1,\cdots,Z_n)} - i_{X;(Z_1,\cdots,Z_n)}$$

which is converging in L^1x to

$$i_{X;Y|Z_1,\cdots} = i_{X;(Y,Z_1,\cdots)} - i_{X;(Z_1,\cdots)},$$

again invoking the density chain rule. If $I(X;Y|Z_1,\cdots) = \infty$ then quantize Y as $q(Y)$ and note since $q(Y)$ has a finite alphabet that

$$I(X;Y|Z_1,Z_2,\cdots,Z_n) \geq I(X;q(Y)|Z_1,Z_2,\cdots,Z_n) \underset{n\to\infty}{\to} I(X;q(Y)|Z_1,\cdots)$$

and hence

$$\liminf_{N\to\infty} I(X;Y|Z_1,\cdots) \geq I(X;q(Y)|Z_1,\cdots).$$

Since the right-hand term above can be made arbitrarily large, the remaining part of the lemma is proved. □

Lemma 5.6.2: If

$$P_X \times P_{Y_1,Y_2,\cdots} >> P_{X,Y_1,Y_2,\cdots}$$

(e.g., $I(X;(Y_1,Y_2,\cdots)) < \infty$), then with probability 1.

$$\lim_{n\to\infty} \frac{1}{n} i(X;(Y_1,\cdots,Y_n)) = 0.$$

Proof: This is a corollary of Theorem 5.4.1. Let P denote the distribution of $\{X,Y_1,Y_2,\cdots\}$ and let M denote the distribution $P_X \times P_{Y_1,\cdots}$. By assumption $M >> P$. The information density is

$$i(X;(Y_1,\cdots,Y_n)) = \ln \frac{dP_n}{dM_n},$$

where P_n and M_n are the restrictions of P and M to $\sigma(X,Y_1,\cdots Y_n)$. Theorem 5.4.1 can therefore be applied to conclude that P-a.e.

$$\lim_{n\to\infty} \frac{1}{n} \ln \frac{dP_n}{dM_n} = 0,$$

which proves the lemma. □

The lemma has the following immediate corollary.

Corollary 5.6.2: If $\{X_n\}$ is a process with the property that

$$I(X_0;X_{-1},X_{-2},\cdots) < \infty,$$

that is, there is a finite amount of information between the zero time sample and the infinite past, then

$$\lim_{n\to\infty} \frac{1}{n} i(X_0;X_{-1},\cdots,X_{-n}) = 0.$$

If the process is stationary, then also

$$\lim_{n\to\infty} \frac{1}{n} i(X_n;X^n) = 0.$$

Chapter 6

Information Rates II

6.1 Introduction

In this chapter we develop general definitions of information rate for processes with standard alphabets and we prove a mean ergodic theorem for information densities. The L^1 results are extensions of the results of Moy [105] and Perez [123] for stationary processes, which in turn extended the Shannon-McMillan theorem from entropies of discrete alphabet processes to information densities. (See also Kieffer [85].) We also relate several different measures of information rate and consider the mutual information between a stationary process and its ergodic component function. In the next chapter we apply the results of Chapter 5 on divergence to the definitions of this chapter for limiting information and entropy rates to obtain a number of results describing the behavior of such rates. In Chapter 8 almost everywhere ergodic theorems for relative entropy and information densities are proved.

6.2 Information Rates for General Alphabets

Suppose that we are given a pair random process $\{X_n, Y_n\}$ with distribution p. The most natural definition of the information rate between the two processes is the extension of the definition for the finite alphabet case:

$$\bar{I}(X;Y) = \limsup_{n\to\infty} \frac{1}{n} I(X^n;Y^n).$$

This was the first general definition of information rate and it is due to Dobrushin [32]. While this definition has its uses, it also has its problems.

Another definition is more in the spirit of the definition of information itself: We formed the general definitions by taking a supremum of the finite alphabet definitions over all finite alphabet codings or quantizers. The above definition takes the limit of such suprema. An alternative definition is to instead reverse the order and take the supremum of the limit and hence the supremum of the information rate over all finite alphabet codings of the process. This will provide a definition of information rate similar to the definition of the entropy of a dynamical system. There is a question as to what kind of codings we permit, that is, do the quantizers quantize individual outputs or long sequences of outputs. We shall shortly see that it makes no difference. Suppose that we have a pair random process $\{X_n, Y_n\}$ with standard alphabets A_X and A_Y and suppose that $f : A_X^\infty \to A_f$ and $g : A_Y^\infty \to A_g$ are stationary codings of the X and Y sequence spaces into a finite alphabet. We will call such finite alphabet stationary mappings *sliding block codes* or *stationary digital codes.* Let $\{f_n, g_n\}$ be the induced output process, that is, if T denotes the shift (on any of the sequence spaces) then $f_n(x,y) = f(T^n x)$ and $g_n(x,y) = g(T^n y)$. Recall that $f(T^n(x,y)) = f_n(x,y)$, that is, shifting the input n times results in the output being shifted n times.

Since the new process $\{f_n, g_n\}$ has a finite alphabet, its mutual information rate is defined. We now define the information rate for general alphabets as follows:

$$I^*(X;Y) = \sup_{\text{sliding block codes } f,g} \bar{I}(f;g)$$

$$= \sup_{\text{sliding block codes } f,g} \limsup_{n\to\infty} \frac{1}{n} I(f^n; g^n).$$

We now focus on AMS processes, in which case the information rates for finite alphabet processes (e.g., quantized processes) is given by the limit, that is,

$$I^*(X;Y) = \sup_{\text{sliding block codes } f,g} \bar{I}(f;g)$$

$$= \sup_{\text{sliding block codes } f,g} \lim_{n\to\infty} \frac{1}{n} I(f^n; g^n).$$

The following lemma shows that for AMS sources I^* can also be evaluated by constraining the sliding block codes to be scalar quantizers.

Lemma 6.2.1: Given an AMS pair random process $\{X_n, Y_n\}$ with standard alphabet,

$$I^*(X;Y) = \sup_{q,r} \bar{I}(q(X); r(Y)) = \sup_{q,r} \limsup_{n\to\infty} \frac{1}{n} I(q(X)^n; r(Y)^n),$$

where the supremum is over all quantizers q of A_X and r of A_Y and where $q(X)^n = q(X_0), \cdots, q(X_{n-1})$.

Proof: Clearly the right hand side above is less than I^* since a scalar quantizer is a special case of a stationary code. Conversely, suppose that f and g are sliding block codes such that $\bar{I}(f;g) \geq I^*(X;Y) - \epsilon$. Then from Corollary 4.3.1 there are quantizers q and r and codes f' and g' depending only on the quantized processes $q(X_n)$ and $r(Y_n)$ such that $\bar{I}(f';g') \geq \bar{I}(f;g) - \epsilon$. From Lemma 4.3.3, however, $\bar{I}(q(X);r(Y)) \geq \bar{I}(f';g')$ since f' and g' are stationary codings of the quantized processes. Thus $\bar{I}(q(X);r(Y)) \geq I^*(X;Y) - 2\epsilon$, which proves the lemma. □

Corollary 6.2.1:

$$I^*(X;Y) \leq \bar{I}(X;Y).$$

If the alphabets are finite, then the two rates are equal.

Proof: The inequality follows from the lemma and the fact that

$$I(X^n;Y^n) \geq I(q(X)^n;r(Y)^n)$$

for any scalar quantizers q and r (where $q(X)^n$ is $q(X_0), \cdots, q(X_{n-1})$). If the alphabets are finite, then the identity mappings are quantizers and yield $I(X^n;Y^n)$ for all n. □

Pinsker [125] introduced the definition of information rate as a supremum over all scalar quantizers and hence we shall refer to this information rate as the Pinsker rate. The Pinsker definition has the advantage that we can use the known properties of information rates for finite alphabet processes to infer those for general processes, an attribute the first definition lacks.

Corollary 6.2.2: Given a standard alphabet pair process alphabet $A_X \times A_Y$ there is a sequence of scalar quantizers q_m and r_m such that for any AMS pair process $\{X_n, Y_n\}$ having this alphabet (that is, for any process distribution on the corresponding sequence space)

$$I(X^n;Y^n) = \lim_{m\to\infty} I(q_m(X)^n;r_m(Y)^n)$$

$$I^*(X;Y) = \lim_{m\to\infty} \bar{I}(q_m(X);r_m(Y)).$$

Furthermore, the above limits can be taken to be increasing by using finer and finer quantizers. *Comment:* It is important to note that the same sequence of quantizers gives both of the limiting results.

Proof: The first result is Lemma 5.5.5. The second follows from the previous lemma. □

Observe that

$$I^*(X;Y) = \lim_{m\to\infty} \limsup_{n\to\infty} \frac{1}{n} I(q_m(X); r_m(Y))$$

whereas

$$\bar{I}(X;Y) = \limsup_{n\to\infty} \lim_{m\to\infty} \frac{1}{n} I(q_m(X); r_m(Y)).$$

Thus the two notions of information rate are equal if the two limits can be interchanged. We shall later consider conditions under which this is true and we shall see that equality of these two rates is important for proving ergodic theorems for information densities.

Lemma 6.2.2: Suppose that $\{X_n, Y_n\}$ is an AMS standard alphabet random process with distribution p and stationary mean $\bar{p}$. Then

$$I_p^*(X;Y) = I_{\bar{p}}^*(X;Y).$$

I_p^* is an affine function of the distribution p. If $\bar{p}$ has ergodic decomposition $\bar{p}_{xy}$, then

$$I_p^*(X;Y) = \int d\bar{p}(x,y) I^*{}_{\bar{p}_{xy}}(X;Y).$$

If f and g are stationary codings of X and Y, then

$$I_p^*(f;g) = \int d\bar{p}(x,y) I_{\bar{p}_{xy}}^*(f;g).$$

Proof: For any scalar quantizers q and r of X and Y we have that $\bar{I}_p(q(X); r(Y)) = \bar{I}_{\bar{p}}(q(X); r(Y))$. Taking a limit with ever finer quantizers yields the first equality. The fact that I^* is affine follows similarly. Suppose that $\bar{p}$ has ergodic decomposition $\bar{p}_{xy}$. Define the induced distributions of the quantized process by m and m_{xy}, that is, $m(F) = \bar{p}(x, y : \{q(x_i), r(y_i); i \in \mathcal{T}\} \in F)$ and similarly for m_{xy}. It is easy to show that m is stationary (since it is a stationary coding of a stationary process), that the m_{xy} are stationary ergodic (since they are stationary codings of stationary ergodic processes), and that the m_{xy} form an ergodic decomposition of m. If we let X'_n, Y'_n denote the coordinate functions on the quantized output sequence space (that is, the processes $\{q(X_n), r(Y_n)\}$ and $\{X'_n, Y'_n\}$ are equivalent), then using the ergodic decomposition of mutual information for finite alphabet processes (Lemma 4.3.1) we have that

$$\bar{I}_p(q(X); r(Y)) = \bar{I}_m(X';Y') = \int \bar{I}_{m_{x'y'}}(X';Y')\, dm(x',y')$$

$$= \int \bar{I}_{\bar{p}_{xy}}(q(X); r(Y))\, d\bar{p}(x,y).$$

Replacing the quantizers by the sequence q_m, r_m the result then follows by taking the limit using the monotone convergence theorem. The result for stationary codings follows similarly by applying the previous result to the induced distributions and then relating the equation to the original distributions. □

The above properties are not known to hold for $\bar{I}$ in the general case. Thus although $\bar{I}$ may appear to be a more natural definition of mutual information rate, I^* is better behaved since it inherits properties from the discrete alphabet case. It will be of interest to find conditions under which the two rates are the same, since then $\bar{I}$ will share the properties possessed by I^*. The first result of the next section adds to the interest by demonstrating that when the two rates are equal, a mean ergodic theorem holds for the information densities.

6.3 A Mean Ergodic Theorem for Densities

Theorem 6.3.1: Given an AMS pair process $\{X_n, Y_n\}$ with standard alphabets, assume that for all n

$$P_{X^n} \times P_{Y^n} >> P_{X^nY^n}$$

and hence that the information densities

$$i_{X^n;Y^n} = \ln \frac{dP_{X^n,Y^n}}{d(P_{X^n} \times P_{Y^n})}$$

are well defined. For simplicity we abbreviate $i_{X^n;Y^n}$ to i_n when there is no possibility of confusion. If the limit

$$\lim_{n\to\infty} \frac{1}{n} I(X^n;Y^n) = \bar{I}(X;Y)$$

exists and

$$\bar{I}(X;Y) = I^*(X;Y) < \infty,$$

then $n^{-1} i_n(X^n;Y^n)$ converges in L^1 to an invariant function $i(X;Y)$. If the stationary mean of the process has an ergodic decomposition $\bar{p}_{xy}$, then the limiting density is $I^*{}_{\bar{p}_{xy}}(X;Y)$, the information rate of the ergodic component in effect.

Proof: Let q_m and r_m be asymptotically accurate quantizers for A_X and A_Y. Define the discrete approximations $\hat{X}_n = q_m(X_n)$ and $\hat{Y}_n = r_m(Y_n)$. Observe that $P_{X^n} \times P_{Y^n} >> P_{X^nY^n}$ implies that $P_{\hat{X}^n} \times P_{\hat{Y}^n} >> P_{\hat{X}^n\hat{Y}^n}$

and hence we can define the information densities of the quantized vectors by

$$\hat{i}_n = \ln \frac{dP_{\hat{X}_n \hat{Y}^n}}{d(P_{\hat{X}^n} \times P_{\hat{Y}^n})}.$$

For any m we have that

$$\int |\frac{1}{n} i_n(x^n; y^n) - I^*{}_{\bar{p}_{xy}}(X;Y)|\, dp(x,y) \leq$$

$$\int |\frac{1}{n} i_n(x^n; y^n) - \frac{1}{n}\hat{i}_n(q_m(x)^n; r_m(y)^n)|\, dp(x,y) +$$

$$\int |\frac{1}{n}\hat{i}_n(q_m(x)^n; r_m(y)^n) - \bar{I}_{\bar{p}_{xy}}(q_m(X); r_m(Y))|\, dp(x,y) +$$

$$\int |\bar{I}_{\bar{p}_{xy}}(q_m(X); r_m(Y)) - I^*{}_{\bar{p}_{xy}}(X;Y)|\, dp(x,y), \tag{6.3.1}$$

where

$$q_m(x)^n = (q_m(x_0), \cdots, q_m(x_{n-1})),$$
$$r_m(y)^n = (r_m(y_0), \cdots, r_m(y_{n-1})),$$

and $\bar{I}_p(q_m(X); r_m(Y))$ denotes the information rate of the process $\{q_m(X_n), r_m(Y_n)$ $n = 0, 1, \cdots,\}$ when p is the process measure describing $\{X_n, Y_n\}$.

Consider first the right-most term of (6.3.1). Since I^* is the supremum over all quantized versions,

$$\int |\bar{I}_{\bar{p}_{xy}}(q_m(X); r_m(Y)) - I^*{}_{\bar{p}_{xy}}(X;Y)|\, dp(x,y)$$

$$= \int (I^*{}_{\bar{p}_{xy}}(X;Y) - \bar{I}_{\bar{p}_{xy}}(q_m(X); r_m(Y)))\, dp(x,y).$$

Using the ergodic decomposition of I^* (Lemma 6.2.2) and that of $\bar{I}$ for discrete alphabet processes (Lemma 4.3.1) this becomes

$$\int |\bar{I}_{\bar{p}_{xy}}(q_m(X); r_m(Y)) - I^*{}_{\bar{p}_{xy}}(X;Y)|\, dp(x,y)$$

$$= I_p^*(X;Y) - \bar{I}_p(q_m(X); r_m(Y)). \tag{6.3.2}$$

For fixed m the middle term of (6.3.1) can be made arbitrarily small by taking n large enough from the finite alphabet result of Lemma 4.3.1. The first term on the right can be bounded above using Corollary 5.2.6 with $\mathcal{F} = \sigma(q(X)^n; r(Y)^n)$ by

$$\frac{1}{n}(I(X^n; Y^n) - I(\hat{X}^n; \hat{Y}^n) + \frac{2}{e})$$

which as $n \to \infty$ goes to $\bar{I}(X;Y) - \bar{I}(q_m(X);r_m(Y))$. Thus we have for any m that

$$\limsup_{n\to\infty} \int |\frac{1}{n} i_n(x^n;y^n) - I^*_{\bar{p}_{xy}}(X;Y)|\, dp(x,y)$$
$$\leq \bar{I}(X;Y) - \bar{I}(q_m(X);r_m(Y)) + I^*(X;Y) - \bar{I}(q_m(X);r_m(Y))$$

which as $m \to \infty$ becomes $\bar{I}(X;Y) - I^*(X;Y)$, which is 0 by assumption. □

6.4 Information Rates of Stationary Processes

In this section we introduce two more definitions of information rates for the case of stationary two-sided processes. These rates are useful tools in relating the Dobrushin and Pinsker rates and they provide additional interpretations of mutual information rates in terms of ordinary mutual information. The definitions follow Pinsker [125].

Henceforth assume that $\{X_n, Y_n\}$ is a stationary two-sided pair process with standard alphabets. Define the sequences $y = \{y_i; i \in \mathcal{T}\}$ and $Y = \{Y_i; i \in \mathcal{T}\}$

First define

$$\tilde{I}(X;Y) = \limsup_{n\to\infty} \frac{1}{n} I(X^n;Y),$$

that is, consider the per-letter limiting information between n-tuples of X and the entire sequence from Y. Next define

$$I^-(X;Y) = I(X_0;Y|X_{-1},X_{-2},\cdots),$$

that is, the average conditional mutual information between one letter from X and the entire Y sequence given the infinite past of the X process. We could define the first rate for one-sided processes, but the second makes sense only when we can consider an infinite past. For brevity we write $X^- = X_{-1}, X_{-2}, \cdots$ and hence

$$I^-(X;Y) = I(X_0;Y|X^-).$$

Theorem 6.4.1:

$$\tilde{I}(X;Y) \geq \bar{I}(X;Y) \geq I^*(X;Y) \geq I^-(X;Y).$$

If the alphabet of X is finite, then the above rates are all equal.

Comment: We will later see more general sufficient conditions for the equality of the various rates, but the case where one alphabet is finite is simple and important and points out that the rates are all equal in the finite alphabet case.

Proof: We have already proved the middle inequality. The left inequality follows immediately from the fact that $I(X^n;Y) \geq I(X^n;Y^n)$ for all n. The remaining inequality is more involved. We prove it in two steps. First we prove the second half of the theorem, that the rates are the same if X has finite alphabet. We then couple this with an approximation argument to prove the remaining inequality. Suppose now that the alphabet of X is finite. Using the chain rule and stationarity we have that

$$\frac{1}{n}I(X^n;Y^n) = \frac{1}{n}\sum_{i=0}^{n-1} I(X_i;Y^n|X_0,\cdots,X_{i-1})$$

$$= \frac{1}{n}\sum_{i=0}^{n-1} I(X_0;Y^n_{-i}|X_{-1},\cdots,X_{-i}),$$

where Y^n_{-i} is $Y_{-i},\cdots,Y_{-i+n-1}$, that is, the n-vector starting at $-i$. Since X has finite alphabet, each term in the sum is bounded. We can show as in Section 5.5 (or using Kolmogorov's formula and Lemma 5.5.1) that each term converges as $i \to \infty$, $n \to \infty$, and $n-i \to \infty$ to $I(X_0;Y|X_{-1},X_{-2},\cdots)$ or $I^-(X;Y)$. These facts, however, imply that the above Cesàro average converges to the same limit and hence $\bar{I} = I^-$. We can similarly expand $\tilde{I}$ as

$$\frac{1}{n}\sum_{i=0}^{n-1} I(X_i;Y|X_0,\cdots,X_{i-1}) = \frac{1}{n}\sum_{i=0}^{n-1} I(X_0;Y|X_{-1},\cdots,X_{-i}),$$

which converges to the same limit for the same reasons. Thus $\tilde{I} = \bar{I} = I^-$ for stationary processes when the alphabet of X is finite. Now suppose that X has a standard alphabet and let q_m be an asymptotically accurate sequences of quantizers. Recall that the corresponding partitions are increasing, that is, each refines the previous partition. Fix $\epsilon > 0$ and choose m large enough so that the quantizer $\alpha(X_0) = q_m(X_0)$ satisfies

$$I(\alpha(X_0);Y|X^-) \geq I(X_0;Y|X^-) - \epsilon.$$

Observe that so far we have only quantized X_0 and not the past. Since

$$\mathcal{F}_m = \sigma(\alpha(X_0),Y,q_m(X_{-i});\ i = 1,2,\cdots)$$

asymptotically generates

$$\sigma(\alpha(X_0),Y,X_{-i};\ i = 1,2,\cdots),$$

given ϵ we can choose for m large enough (larger than before) a quantizer $\beta(x) = q_m(x)$ such that if we define $\beta(X^-)$ to be $\beta(X_{-1}), \beta(X_{-2}), \cdots$, then

$$|I(\alpha(X_0); (Y, \beta(X^-))) - I(\alpha(X_0); (Y, X^-))| \leq \epsilon$$

and

$$|I(\alpha(X_0); \beta(X^-)) - I(\alpha(X_0); X^-)| \leq \epsilon.$$

Using Kolmogorov's formula this implies that

$$|I(\alpha(X_0); Y|X^-) - I(\alpha(X_0); Y|\beta(X^-))| \leq 2\epsilon$$

and hence that

$$I(\alpha(X_0); Y|\beta(X^-)) \geq I(\alpha(X_0); Y|X^-) - 2\epsilon \geq I(X_0; Y|X^-) - 3\epsilon.$$

But the partition corresponding to β refines that of α and hence increases the information; hence

$$I(\beta(X_0); Y|\beta(X^-)) \geq I(\alpha(X_0); Y|\beta(X^-)) \geq I(X_0; Y|X^-) - 3\epsilon.$$

Since $\beta(X_n)$ has a finite alphabet, however, from the finite alphabet result the left-most term above must be $\bar{I}(\beta(X); Y)$, which can be made arbitrarily close to $I^*(X; Y)$. Since ϵ is arbitrary, this proves the final inequality. □

The following two theorems provide sufficient conditions for equality of the various information rates. The first result is almost a special case of the second, but it is handled separately as it is simpler, much of the proof applies to the second case, and it is not an exact special case of the subsequent result since it does not require the second condition of that result. The result corresponds to condition (7.4.33) of Pinsker [125], who also provides more general conditions. The more general condition is also due to Pinsker and strongly resembles that considered by Barron [9].

Theorem 6.4.2: Given a stationary pair process $\{X_n, Y_n\}$ with standard alphabets, if

$$I(X_0; (X_{-1}, X_{-2}, \cdots)) < \infty,$$

then

$$\tilde{I}(X; Y) = \bar{I}(X; Y) = I^*(X; Y) = I^-(X; Y). \tag{6.4.1}$$

Proof: We have that

$$\frac{1}{n} I(X^n; Y) \leq \frac{1}{n} I(X^n; (Y, X^-))$$

$$= \frac{1}{n} I(X^n; X^-) + \frac{1}{n} I(X^n; Y|X^-), \tag{6.4.2}$$

where, as before, $X^- = \{X_{-1}, X_{-2}, \cdots\}$. Consider the first term on the right. Using the chain rule for mutual information

$$\frac{1}{n} I(X^n; X^-) = \frac{1}{n} \sum_{i=0}^{n-1} I(X_i; X^- | X^i)$$

$$= \frac{1}{n} \sum_{i=0}^{n-1} (I(X_i; (X^i, X^-)) - I(X_i; X^i)). \tag{6.4.3}$$

Using stationarity we have that

$$\frac{1}{n} I(X^n; X^-) = \frac{1}{n} \sum_{i=0}^{n-1} (I(X_0; X^-) - I(X_0; (X_{-1}, \cdots, X_{-i})). \tag{6.4.4}$$

The terms $I(X_0; (X_{-1}, \cdots, X_{-i}))$ are converging to $I(X_0; X^-)$, hence the terms in the sum are converging to 0, i.e.,

$$\lim_{i \to \infty} I(X_i; X^- | X^i) = 0. \tag{6.4.5}$$

The Cesàro mean of (6.4.3) is converging to the same thing and hence

$$\frac{1}{n} I(X^n; X^-) \to 0. \tag{6.4.6}$$

Next consider the term $I(X^n; Y | X^-)$. For any positive integers n,m we have

$$I(X^{n+m}; Y | X^-) = I(X^n; Y | X^-) + I(X_n^m; Y | X^-, X^n), \tag{6.4.7}$$

where $X_n^m = X_n, \cdots, X_{n+m-1}$. From stationarity, however, the rightmost term is just $I(X^m; Y | X^-)$ and hence

$$I(X^{m+n}; Y | X^-) = I(X^n; Y | X^-) + I(X^m; Y | X^-). \tag{6.4.8}$$

This is just a linear functional equation of the form $f(n+m) = f(n) + f(m)$ and the unique solution to such an equation is $f(n) = nf(1)$, that is,

$$\frac{1}{n} I(X^n; Y | X^-) = I(X_0; Y | X^-) = I^-(X; Y). \tag{6.4.9}$$

Taking the limit supremum in (6.4.2) we have shown that

$$\bar{I}(X; Y) \leq I^-(X; Y), \tag{6.4.10}$$

which with Theorem 6.4.1 completes the proof. □

Intuitively, the theorem states that if one of the processes has finite average mutual information between one symbol and its infinite past, then the Dobrushin and Pinsker information rates yield the same value and hence there is an L^1 ergodic theorem for the information density.

To generalize the theorem we introduce a condition that will often be useful when studying asymptotic properties of entropy and information. A stationary process $\{X_n\}$ is said to have the *finite-gap information property* if there exists an integer K such that

$$I(X_K; X^- | X^K) < \infty, \tag{6.4.11}$$

where, as usual, $X^- = (X_{-1}, X_{-2}, \cdots)$. When a process has this property for a specific K, we shall say that it has the K-gap information property. Observe that if a process possesses this property, then it follows from Lemma 5.5.4

$$I(X_K; (X_{-1}, \cdots, X_{-l}) | X^K) < \infty;\ l = 1, 2, \cdots \tag{6.4.12}$$

Since these informations are finite,

$$P_{X^n}^{(K)} >> P_{X^n};\ n = 1, 2, \cdots, \tag{6.4.13}$$

where $P_{X^n}^{(K)}$ is the Kth order Markov approximation to P_{X^n}.

Theorem 6.4.3: Given a stationary standard alphabet pair process $\{X_n, Y_n\}$, if $\{X_n\}$ satisfies the finite-gap information property (6.4.11) and if, in addition,

$$I(X^K; Y) < \infty, \tag{6.4.14}$$

then (6.4.1) holds.

(If $K = 0$ then there is no conditioning and (6.4.14) is trivial, that is, the previous theorem is the special case with $K = 0$.)

Comment: This theorem shows that if there is any finite dimensional future vector $(X_K, X_{K+1}, \cdots, X_{K+N-1})$ which has finite mutual information with respect to the infinite past X^- when conditioned on the intervening gap $(X_0, \cdots, X_{K-1})$, then the various definitions of mutual information are equivalent provided that the mutual information betwen the "gap" X^K and the sequence $\mathbf{Y}$ are finite. Note that this latter condition will hold if, for example, $\tilde{I}(X; Y)$ is finite.

Proof: For $n > K$

$$\frac{1}{n} I(X^n; Y) = \frac{1}{n} I(X^K; Y) + \frac{1}{n} I(X_K^{n-K}; Y | X^K).$$

By assumption the first term on the left will tend to 0 as $n \to \infty$ and hence we focus on the second, which can be broken up analogous to the previous theorem with the addition of the conditioning:

$$\frac{1}{n} I(X_K^{n-K}; Y|X^K) \leq \frac{1}{n} I(X_K^{n-K}; (Y, X^-|X^K))$$

$$= \frac{1}{n} I(X_K^{n-K}; X^-|X^K) + \frac{1}{n} I(X_K^{n-K}; Y|X^-, X^K).$$

Consider first the term

$$\frac{1}{n} I(X_K^{n-K}; X^-|X^K) = \frac{1}{n} \sum_{i=K}^{n-1} I(X_i; X^-|X^i),$$

which is as (6.4.3) in the proof of Theorem 6.4.2 except that the first K terms are missing. The same argument then shows that the limit of the sum is 0. The remaining term is

$$\frac{1}{n} I(X_K^{n-K}; Y|X^-, X^K) = \frac{1}{n} I(X^n; Y|X^-)$$

exactly as in the proof of Theorem 6.4.2 and the same argument then shows that the limit is $I^-(X;Y)$, which completes the proof. □

One result developed in the proofs of Theorems 6.4.2 and 6.4.3 will be important later in its own right and hence we isolate it as a corollary. The result is just (6.4.5), which remains valid under the more general conditions of Theorem 6.4.3, and the fact that the Cesàro mean of converging terms has the same limit.

Corollary 6.4.1: If a process $\{X_n\}$ has the finite-gap information property

$$I(X_K; X^-|X^K) < \infty$$

for some K, then

$$\lim_{n\to\infty} I(X_n; X^-|X^n) = 0$$

and

$$\lim_{n\to\infty} \frac{1}{n} I(X^n; X^-) = 0.$$

The corollary can be interpreted as saying that if a process has the the finite gap information property, then the mutual information between a single sample and the infinite past conditioned on the intervening samples goes to zero as the number of intervening samples goes to infinity. This can be interpreted as a form of asymptotic independence property of the process.

Corollary 6.4.2: If a one-sided stationary source $\{X_n\}$ is such that for some K, $I(X_n; X^{n-K}|X_{n-K}^K)$ is bounded uniformly in n, then it has the finite-gap property and hence

$$\bar{I}(X;Y) = I^*(X;Y).$$

Proof: Simply imbed the one-sided source into a two-sided stationary source with the same probabilities on all finite-dimensional events. For that source

$$I(X_n; X^{n-K}|X_{n-K}^K) = I(X_K; X_{-1}, \cdots, X_{-n-K}|X^K) \underset{n\to\infty}{\to} I(X_K; X^-|X^K).$$

Thus if the terms are bounded, the conditions of Theorem 6.4.2 are met for the two-sided source. The one-sided equality then follows. □

The above results have an information theoretic implication for the ergodic decomposition, which is described in the next theorem.

Theorem 6.4.4: Suppose that $\{X_n\}$ is a stationary process with the finite-gap property (6.4.11). Let ψ be the ergodic component function of Theorem 1.8.3 and suppose that for some n

$$I(X^n; \psi) < \infty. \tag{6.4.15}$$

(This will be the case, for example, if the finite-gap information property holds for 0 gap, that is, $I(X_0; X^-) < \infty$ since ψ can be determined from X^- and information is decreased by taking a function.) Then

$$\lim_{n\to\infty} \frac{1}{n} I(X^n; \psi) = 0.$$

Comment: For discrete alphabet processes this theorem is just the ergodic decomposition of entropy rate in disguise (Theorem 2.4.1). It also follows for finite alphabet processes from Lemma 3.3.1. We shall later prove a corresponding almost everywhere convergence result for the corresponding densities. All of these results have the interpretation that the per-symbol mutual information between the outputs of the process and the ergodic component decreases with time because the ergodic component in effect can be inferred from the process output in the limit of an infinite observation sequence. The finiteness condition on some $I(X^n; \psi)$ is necessary for the nonzero finite-gap case to avoid cases such as where $X_n = \psi$ for all n and hence

$$I(X^n; \psi) = I(\psi; \psi) = H(\psi) = \infty,$$

in which case the theorem does not hold.

Proof:

Define $\psi_n = \psi$ for all n. Since ψ is invariant, $\{X_n, \psi_n\}$ is a stationary process. Since X_n satisfies the given conditions, however, $\bar{I}(X;\psi) = I^*(X;\psi)$. But for any scalar quantizer q, $\bar{I}(q(X);\psi)$ is 0 from Lemma 3.3.1. $I^*(X;\psi)$ is therefore 0 since it is the supremum of $\bar{I}(q(X);\psi)$ over all quantizers q. Thus

$$0 = \bar{I}(X;\psi) = \lim_{n\to\infty} \frac{1}{n} I(X^n;\psi^n) = \lim_{n\to\infty} \frac{1}{n} I(X^n;\psi). \quad \Box$$

Chapter 7

Relative Entropy Rates

7.1 Introduction

This chapter extends many of the basic properties of relative entropy to sequences of random variables and to processes. Several limiting properties of entropy rates are proved and a mean ergodic theorem for relative entropy densities is given. The principal ergodic theorems for relative entropy and information densities in the general case are given in the next chapter.

7.2 Relative Entropy Densities and Rates

Suppose that p and m are two AMS distributions for a random process $\{X_n\}$ with a standard alphabet A. For convenience we assume that the random variables $\{X_n\}$ are coordinate functions of an underlying measurable space $(\Omega, \mathcal{B})$ where Ω is a one-sided or two-sided sequence space and $\mathcal{B}$ is the corresponding σ-field. Thus $x \in \Omega$ has the form $x = \{x_i\}$, where the index i runs from 0 to ∞ for a one-sided process and from $-\infty$ to $+\infty$ for a two-sided process. The random variables and vectors of principal interest are $X_n(x) = x_n$, $X^n(x) = x^n = (x_0, \cdots, x_{n-1})$, and $X_l^k(x) = (x_l, \cdots, x_{l+k-1})$. The process distributions p and m are both probability measures on the measurable space $(\Omega, \mathcal{B})$.

For $n = 1, 2, \cdots$ let M_{X^n} and P_{X^n} be the vector distributions induced by p and m. We assume throughout this section that $M_{X^n} >> P_{X^n}$ and hence that the Radon-Nikodym derivatives $f_{X^n} = dP_{X^n}/dM_{X^n}$ and the entropy densities $h_{X^n} = \ln f_{X^n}$ are well defined for all $n = 1, 2, \cdots$ Strictly speaking, for each n the random variable f_{X^n} is defined on the measurable space $(A^n, \mathcal{B}_{A^n})$ and hence f_{X^n} is defined on a different space for each n. When considering convergence of relative entropy densities, it is necessary to consider a sequence of random variables defined on a common measurable space, and hence two notational modifications are introduced: The random variables $f_{X^n}(X^n) : \Omega \to [0, \infty)$ are defined by

$$f_{X^n}(X^n)(x) \equiv f_{X^n}(X^n(x)) = f_{X^n}(x^n)$$

for $n = 1, 2, \cdots$. Similarly the entropy densities can be defined on the common space $(\Omega, \mathcal{B})$ by

$$h_{X^n}(X^n) = \ln f_{X^n}(X^n).$$

The reader is warned of the potentially confusing dual use of X^n in this notation: the subscript is the name of the random variable X^n and the argument is the random variable X^n itself. To simplify notation somewhat, we will often abbreviate the previous (unconditional) densities to

$$f_n = f_{X^n}(X^n);\ h_n = h_{X^n}(X^n).$$

For $n = 1, 2, \cdots$ define the relative entropy by

$$H_{p||m}(X^n) = D(P_{X^n}||M_{X^n}) = E_{P_{X^n}} h_{X^n} = E_p h_{X^n}(X^n).$$

Define the *relative entropy rate* by

$$\bar{H}_{p||m}(X) = \limsup_{n \to \infty} \frac{1}{n} H_{p||m}(X^n).$$

Analogous to Dobrushin's definition of information rate, we also define

$$H^*{}_{p||m}(X) = \sup_q \bar{H}_{p||m}(q(X)),$$

where the supremum is over all scalar quantizers q.

Define as in Chapter 5 the conditional densities

$$f_{X_n|X^n} = \frac{f_{X^{n+1}}}{f_{X^n}} = \frac{dP_{X^{n+1}}/dM_{X^{n+1}}}{dP_{X^n}/dM_{X^n}} = \frac{dP_{X_n|X^n}}{dM_{X_n|X^n}} \tag{7.2.1}$$

provided $f_{X^n} \neq 0$ and $f_{X_n|X^n} = 1$ otherwise. As for unconditional densities we change the notation when we wish to emphasize that the densities can all be defined on a common underlying sequence space. For example, we follow the notation for ordinary conditional probability density functions and define the random variables

$$f_{X_n|X^n}(X_n|X^n) = \frac{f_{X^{n+1}}(X^{n+1})}{f_{X^n}(X^n)}$$

and

$$h_{X_n|X^n}(X_n|X^n) = \ln f_{X_n|X^n}(X_n|X^n)$$

on $(\Omega, \mathcal{B})$. These densities will not have a simple abbreviation as do the unconditional densities.

Define the conditional relative entropy

$$H_{p||m}(X_n|X^n) = E_{P_{X^n}}(\ln f_{X_n|X^n}) = \int dp \ln f_{X_n|X^n}(X_n|X^n). \quad (7.2.2)$$

All of the above definitions are immediate applications of definitions of Chapter 5 to the random variables X_n and X^n. The difference is that these are now defined for all samples of a random process, that is, for all $n = 1, 2, \cdots$. The focus of this chapter is the interrelations of these entropy measures and on some of their limiting properties for large n.

For convenience define

$$D_n = H_{p||m}(X_n|X^n);\ n = 1, 2, \cdots,$$

and $D_0 = H_{p||m}(X_0)$. From Theorem 5.3.1 this quantity is nonnegative and

$$D_n + D(P_{X^n}||M_{X^n}) = D(P_{X^{n+1}}||M_{X^{n+1}}).$$

If $D(P_{X^n}||M_{X^n}) < \infty$, then also

$$D_n = D(P_{X^{n+1}}||M_{X^{n+1}}) - D(P_{X^n}||M_{X^n}).$$

We can write D_n as a single divergence if we define as in Theorem 5.3.1 the distribution $S_{X^{n+1}}$ by

$$S_{X^{n+1}}(F \times G) = \int_F M_{X_n|X^n}(F|x^n)\, dP_{X^n}(x^n);\ F \in \mathcal{B}_A;\ G \in \mathcal{B}_{A^n}. \quad (7.2.3)$$

Recall that $S_{X^{n+1}}$ combines the distribution P_{X^n} on X^n with the conditional distribution $M_{X_n|X^n}$ giving the conditional probability under M for X_n given X^n. We shall abbreviate this construction by

$$S_{X^{n+1}} = \overline{M_{X_n|X^n} P_{X^n}}. \quad (7.2.4)$$

Then

$$D_n = D(P_{X^{n+1}}||S_{X^{n+1}}). \quad (7.2.5)$$

Note that $S_{X^{n+1}}$ is not in general a consistent family of measures in the sense of the Kolmogorov extension theorem since its form changes with n, the first n samples being chosen according to p and the final sample being chosen using the conditional distribution induced by m given the first n samples. Thus, in particular, we cannot infer that there is a process distribution s which has $S_{X^n};$, $n = 1, 2, \cdots$ as its vector distributions.

We immediately have a chain rule for densities

$$f_{X^n} = \prod_{i=0}^{n-1} f_{X_i|X^i} \tag{7.2.6}$$

and a corresponding chain rule for conditional relative entropies similar to that for ordinary entropies:

$$D(P_{X^n}||M_{X^n}) = H_{p||m}(X^n) = \sum_{i=0}^{n-1} H_{p||m}(X_i|X^i) = \sum_{i=0}^{n-1} D_i. \tag{7.2.7}$$

7.3 Markov Dominating Measures

The evaluation of relative entropy simplifies for certain special cases and reduces to a mutual information when the dominating measure is a Markov approximation of the dominated measure. The following lemma is an extension to sequences of the results of Corollary 5.5.2 and Lemma 5.5.4.

Theorem 7.3.1: Suppose that p is a process distribution for a standard alphabet random process $\{X_n\}$ with induced vector distributions P_{X^n}; $n = 1, 2, \cdots$. Suppose also that there exists a process distribution m with induced vector distributions M_{X^n} such that

(a) under m $\{X_n\}$ is a k-step Markov source, that is, for all $n \geq k$, $X^{n-k} \to X^k_{n-k} \to X_n$ is a Markov chain or, equivalently,

$$M_{X_n|X^n} = M_{X_n|X^k_{n-k}},$$

and

(b) $M_{X^n} >> P_{X^n}$, $n = 1, 2, \cdots$ so that the densities

$$f_{X^n} = \frac{dP_{X^n}}{dM_{X^n}}$$

are well defined.

Suppose also that $p^{(k)}$ is the k-step Markov approximation to p, that is, the source with induced vector distributions $P^{(k)}_{X^n}$ such that

$$P^{(k)}_{X^k} = P_{X^k}$$

and for all $n \geq k$

$$P^{(k)}_{X_n|X^n} = P_{X_n|X^k_{n-k}};$$

that is, $p^{(k)}$ is a k-step Markov process having the same initial distribution and the same kth order conditional probabilities as p. Then for all $n \geq k$

$$M_{X^n} >> P^{(k)}_{X^n} >> P_{X^n} \tag{7.3.1}$$

and

$$\frac{dP^{(k)}_{X^n}}{dM_{X^n}} = f^{(k)}_{X^n} \equiv f_{X^k} \prod_{l=k}^{n-1} f_{X_l|X^k_{l-k}}, \tag{7.3.2}$$

$$\frac{dP_{X^n}}{dP^{(k)}_{X^n}} = \frac{f_{X^n}}{f^{(k)}_{X^n}}. \tag{7.3.3}$$

Furthermore

$$h_{X_n|X^n} = h_{X_n|X^k_{n-k}} + i_{X_n;X^{n-k}|X^k_{n-k}} \tag{7.3.4}$$

and hence

$$\begin{aligned} D_n &= H_{p||m}(X_n|X^n) \\ &= I_p(X_n; X^{n-k}|X^k_{n-k}) + H_{p||m}(X_n|X^k_{n-k}). \end{aligned} \tag{7.3.5}$$

Thus

$$h_{X^n} = h_{X^k} + \sum_{l=k}^{n-1} \left(h_{X_l|X^k_{l-k}} + i_{X_l;X^{l-k}|X^k_{l-k}} \right) \tag{7.3.6}$$

and hence

$$D(P_{X^n}||M_{X^n}) = H_{p||m}(X^k) + \sum_{l=k}^{n-1} (I_p(X_l; X^{l-k}|X^k_{l-k}) + H_{p||m}(X_l|X^k_{l-k})). \tag{7.3.7}$$

If $m = p^{(k)}$, then for all $n \geq k$ we have that $h_{X_n|X^k_{n-k}} = 0$ and hence

$$H_{p||p^{(k)}}(X_n|X^k_{n-k}) = 0 \tag{7.3.8}$$

and

$$D_n = I_p(X_n; X^{n-k}|X^k_{n-k}), \tag{7.3.9}$$

and hence

$$D(P_{X^n}||P^{(k)}_{X^n}) = \sum_{l=k}^{n-1} I_p(X_l; X^{l-k}|X^k_{l-k}). \tag{7.3.10}$$

Proof: If $n = k + 1$, then the results follow from Corollary 5.3.3 and Lemma 5.5.4 with $X = X_n$, $Z = X^k$, and $Y = X_k$. Now proceed by induction and assume that the results hold for n. Consider the distribution $Q_{X^{(n+1)}}$ specified by $Q_{X^n} = P_{X^n}$ and $Q_{X_n|X^n} = P_{X_n|X^k_{n-k}}$. In other words,

$$Q_{X^{n+1}} = \overline{P_{X_n|X^k_{n-k}} P_{X^n}}$$

Application of Corollary 5.3.1 with $Z = X^{n-k}$, $Y = X^k_{n-k}$, and $X = X_n$ implies that $M_{X^{n+1}} >> Q_{X^{n+1}} >> P_{X^{n+1}}$ and that

$$\frac{dP_{X^{n+1}}}{dQ_{X^{n+1}}} = \frac{f_{X_n|X^n}}{f_{X_n|X^k_{n-k}}}.$$

This means that we can write

$$P_{X^{n+1}}(F) = \int_F \frac{dP_{X^{n+1}}}{dQ_{X^{n+1}}} dQ_{X^{n+1}} = \int_F \frac{dP_{X^{n+1}}}{dQ_{X^{n+1}}} dQ_{X_n|X^n} \, dQ_{X^n}$$

$$= \int_F \frac{dP_{X^{n+1}}}{dQ_{X^{n+1}}} dP_{X_n|X^k_{n-k}} \, dP_{X^n}.$$

From the induction hypothesis we can express this as

$$P_{X^{n+1}}(F) = \int_F \frac{dP_{X^{n+1}}}{dQ_{X^{n+1}}} \frac{dP_{X^n}}{dP^{(k)}_{X^n}} dP_{X_n|X^k_{n-k}} \, dP^{(k)}_{X^n}$$

$$= \int_F \frac{dP_{X^{n+1}}}{dQ_{X^{n+1}}} \frac{dP_{X^n}}{dP^{(k)}_{X^n}}, dP^{(k)}_{X^{n+1}},$$

proving that $P^{(k)}_{X^{n+1}} >> P_{X^{n+1}}$ and that

$$\frac{dP_{X^{n+1}}}{dP^{(k)}_{X^{n+1}}} = \frac{dP_{X^{n+1}}}{dQ_{X^{n+1}}} \frac{dP_{X^n}}{dP^{(k)}_{X^n}} = \frac{f_{X_n|X^n}}{f_{X_n|X^k_{n-k}}} \frac{dP_{X^n}}{dP^{(k)}_{X^n}}.$$

This proves the right hand part of (7.3.2) and (7.3.3).

Next define the distribution $\hat{P}_{X^n}$ by

$$\hat{P}_{X^n}(F) = \int_F f^{(k)}_{X^n} \, dM_{X^n},$$

where $f^{(k)}_{X^n}$ is defined in (7.3.2). Proving that $\hat{P}_{X^n} = P^{(k)}_{X^n}$ will prove both the left hand relation of (7.3.1) and (7.3.2). Clearly

$$\frac{d\hat{P}_{X^n}}{dM_{X^n}} = f^{(k)}_{X^n}$$

and from the definition of $f^{(k)}$ and conditional densities

$$f^{(k)}_{X_n|X^n} = f^{(k)}_{X_n|X^k_{n-k}}. \tag{7.3.11}$$

From Corollary 5.3.1 it follows that $X^{n-k} \to X^k_{n-k} \to X_n$ is a Markov chain. Since this is true for any $n \geq k$, $\hat{P}_{X^n}$ is the distribution of a k-step Markov process. By construction we also have that

$$f^{(k)}_{X_n|X^k_{n-k}} = f_{X_n|X^k_{n-k}} \tag{7.3.12}$$

and hence from Theorem 5.3.1

$$P^{(k)}_{X_n|X^k_{n-k}} = P_{X_n|X^k_{n-k}}.$$

Since also $f^{(k)}_{X^k} = f_{X^k}$, $\hat{P}_{X^n} = P^{(k)}_{X^n}$ as claimed. This completes the proof of (7.3.1)–(7.3.3). Eq. (7.3.4) follows since

$$f_{X_n|X^n} = f_{X_n|X^k_{n-k}} \times \frac{f_{X_n|X^n}}{f_{X_n|X^k_{n-k}}}.$$

Eq. (7.3.5) then follows by taking expectations. Eq. (7.3.6) follows from (7.3.4) and

$$f_{X^n} = f_{X^k} \prod_{l=k}^{n-1} f_{X_l|X^l},$$

whence (7.3.7) follows by taking expectations. If $m = p^{(k)}$, then the claims follow from (5.3.11)–(5.3.12). □

Corollary 7.3.1: Given a stationary source p, suppose that for some K there exists a K-step Markov source m with distributions $M_{X^n} >> P_{X^n}$, $n = 1, 2, \cdots$. Then for all $k \geq K$ (7.3.1)–(7.3.3) hold.

Proof: If m is a K-step Markov source with the property $M_{X^n} >> P_{X^n}$, $n = 1, 2, \cdots$, then it is also a k-step Markov source with this property for all $k \geq K$. The corollary then follows from the theorem. □

Comment: The corollary implies that if *any* K-step Markov source dominates p on its finite dimensional distributions, then for *all* $k \geq K$ the k-step Markov approximations $p^{(k)}$ also dominate p on its finite dimensional distributions.

The following variational corollary follows from Theorem 7.3.1.

Corollary 7.3.2: For a fixed k let Let $\mathcal{M}^k$ denote the set of all k-step Markov distributions. Then $\inf_{M \in \mathcal{M}^k} D(P_{X^n}||M)$ is attained by $P^{(k)}$, and

$$\inf_{M \in \mathcal{M}^k} D(P_{X^n}||M) = D(P_{X^n}||P^{(k)}_{X^n}) = \sum_{l=k}^{n-1} I_p(X_l; X^{l-k}|X^k_{l-k}).$$

Since the divergence can be thought of as a distance between probability distributions, the corollary justifies considering the k-step Markov process with the same kth order distributions as the k-step Markov *approximation* or *model* for the original process: It is the minimum divergence distribution meeting the k-step Markov requirement.

7.4 Stationary Processes

Several of the previous results simplify when the processes m and p are both stationary. We can consider the processes to be two-sided since given a stationary one-sided process, there is always a stationary two-sided process with the same probabilities on all positive time events. When both processes are stationary, the densities $f_{X_m^n}$ and f_{X^n} satisfy

$$f_{X_m^n} = \frac{dP_{X_m^n}}{dM_{X_m^n}} = f_{X^n}T^m = \frac{dP_{X^n}}{dM_{X^n}}T^m,$$

and have the same expectation for any integer m. Similarly the conditional densities $f_{X_n|X^n}$, $f_{X_k|X_{k-n}^n}$, and $f_{X_0|X_{-1},X_{-2},\cdots,X_{-n}}$ satisfy

$$f_{X_n|X^n} = f_{X_k|X_{k-n}^n}T^{n-k} = f_{X_0|X_{-1},X_{-2},\cdots,X_{-n}}T^n \tag{7.4.1}$$

for any k and have the same expectation. Thus

$$\frac{1}{n}H_{p||m}(X^n) = \frac{1}{n}\sum_{i=0}^{n-1} H_{p||m}(X_0|X_{-1},\cdots,X_{-i}). \tag{7.4.2}$$

Using the construction of Theorem 5.3.1 we have also that

$$D_i = H_{p||m}(X_i|X^i) = H_{p||m}(X_0|X_{-1},\cdots,X_{-i})$$
$$= D(P_{X_0,X_{-1},\cdots,X_{-i}}||S_{X_0,X_{-1},\cdots,X_{-i}}),$$

where now

$$S_{X_0,X_{-1},\cdots,X_{-i}} = \overline{M_{X_0|X_{-1},\cdots,X_{-i}}P_{X_{-1},\cdots,X_{-i}}}; \tag{7.4.3}$$

that is,

$$S_{X_0,X_{-1},\cdots,X_{-i}}(F\times G) = \int_F M_{X_0|X_{-1},\cdots,X_{-i}}(F|x^i)\,dP_{X_{-1},\cdots,X_{-i}}(x^i);$$
$$F\in\mathcal{B}_A; G\in\mathcal{B}_{A^i}.$$

As before the S_{X^n} distributions are not in general consistent. For example, they can yield differing marginal distributions S_{X_0}. As we saw in the finite case, general conclusions about the behavior of the limiting conditional relative entropies cannot be drawn for arbitrary reference measures. If, however, we assume as in the finite case that the reference measures are Markov, then we can proceed.

Suppose now that under m the process is a k-step Markov process. Then for any $n \geq k$ $(X_{-n}, \cdots, X_{-k-2}, X_{-k-1}) \to X^k_{-k} \to X_0$ is a Markov chain under m and Lemma 5.5.4 implies that

$$H_{p||m}(X_0|X_{-1}, \cdots, X_{-n}) = H_{p||m}(X_k|X^k) + I_p(X_k; (X_{-1}, \cdots, X_{-n})|X^k) \tag{7.4.4}$$

and hence from (7.4.2)

$$\bar{H}_{p||m}(X) = H_{p||m}(X_k|X^k) + I_p(X_k; X^-|X^k). \tag{7.4.5}$$

We also have, however, that $X^- \to X^k \to X_k$ is a Markov chain under m and hence a second application of Lemma 5.5.4 implies that

$$H_{p||m}(X_0|X^-) = H_{p||m}(X_k|X^k) + I_p(X_k; X^-|X^k). \tag{7.4.6}$$

Putting these facts together and using (7.2.2) yields the following lemma.

Lemma 7.4.1: Let $\{X_n\}$ be a two-sided process with a standard alphabet and let p and m be stationary process distributions such that $M_{X^n} >> P_{X^n}$ all n and m is kth order Markov. Then the relative entropy rate exists and

$$\begin{aligned}\bar{H}_{p||m}(X) &= \lim_{n\to\infty} \frac{1}{n} H_{p||m}(X^n) \\ &= \lim_{n\to\infty} H_{p||m}(X_0|X_{-1}, \cdots, X_{-n}) = H_{p||m}(X_0|X^-) \\ &= H_{p||m}(X_k|X^k) + I_p(X_k; X^-|X^k) \\ &= E_p[\ln f_{X_k|X^k}(X_k|X^k)] + I_p(X_k; X^-|X^k).\end{aligned} \tag{7.4.7}$$

Corollary 7.4.1: Given the assumptions of Lemma 7.4.1,

$$H_{p||m}(X^N|X^-) = NH_{p||m}(X_0|X^-).$$

Proof: From the chain rule for conditional relative entropy (equation (7.2.7),

$$H_{p||m}(X^N|X^-) = \sum_{l=0}^{n-1} H_{p||m}(X_l|X^l, X^-).$$

Stationarity implies that each term in the sum equals $H_{p\|m}(X_0|X^-)$, proving the corollary. □

The next corollary extends Corollary 7.3.1 to processes.

Corollary 7.4.2: Given k and $n \geq k$, let $\mathcal{M}^k$ denote the class of all k-step stationary Markov process distributions. Then

$$\inf_{m \in \mathcal{M}^k} \bar{H}_{p\|m}(X) = \bar{H}_{p\|p^{(k)}}(X) = I_p(X_k; X^-|X^k).$$

Proof: Follows from (7.4.4) and Theorem 7.3.1. □

This result gives an interpretation of the finite-gap information property (6.4.11): If a process has this property, then there exists a k-step Markov process which is only a finite "distance" from the given process in terms of limiting per-symbol divergence. If any such process has a finite distance, then the k-step Markov approximation also has a finite distance. Furthermore, we can apply Corollary 6.4.1 to obtain the generalization of the finite alphabet result of Theorem 2.6.2

Corollary 7.4.3: Given a stationary process distribution p which satisfies the finite-gap information property,

$$\inf_k \inf_{m \in \mathcal{M}^k} \bar{H}_{p\|m}(X) = \inf_k \bar{H}_{p\|p^{(k)}}(X) = \lim_{k \to \infty} \bar{H}_{p\|p^{(k)}}(X) = 0.$$

Lemma 7.4.1 also yields the following approximation lemma.

Corollary 7.4.4: Given a process $\{X_n\}$ with standard alphabet A let p and m be stationary measures such that $P_{X^n} << M_{X^n}$ for all n and m is kth order Markov. Let q_k be an asymptotically accurate sequence of quantizers for A. Then

$$\bar{H}_{p\|m}(X) = \lim_{k \to \infty} \bar{H}_{p\|m}(q_k(X)),$$

that is, the divergence rate can be approximated arbitrarily closely by that of a quantized version of the process. Thus, in particular,

$$\bar{H}_{p\|m}(X) = H^*_{p\|m}(X).$$

Proof: This follows from Corollary 5.2.3 by letting the generating σ-fields be $\mathcal{F}_n = \sigma(q_n(X_i);\ i = 0, -1, \cdots)$ and the representation of conditional relative entropy as an ordinary divergence. □

Another interesting property of relative entropy rates for stationary processes is that we can "reverse time" when computing the rate in the sense of the following lemma.

Lemma 7.4.2: Let $\{X_n\}$, p, and m be as in Lemma 7.4.1. If either $\bar{H}_{p||m}(X) < \infty$ or $H_{P||M}(X_0|X^-) < \infty$, then

$$H_{p||m}(X_0|X_{-1},\cdots,X_{-n}) = H_{p||m}(X_0|X_1,\cdots,X_n)$$

and hence

$$H_{p||m}(X_0|X_1,X_2,\cdots) = H_{p||m}(X_0||X_{-1},X_{-2},\cdots) = \bar{H}_{p||m}(X) < \infty.$$

Proof: If $\bar{H}_{p||m}(X)$ is finite, then so must be the terms $H_{p||m}(X^n) = D(P_{X^n}||M_{X^n})$ (since otherwise all such terms with larger n would also be infinite and hence $\bar{H}$ could not be finite). Thus from stationarity

$$H_{p||m}(X_0|X_{-1},\cdots,X_{-n}) = H_{p||m}(X_n|X^n)$$

$$= D(P_{X^{n+1}}||M_{X^{n+1}}) - D(P_{X^n}||M_{X^n})$$

$$D(P_{X^{n+1}}||M_{X^{n+1}}) - D(P_{X_1^n}||M_{X_1^n}) = H_{p||m}(X_0|X_1,\cdots,X_n)$$

from which the results follow. If on the other hand the conditional relative entropy is finite, the results then follow as in the proof of Lemma 7.4.1 using the fact that the joint relative entropies are arithmetic averages of the conditional relative entropies and that the conditional relative entropy is defined as the divergence between the P and S measures (Theorem 5.3.2). □

7.5 Mean Ergodic Theorems for Relative Entropy Densities

In this section we state and prove some preliminary ergodic theorems for relative entropy densities analogous to those first developed for entropy densities in Chapter 3 and for information densities in Section 6.3. In particular, we show that an almost everywhere ergodic theorem for finite alphabet processes follows easily from the sample entropy ergodic theorem and that an approximation argument then yields an L^1 ergodic theorem for stationary sources. The results involve little new and closely parallel those for mutual information densities and therefore the details are skimpy. The results are given for completeness and because the L^1 results yield the byproduct that relative entropies are uniformly integrable, a fact which does not follow as easily for relative entropies as it did for entropies.

Finite Alphabets

Suppose that we now have two process distributions p and m for a random process $\{X_n\}$ with finite alphabet. Let P_{X^n} and M_{X^n} denote the induced nth order distributions and p_{X^n} and m_{X^n} the corresponding probability mass functions (pmf's). For example, $p_{X^n}(a^n) = P_{X^n}(\{x^n : x^n = a^n\}) = p(\{x : X^n(x) = a^n\})$. We assume that $P_{X^n} << M_{X^n}$. In this case the relative entropy density is given simply by

$$h_n(x) = h_{X^n}(X^n)(x) = \ln \frac{p_{X^n}(x^n)}{m_{X^n}(x^n)},$$

where $x^n = X^n(x)$.

The following lemma generalizes Theorem 3.1.1 from entropy densities to relative entropy densities for finite alphabet processes. Relative entropies are of more general interest than ordinary entropies because they generalize to continuous alphabets in a useful way while ordinary entropies do not.

Lemma 7.5.1: Suppose that $\{X_n\}$ is a finite alphabet process and that p and m are two process distributions with $M_{X^n} >> P_{X^n}$ for all n, where p is AMS with stationary mean $\bar{p}$, m is a kth order Markov source with stationary transitions, and $\{\bar{p}_x\}$ is the ergodic decomposition of the stationary mean of p. Assume also that $M_{X^n} >> \bar{P}_{X^n}$ for all n. Then

$$\lim_{n\to\infty} \frac{1}{n} h_n = h;\ p-\text{a.e. and in } L^1(p),$$

where $h(x)$ is the invariant function defined by

$$\begin{aligned} h(x) &= -\bar{H}_{\bar{p}_x}(X) - E_{\bar{p}_x} \ln m(X_k|X^k) \\ &= \lim_{n\to\infty} \frac{1}{n} H_{\bar{p}_x||m}(X^n) = \bar{H}_{\bar{p}_x||m}(X), \end{aligned} \tag{7.5.1}$$

where

$$m(X_k|X^k)(x) \equiv \frac{m_{X^{k+1}}(x^{k+1})}{m_{X^k}(x^k)} = M_{X_k|X^k}(x_k|x^k).$$

Furthermore,

$$E_p h = \bar{H}_{p||m}(X) = \lim_{n\to\infty} \frac{1}{n} H_{p||m}(X^n), \tag{7.5.2}$$

that is, the relative entropy rate of an AMS process with respect to a Markov process with stationary transitions is given by the limit. Lastly,

$$\bar{H}_{p||m}(X) = \bar{H}_{\bar{p}||m}(X); \tag{7.5.3}$$

that is, the relative entropy rate of the AMS process with respect to m is the same as that of its stationary mean with respect to m.

Proof: We have that

$$\frac{1}{n}h(X^n) = \frac{1}{n}\ln p(X^n) - \frac{1}{n}\ln m(X^k) + \frac{1}{n}\sum_{i=k}^{n-1}\ln m(X_i|X_{i-k}^k)$$

$$= \frac{1}{n}\ln p(X^n) - \frac{1}{n}\ln m(X^k) - \frac{1}{n}\sum_{i=k}^{n-1}\ln m(X_k|X^k)T^{i-k}, \tag{7.5.4}$$

where T is the shift transformation, $p(X^n)$ is an abbreviation for $P_{X^n}(X^n)$, and $m(X_k|X^k) = M_{X_k|X^k}(X_k|X^k)$. From Theorem 3.1.1 the first term converges to $-\bar{H}_{\bar{p}_x}(X)p$-a.e. and in $L^1(p)$.

Since $M_{X^k} >> P_{X^k}$, if $M_{X^k}(F) = 0$, then also $P_{X^k}(F) = 0$. Thus P_{X^k} and hence also p assign zero probability to the event that $M_{X^k}(X^k) = 0$. Thus with probability one under p, $\ln m(X^k)$ is finite and hence the second term in (7.5.4) converges to 0 p-a.e. as $n \to \infty$.

Define α as the minimum nonzero value of the conditional probability $m(x_k|x^k)$. Then with probability 1 under M_{X^n} and hence also under P_{X^n} we have that

$$\frac{1}{n}\sum_{i=k}^{n-1}\ln\frac{1}{m(X_i|X_{i-k}^k)} \leq \ln\frac{1}{\alpha}$$

since otherwise the sequence X^n would have 0 probability under M_{X^n} and hence also under P_{X^n} and $0\ln 0$ is considered to be 0. Thus the rightmost term of (7.5.4) is uniformly integrable with respect to p and hence from Theorem 1.8.3 this term converges to $E_{\bar{p}_x}(\ln m(X_k|X^k))$. This proves the leftmost equality of (7.5.1).

Let $\bar{p}_{X^n|x}$ denote the distribution of X^n under the ergodic component $\bar{p}_x$. Since $M_{X^n} >> \bar{P}_{X^n}$ and $\bar{P}_{X^n} = \int d\bar{p}(x)\bar{p}_{X^n|x}$, if $M_{X^n}(F) = 0$, then $\bar{p}_{X^n|x}(F) = 0$ p-a.e. Since the alphabet of X_n if finite, we therefore also have with probability one under $\bar{p}$ that $M_{X^n} >> \bar{p}_{X^n|x}$ and hence

$$H_{\bar{p}_x||m}(X^n) = \sum_{a^n}\bar{p}_{X^n|x}(a^n)\ln\frac{\bar{p}_{X^n|x}(a^n)}{M_{X^n}(a^n)}$$

is well defined for $\bar{p}$-almost all x. This expectation can also be written as

$$H_{\bar{p}_x||m}(X^n) = -H_{\bar{p}_x}(X^n) - E_{\bar{p}_x}[\ln m(X^k) + \sum_{i=k}^{n-1}\ln m(X_k|X^k)T^{i-k}]$$

$$= -H_{\bar{p}_x}(X^n) - E_{\bar{p}_x}[\ln m(X^k)] - (n-k)E_{\bar{p}_x}[\ln m(X_k|X^k)],$$

where we have used the stationarity of the ergodic components. Dividing by n and taking the limit as $n \to \infty$, the middle term goes to zero as previously and the remaining limits prove the middle equality and hence the rightmost inequality in (7.5.1).

Equation (7.5.2) follows from (7.5.1) and $L^1(p)$ convergence, that is, since $n^{-1}h_n \to h$, we must also have that $E_p(n^{-1}h_n(X^n)) = n^{-1}H_{p||m}(X^n)$ converges to $E_p h$. Since the former limit is $\bar{H}_{p||m}(X)$, (7.5.2) follows. Since $\bar{p}_x$ is invariant (Theorem 1.8.2) and since expectations of invariant functions are the same under an AMS measure and its stationary mean (Lemma 6.3.1 of [50]), application of the previous results of the lemma to both p and $\bar{p}$ proves that

$$\bar{H}_{p||m}(X) = \int dp(x)\bar{H}_{\bar{p}_x||m}(X) = \int d\bar{p}(x)\bar{H}_{\bar{p}_x||m}(X) = \bar{H}_{\bar{p}||m}(X),$$

which proves (7.5.4) and completes the proof of the lemma. □

Corollary 7.5.1: Given p and m as in the Lemma, then the relative entropy rate of p with respect to m has an ergodic decomposition, that is,

$$\bar{H}_{p||m}(X) = \int dp(x)\bar{H}_{\bar{p}_x||m}(X).$$

Proof: This follows immediately from (7.5.1) and (7.5.2). □

Standard Alphabets

We now drop the finite alphabet assumption and suppose that $\{X_n\}$ is a standard alphabet process with process distributions p and m, where p is stationary, m is kth order Markov with stationary transitions, and $M_{X^n} >> P_{X^n}$ are the induced vector distributions for $n = 1, 2, \cdots$. Define the densities f_n and entropy densities h_n as previously.

As an easy consequence of the development to this point, the ergodic decomposition for divergence rate of finite alphabet processes combined with the definition of H^* as a supremum over rates of quantized processes yields an extension of Corollary 6.2.1 to divergences. This yields other useful properties as summarized in the following corollary.

Corollary 7.5.1: Given a standard alphabet process $\{X_n\}$ suppose that p and m are two process distributions such that p is AMS and m is kth order Markov with stationary transitions and $M_{X^n} >> P_{X^n}$ are the induced vector distributions. Let $\bar{p}$ denote the stationary mean of p and let $\{\bar{p}_x\}$ denote the ergodic decomposition of the stationary mean $\bar{p}$. Then

$$H^*_{p||m}(X) = \int dp(x)H^*_{\bar{p}_x||m}(X). \tag{7.5.5}$$

In addition,

$$H^*_{p||m}(X) = H^*_{\bar{p}||m}(X) = \bar{H}_{\bar{p}||m}(X) = \bar{H}_{p||m}(X); \tag{7.5.6}$$

that is, the two definitions of relative entropy rate yield the same values for AMS p and stationary transition Markov m and both rates are the same as the corresponding rates for the stationary mean. Thus relative entropy rate has an ergodic decomposition in the sense that

$$\bar{H}_{p||m}(X) = \int dp(x)\bar{H}_{\bar{p}_x||m}(X). \tag{7.5.7}$$

Comment: Note that the extra technical conditions of Theorem 6.4.2 for equality of the analogous mutual information rates $\bar{I}$ and I^* are not needed here. Note also that only the ergodic decomposition of the stationary mean $\bar{p}$ of the AMS measure p is considered and not that of the Markov source m.

Proof: The first statement follows as previously described from the finite alphabet result and the definition of H^*. The left-most and right-most equalities of (7.5.6) both follow from the previous lemma. The middle equality of (7.5.6) follows from Corollary 7.4.2. Eq. (7.5.7) then follows from (7.5.5) and (7.5.6). □

Theorem 7.5.1: Given a standard alphabet process $\{X_n\}$ suppose that p and m are two process distributions such that p is AMS and m is kth order Markov with stationary transitions and $M_{X^n} >> P_{X^n}$ are the induced vector distributions. Let $\{\bar{p}_x\}$ denote the ergodic decomposition of the stationary mean $\bar{p}$. If

$$\lim_{n\to\infty} \frac{1}{n} H_{p||m}(X^n) = \bar{H}_{p||m}(X) < \infty,$$

then there is an invariant function h such that $n^{-1}h_n \to h$ in $L^1(p)$ as $n \to \infty$. In fact,

$$h(x) = \bar{H}_{\bar{p}_x||m}(X),$$

the relative entropy rate of the ergodic component $\bar{p}_x$ with respect to m. Thus, in particular, under the stated conditions the relative entropy densities h_n are uniformly integrable with respect to p.

Proof: The proof exactly parallels that of Theorem 6.3.1, the mean ergodic theorem for information densities, with the relative entropy densities replacing the mutual information densities. The density is approximated by that of a quantized version and the integral bounded above using the triangle inequality. One term goes to zero from the finite alphabet case.

Since $\bar{H} = H^*$ (Corollary 7.5.1) the remaining terms go to zero because the relative entropy rate can be approximated arbitrarily closely by that of a quantized process. □

It should be emphasized that although Theorem 7.5.1 and Theorem 6.3.1 are similar in appearance, neither result directly implies the other. It is true that mutual information can be considered as a special case of relative entropy, but given a pair process $\{X_n, Y_n\}$ we cannot in general find a kth order Markov distribution m for which the mutual information rate $\bar{I}(X;Y)$ equals a relative entropy rate $\bar{H}_{p\|m}$. We will later consider conditions under which convergence of relative entropy densities does imply convergence of information densities.

Chapter 8

Ergodic Theorems for Densities

8.1 Introduction

This chapter is devoted to developing ergodic theorems first for relative entropy densities and then information densities for the general case of AMS processes with standard alphabets. The general results were first developed by Barron [9] using the martingale convergence theorem and a new martingale inequality. The similar results of Algoet and Cover [7] can be proved without direct recourse to martingale theory. They infer the result for the stationary Markov approximation and for the infinite order approximation from the ordinary ergodic theorem. They then demonstrate that the growth rate of the true density is asymptotically sandwiched between that for the kth order Markov approximation and the infinite order approximation and that no gap is left between these asymptotic upper and lower bounds in the limit as $k \to \infty$. They use martingale theory to show that the values between which the limiting density is sandwiched are arbitrarily close to each other, but we shall see that this is not necessary and this property follows from the results of Chapter 6.

8.2 Stationary Ergodic Sources

Theorem 8.2.1: Given a standard alphabet process $\{X_n\}$, suppose that p and m are two process distributions such that p is stationary ergodic and m is a K-step Markov source with stationary transition probabilities. Let $M_{X^n} >> P_{X^n}$ be the vector distributions induced by p and m. As before let

$$h_n = \ln f_{X^n}(X^n) = \ln \frac{dP_{X^n}}{dM_{X^n}}(X^n).$$

Then with probability one under p

$$\lim_{n\to\infty} \frac{1}{n} h_n = \bar{H}_{p\|m}(X).$$

Proof: Let $p^{(k)}$ denote the k-step Markov approximation of p as defined in Theorem 7.3.1, that is, $p^{(k)}$ has the same kth order conditional probabilities and k-dimensional initial distribution. From Corollary 7.3.1, if $k \geq K$, then (7.3.1)–(7.3.3) hold. Consider the expectation

$$E_p\left(\frac{f^{(k)}_{X^n}(X^n)}{f_{X^n}(X^n)}\right) = E_{P_{X^n}}\left(\frac{f^{(k)}_{X^n}}{f_{X^n}}\right) = \int \left(\frac{f^{(k)}_{X^n}}{f_{X^n}}\right) dP_{X^n}.$$

Define the set $A_n = \{x^n : \ f_{X^n} > 0\}$; then $P_{X^n}(A_n) = 1$. Use the fact that $f_{X^n} = dP_{X^n}/dM_{X^n}$ to write

$$E_P\left(\frac{f^{(k)}_{X^n}(X^n)}{f_{X^n}(X^n)}\right) = \int_{A_n} \left(\frac{f^{(k)}_{X^n}}{f_{X^n}}\right) f_{X^n}\, dM_{X^n}$$

$$= \int_{A_n} f^{(k)}_{X^n}\, dM_{X^n}.$$

From Corollary 7.3.1,

$$f^{(k)}_{X^n} = \frac{dP^{(k)}_{X^n}}{dM_{X^n}}$$

and therefore

$$E_p\left(\frac{f^{(k)}_{X^n}(X^n)}{f_{X^n}(X^n)}\right) = \int_{A_n} \frac{dP^{(k)}_{X^n}}{dM_{X^n}} dM_{X^n} = P^{(k)}_{X^n}(A_n) \leq 1.$$

Thus we can apply Lemma 5.4.2 to the sequence $f^{(k)}_{X^n}(X^n)/f_{X^n}(X^n)$ to conclude that with p-probability 1

$$\lim_{n\to\infty} \frac{1}{n} \ln \frac{f^{(k)}_{X^n}(X^n)}{f_{X^n}(X^n)} \leq 0$$

and hence

$$\lim_{n\to\infty} \frac{1}{n} \ln f^{(k)}_{X^n}(X^n) \leq \liminf_{n\to\infty} \frac{1}{n} f_{X^n}(X^n). \tag{8.2.1}$$

The left-hand limit is well defined by the usual ergodic theorem:

$$\lim_{n\to\infty} \frac{1}{n} \ln f_{X^n}^{(k)}(X^n) = \lim_{n\to\infty} \frac{1}{n} \sum_{l=k}^{n-1} \ln f_{X_l|X_{l-k}^k}(X_l|X_{l-k}^k) + \lim_{n\to\infty} \frac{1}{n} \ln f_{X^k}(X^k).$$

Since $0 < f_{X^k} < \infty$ with probability 1 under M_{X^k} and hence also under P_{X^k}, then $0 < f_{X^k}(X^k) < \infty$ under p and therefore $n^{-1} \ln f_{X^k}(X^k) \to 0$ as $n \to \infty$ with probability one. Furthermore, from the ergodic theorem for stationary and ergodic processes (e.g., Theorem 7.2.1 of [50]), since p is stationary ergodic we have with probability one under p using (7.4.1) and Corollary 7.4.1 that

$$\lim_{n\to\infty} \frac{1}{n} \sum_{l=k}^{n-1} \ln f_{X_l|X_{l-k}^k}(X_l|X_{l-k}^k)$$

$$= \lim_{n\to\infty} \frac{1}{n} \sum_{l=k}^{n-1} \ln f_{X_0|X_{-1},\cdots,X_{-k}}(X_0|$$

$$X_{-1},\cdots,X_{-k})T^l = E_p \ln f_{X_0|X_{-1},\cdots,X_{-k}}(X_0|X_{-1},\cdots,X_{-k})$$
$$= H_{p\|m}(X_0|X_{-1},\cdots,X_{-k}) = \bar{H}_{p^{(k)}\|m}(X).$$

Thus with (8.2.1) we now have that

$$\liminf_{n\to\infty} \frac{1}{n} \ln f_{X^n}(X^n) \geq H_{p\|m}(X_0|X_{-1},\cdots,X_{-k}) \tag{8.2.2}$$

for any positive integer k. Since m is Kth order Markov, Lemma 7.4.1 and the above imply that

$$\liminf_{n\to\infty} \frac{1}{n} \ln f_{X^n}(X^n) \geq H_{p\|m}(X_0|X^-) = \bar{H}_{p\|m}(X), \tag{8.2.3}$$

which completes half of the sandwich proof of the theorem.

If $\bar{H}_{p\|m}(X) = \infty$, the proof is completed with (8.2.3). Hence we can suppose that $\bar{H}_{p\|m}(X) < \infty$. From Lemma 7.4.1 using the distribution $S_{X_0,X_{-1},X_{-2},\cdots}$ constructed there, we have that

$$D(P_{X_0,X_{-1},\cdots}\|S_{X_0,X_{-1},\cdots}) = H_{p\|m}(X_0|X^-) = \int dP_{X_0,X^-} \ln f_{X_0|X^-}$$

where

$$f_{X_0|X^-} = \frac{dP_{X_0,X_{-1},\cdots}}{dS_{X_0,X_{-1},\cdots}}.$$

It should be pointed out that we have not (and will not) prove that $f_{X_0|X_{-1},\cdots,X_{-n}} \to f_{X_0|X^-}$; the convergence of conditional probability densities which follows from the martingale convergence theorem and the result about which most generalized Shannon-McMillan-Breiman theorems are built. (See, e.g., Barron [9].) We have proved, however, that the expectations converge (Lemma 7.4.1), which is what is needed to make the sandwich argument work.

For the second half of the sandwich proof we construct a measure Q which will be dominated by p on semi-infinite sequences using the above conditional densities given the infinite past. Define the semi-infinite sequence $X_n^- = \{\cdots, X_{n-1}\}$ for all nonnegative integers n. Let $\mathcal{B}_k^n = \sigma(X_k^n)$ and $\mathcal{B}_k^- = \sigma(X_k^-) = \sigma(\cdots, X_{k-1})$ be the σ-fields generated by the finite dimensional random vector X_k^n and the semi-infinite sequence X_k^-, respectively. Let Q be the process distribution having the same restriction to $\sigma(X_k^-)$ as does p and the same restriction to $\sigma(X_0, X_1, \cdots)$ as does p, but which makes X^- and X_k^n conditionally independent given X^k for any n; that is,

$$Q_{X_k^-} = P_{X_k^-},$$

$$Q_{X_k, X_{k+1}, \cdots} = P_{X_k, X_{k+1}, \cdots},$$

and $X^- \to X^k \to X_k^n$ is a Markov chain for all positive integers n so that

$$Q(X_k^n \in F|X_k^-) = Q(X_k^n \in F|X^k).$$

The measure Q is a (nonstationary) k-step Markov approximation to P in the sense of Section 5.3 and

$$Q = P_{X^- \times (X_k, X_{k+1}, \cdots)|X^k}$$

(in contrast to $P = P_{X^- X^k X_k^\infty}$). Observe that $X^- \to X^k \to X_k^n$ is a Markov chain under both Q and m.

By assumption,

$$H_{p||m}(X_0|X^-) < \infty$$

and hence from Corollary 7.4.1

$$H_{p||m}(X_k^n|X_k^-) = nH_{p||m}(X_k^n|X_k^-) < \infty$$

and hence from Theorem 5.3.2 the density $f_{X_k^n|X_k^-}$ is well-defined as

$$f_{X_k^n|X_k^-} = \frac{dS_{X_{n+k}^-}}{P_{X_{n+k}^-}} \tag{8.2.4}$$

where

$$S_{X^-_{n+k}} = \overline{M_{X^n_k|X^k} P_{X^-_k}}, \qquad (8.2.5)$$

and

$$\int dP_{X^-_{n+k}} \ln f_{X^n_k|X^-_k} = D(P_{X^-_{n+k}} \| S_{X^-_{n+k}})$$
$$= nH_{p\|m}(X^n_k|X^-_k) < \infty. \qquad (8.2.6)$$

Thus, in particular,

$$S_{X^-_{n+k}} >> P_{X^-_{n+k}}.$$

Consider now the sequence of ratios of conditional densities

$$\zeta_n = \frac{f_{X^n_k|X^k}(X^{n+k})}{f_{X^n_k|X^-_k}(X^-_{n+k})}$$

We have that

$$\int dp\zeta_n = \int_{G_n} \zeta_n$$

where

$$G_n = \{x : f_{X^n_k|X^-_k}(\mathbf{x}^-_{n+k}) > 0\}$$

since G_n has probability 1 under p (or else (8.2.6) would be violated). Thus

$$\int dp\zeta_n = \int dP_{X^-_{n+k}} \left(\frac{f_{X^n_k|X^k}(X^{n+k})}{f_{X^n_k|X^-_k}} 1_{\{f_{X^n_k|X^-_k}>0\}} \right)$$

$$= \int dS_{X^-_{n+k}} f_{X^n_k|X^-_k} \left(\frac{f_{X^n_k|X^k}(X^{n+k})}{f_{X^n_k|X^-_k}} 1_{\{f_{X^n_k|X^-_k}>0\}} \right)$$

$$= \int dS_{X^-_{n+k}} f_{X^n_k|X^k}(X^{n+k}) 1_{\{f_{X^n_k|X^-_k}>0\}} \le \int dS_{X^-_{n+k}} f_{X^n_k|X^k}(X^{n+k}).$$

Using the definition of the measure S and iterated expectation we have that

$$\int dp\zeta_n \le \int dM_{X^n_k|X^-_k} dP_{X^-_k} f_{X^n_k|X^k}(X^{n+k}).$$

$$= \int dM_{X^n_k|X^k} dP_{X^-_k} f_{X^n_k|X^k}(X^{n+k}).$$

Since the integrand is now measurable with respect to $\sigma(X^{n+k})$, this reduces to

$$\int dp\zeta_n \le \int dM_{X^n_k|X^k} dP_{X^k} f_{X^n_k|X^k}.$$

Applying Lemma 5.3.2 we have

$$\int dp\zeta_n \le \int dM_{X_k^n|X^k} dP_{X^k} \frac{dP_{X_k^n|X^k}}{dM_{X_k^n|X^k}}$$

$$= \int dP_{X^k} dP_{X_k^n|X^k} = 1.$$

Thus

$$\int dp\zeta_n \le 1$$

and we can apply Lemma 5.4.1 to conclude that p-a.e.

$$\limsup_{n\to\infty} \zeta_n = \limsup_{n\to\infty} \frac{1}{n} \ln \frac{f_{X_k^n|X^k}}{f_{X_k^n|X_k^-}} \le 0. \tag{8.2.7}$$

Using the chain rule for densities,

$$\frac{f_{X_k^n|X^k}}{f_{X_k^n|X_k^-}} = \frac{f_{X^n}}{f_{X^k}} \times \frac{1}{\prod_{l=k}^{n-1} f_{X_l|X_l^-}}.$$

Thus from (8.2.7)

$$\limsup_{n\to\infty} \left(\frac{1}{n} \ln f_{X^n} - \frac{1}{n} \ln f_{X^k} - \frac{1}{n} \sum_{l=k}^{n-1} \ln f_{X_l|X_l^-} \right) \le 0.$$

Invoking the ergodic theorem for the rightmost terms and the fact that the middle term converges to 0 almost everywhere since $\ln f_{X^k}$ is finite almost everywhere implies that

$$\limsup_{n\to\infty} \frac{1}{n} \ln f_{X^n} \le E_p(\ln f_{X_k|X_k^-}) = E_p(\ln f_{X_0|X^-})$$

$$= \bar{H}_{p||m}(X). \tag{8.2.8}$$

Combining this with (8.2.3) completes the sandwich and proves the theorem. □

8.3 Stationary Nonergodic Sources

Next suppose that the source p is stationary with ergodic decomposition $\{p_\lambda;\ \lambda \in \Lambda\}$ and ergodic component function ψ as in Theorem 1.8.3.

We first require some technical details to ensure that the various Radon-Nikodym derivatives are well defined and that the needed chain rules for densities hold.

Lemma 8.3.1: Given a stationary source $\{X_n\}$, let $\{p_\lambda;\ \lambda \in \Lambda\}$ denote the ergodic decomposition and ψ the ergodic component function of Theorem 1.8.3. Let P_ψ denote the induced distribution of ψ. Let P_{X^n} and $P^\lambda_{X^n}$ denote the induced marginal distributions of p and p_λ. Assume that $\{X_n\}$ has the finite-gap information property of (6.4.11); that is, there exists a K such that

$$I_p(X_K; X^- | X^K) < \infty, \tag{8.3.1}$$

where $X^- = (X_{-1}, X_{-2}, \cdots)$. We also assume that for some n

$$I(X^n; \psi) < \infty. \tag{8.3.2}$$

This will be the case, for example, if (8.3.1) holds for $K = 0$. Let m be a K-step Markov process such that $M_{X^n} >> P_{X^n}$ for all n. (Observe that such a process exists since from (8.3.1) the Kth order Markov approximation $p^{(K)}$ suffices.) Define $M_{X^n,\psi} = M_{X^n} \times P_\psi$. Then

$$M_{X^n,\psi} >> P_{X^n} \times P_\psi >> P_{X^n,\psi}, \tag{8.3.3}$$

and with probability 1 under p

$$M_{X^n} >> P_{X^n} >> P^\psi_{X^n}.$$

Lastly,

$$\frac{dP^\psi_{X^n}}{dM_{X^n}} = f_{X^n|\psi} = \frac{dP_{X^n,\psi}}{d(M_{X^n} \times P_\psi)}. \tag{8.3.4}$$

and therefore

$$\frac{dP^\psi_{X^n}}{dP_{X^n}} = \frac{dP^\psi_{X^n}/dM_{X^n}}{dP_{X^n}/dM_{X^n}} = \frac{f_{X^n|\psi}}{f_{X^n}}. \tag{8.3.5}$$

Proof: From Theorem 6.4.4 the given assumptions ensure that

$$\lim_{n\to\infty} \frac{1}{n} E_p i(X^n; \psi) = \lim_{n\to\infty} \frac{1}{n} I(X^n; \psi) = 0 \tag{8.3.6}$$

and hence $P_{X^n} \times P_\psi >> P_{X^n,\psi}$ (since otherwise $I(X^n; \psi)$ would be infinite for some n and hence infinite for all larger n since it is increasing with n). This proves the right-most absolute continuity relation of (8.3.3). This in turn implies that $M_{X^n} \times P_\psi >> P_{X^n,\psi}$. The lemma then follows from Theorem 5.3.1 with $X = X^n$, $Y = \psi$ and the chain rule for Radon-Nikodym derivatives. □

We know that the source will produce with probability one an ergodic component p_λ and hence Theorem 8.2.1 will hold for this ergodic component. In other words, we have for all λ that

$$\lim_{n\to\infty} \frac{1}{n} \ln f_{X^n|\psi}(X^n|\lambda) = \bar{H}_{p_\lambda}(X);\ p_\lambda - \text{a.e.}$$

This implies that

$$\lim_{n\to\infty} \frac{1}{n} \ln f_{X^n|\psi}(X^n|\psi) = \bar{H}_{p_\psi}(X);\ p - \text{a.e.} \tag{8.3.7}$$

Making this step precise generalizes Lemma 3.3.1.

Lemma 8.3.2: Suppose that $\{X_n\}$ is a stationary not necessarily ergodic source with ergodic component function ψ. Then (8.3.7) holds.

Proof: The proof parallels that for Lemma 3.3.1. Observe that if we have two random variables U, V ($U = X_0, X_1, \cdots$ and $Y = \psi$ above) and a sequence of functions $g_n(U, V)$ ($n^{-1} f_{X^n|\psi}(X^n|\psi)$) and a function $g(V)$ ($\bar{H}_{p_\psi}(X)$) with the property

$$\lim_{n\to\infty} g_n(U, v) = g(v), P_{U|V=v} - \text{a.e.},$$

then also

$$\lim_{n\to\infty} g_n(U, V) = g(V);\ P_{UV} - \text{a.e.}$$

since defining the (measurable) set $G = \{u, v : \lim_{n\to\infty} g_n(u, v) = g(v)\}$ and its section $G_v = \{u : (u, v) \in G\}$, then from (1.6.8)

$$P_{UV}(G) = \int P_{U|V}(G_v|v) dP_V(v) = 1$$

if $P_{U|V}(G_v|v) = 1$ with probability 1. □

It is not, however, the relative entropy density using the distribution of the ergodic component that we wish to show converges. It is the original sample density f_{X^n}. The following lemma shows that the two sample entropies converge to the same thing. The lemma generalizes Lemma 3.3.1 and is proved by a sandwich argument analogous to Theorem 8.2.1. The result can be viewed as an almost everywhere version of (8.3.6).

Theorem 8.3.1: Given a stationary source $\{X_n\}$, let $\{p_\lambda;\ \lambda \in \Lambda\}$ denote the ergodic decomposition and ψ the ergodic component function of Theorem 1.8.3. Assume that the finite-gap information property (8.3.1) is satisfied and that (8.3.2) holds for some n. Then

$$\lim_{n\to\infty} \frac{1}{n} i(X^n; \psi) = \lim_{n\to\infty} \frac{1}{n} \ln \frac{f_{X^n|\psi}}{f_{X^n}} = 0;\ p - \text{a.e.}$$

Proof: From Theorem 5.4.1 we have immediately that

$$\liminf_{n\to\infty} i_n(X^n;\psi) \geq 0, \tag{8.3.8}$$

which provides half of the sandwich proof.

To develop the other half of the sandwich, for each $k \geq K$ let $p^{(k)}$ denote the k-step Markov approximation of p. Exactly as in the proof of Theorem 8.2.1, it follows that (8.2.1) holds. Now, however, the Markov approximation relative entropy density converges instead as

$$\lim_{n\to\infty} \frac{1}{n} \ln f_{X^n}^{(k)}(X^n) = \lim_{n\to\infty} \frac{1}{n} \sum_{l=k}^{\infty} f_{X_k|X^k}(X_k|X^k)T^k = E_{p_\psi} f_{X_k|X^k}(X_k|X^k).$$

Combining this with (8.3.7 we have that

$$\limsup_{n\to\infty} \frac{1}{n} \ln \frac{f_{X^n|\psi}(X^n|\psi)}{f_{X^n}(X^n)} \leq \bar{H}_{p_\psi||m}(X) - E_{p_\psi} f_{X_k|X^k}(X_k|X^k).$$

From Lemma 7.4.1, the right hand side is just $I_{p_\psi}(X_k;X^-|X^k)$ which from Corollary 7.4.2 is just $\bar{H}_{p||p^{(k)}}(X)$. Since the bound holds for all k, we have that

$$\limsup_{n\to\infty} \frac{1}{n} \ln \frac{f_{X^n|\psi}(X^n|\psi)}{f_{X^n}(X^n)} \leq \inf_k \bar{H}_{p_\psi||p^{(k)}}(X) \equiv \zeta.$$

Using the ergodic decompostion of relative entropy rate (Corollary 7.5.1) that and the fact that Markov approximations are asymptotically accurate (Corollary 7.4.3) we have further that

$$\int dP_\psi \zeta = \int dP_\psi \inf_k \bar{H}_{p_\psi||p^{(k)}}(X)$$

$$\leq \inf_k \int dP_\psi \bar{H}_{p_\psi||p^{(k)}}(X) = \inf_k \bar{H}_{p||p^{(k)}}(X) = 0$$

and hence $\zeta = 0$ with P_ψ probability 1. Thus

$$\limsup_{n\to\infty} \frac{1}{n} \ln \frac{f_{X^n|\psi}(X^n|\psi)}{f_{X^n}(X^n)} \leq 0, \tag{8.3.9}$$

which with (8.3.8) completes the sandwich proof. □

Simply restating the theorem yields and using (8.3.7) the ergodic theorem for relative entropy densities in the general stationary case.

Corollary 8.3.1: Given the assumptions of Theorem 8.3.1,

$$\lim_{n\to\infty} \frac{1}{n} \ln f_{X^n}(X^n) = \bar{H}_{p_\psi||m}(X), p-\text{a.e.}$$

The corollary states that the sample relative entropy density of a process satisfying (8.3.1) converges to the conditional relative entropy rate with respect to the underlying ergodic component. This is a slight extension and elaboration of Barron's result [9] which made the stronger assumption that $H_{p||m}(X_0|X^-) = \bar{H}_{p||m}(X) < \infty$. From Corollary 7.4.3 this condition is sufficient but not necessary for the finite-gap information property of (8.3.1). In particular, the finite gap information property implies that

$$\bar{H}_{p||p^{(k)}}(X) = I_p(X_k; X^-|X^k) < \infty,$$

but it need not be true that $\bar{H}_{p||m}(X) < \infty$. In addition, Barron [9] and Algoet and Cover [7] do not characterize the limiting density as the entropy rate of the ergodic component, instead they effectively show that the limit is $E_{p_\psi}(\ln f_{X_0|X^-}(X_0|X^-))$. This, however, is equivalent since it follows from the ergodic decomposition (see specifically Lemma 8.6.2 [50]) that $f_{X_0|X^-} = f_{X_0|X^-,\psi}$ with probability one since the ergodic component ψ can be determined from the infinite past X^-.

8.4 AMS Sources

The following lemma is a generalization of Lemma 3.4.1. The result is due to Barron [9], who proved it using martingale inequalities and convergence results.

Lemma 8.4.1: Let $\{X_n\}$ be an AMS source with the property that for every integer k there exists an integer $l = l(k)$ such that

$$I_p(X^k; (X_{k+l}, X_{k+l+1}, \cdots)|X_k^l). < \infty. \tag{8.4.1}$$

Then

$$\lim_{n\to\infty} \frac{1}{n} i(X^k; (X_k + l, \cdots, X_{n-1})|X_k^l) = 0;\ p-\text{a.e.}$$

Proof: By assumption

$$I_p(X^k; (X_{k+l}, X_{k+l+1}, \cdots)|X_k^l) =$$

$$E_p \ln \frac{f_{X^k|X_k,X_{k+1},\cdots}(X^k|X_k, X_{k+1}, \cdots)}{f_{X^k|X_k^l}(X^k|X_k^l)} < \infty.$$

This implies that

$$P_{X^k \times (X_k+l,\cdots)|X_k^l} >> P_{X_0,X_1,\cdots}.$$

with

$$\frac{dP_{X_0,X_1,\ldots}}{dP_{X^k\times(X_k+l,\cdots)|X_k^l}} = \frac{f_{X^k|X_k,X_k+1,\cdots}(X^k|X_k,X_k+1,\cdots)}{f_{X^k|X_k^l}(X^k|X_k^l)}.$$

Restricting the measures to X^n for $n > k + l$ yields

$$\frac{dP_{X^n}}{dP_{X^k\times(X_k+l,\cdots,X^n)|X_k^l}} = \frac{f_{X^k|X_k,X_k+1,\cdots,X_n}(X^k|X_k,X_k+1,\cdots)}{f_{X^k|X_k^l}(X^k|X_k^l)}$$

$$= i(X^k;(X_k+l,\cdots,X_n)|X_k^l).$$

With this setup the lemma follows immediately from Theorem 5.4.1. □

The following lemma generalizes Lemma 3.4.2 and will yield the general theorem. The lemma was first proved by Barron [9] using martingale inequalities.

Theorem 8.4.1: Suppose that p and m are distributions of a standard alphabet process $\{X_n\}$ such that p is AMS and m is k-step Markov. Let $\bar{p}$ be a stationary measure that asymptotically dominates p (e.g., the stationary mean). Suppose that P_{X^n}, $\bar{P}_{X^n}$, and M_{X^n} are the distributions induced by p, $\bar{p}$, and m and that M_{X^n} dominates both P_{X^n} and $\bar{P}_{X^n}$ for all n and that f_{X^n} and $\bar{f}_{X^n}$ are the corresponding densities. If there is an invariant function h such that

$$\lim_{n\to\infty}\frac{1}{n}\ln \bar{f}_{X^n}(X^n) = h;\ \bar{p}-\text{a.e.}$$

then also

$$\lim_{n\to\infty}\frac{1}{n}\ln f_{X^n}(X^n) = h;\ p-\text{a.e.}$$

Proof: For any k and $n \geq k$ we can write using the chain rule for densities

$$\frac{1}{n}\ln f_{X^n} - \frac{1}{n}\ln f_{X_k^{n-k}} = \frac{1}{n}\ln f_{X^k|X_k^{n-k}}.$$

Since for $k \leq l < n$

$$\frac{1}{n}\ln f_{X^k|X_k^{n-k}} = \frac{1}{n}\ln f_{X^k|X_k^l} + \frac{1}{n}i(X^k;(X_{k+l},\cdots,X_{n-1})|X_k^l),$$

Lemma 8.4.1 and the fact that densities are finite with probability one implies that

$$\lim_{n\to\infty}\frac{1}{n}\ln f_{X^k|X_k^{n-k}} = 0;\ p-\text{a.e.}$$

This implies that there is a subsequence $k(n)\to\infty$ such that

$$\frac{1}{n}\ln f_{X^n}(X^n) - \frac{1}{n}\ln f_{X_{k(n)}^{n-k(n)}}(X_{k(n)}^{n-k(n)});\ \to 0, p-\text{a.e.}$$

To prove this, for each k chose $N(k)$ large enough so that

$$p(|\frac{1}{N}(k)\ln f_{X^k|X_k^{N(k)-k}}(X^k|X_k^{N(k)-k})|>2^{-k})\leq 2^{-k}$$

and then let $k(n)=k$ for $N(k)\leq n<N(k+1)$. Then from the Borel-Cantelli lemma we have for any ϵ that

$$p(|\frac{1}{N(k)}\ln f_{X^k|X_k^{N(k)-k}}(X^k|X_k^{N(k)-k})|>\epsilon \text{ i.o.})=0$$

and hence

$$\lim_{n\to\infty}\frac{1}{n}\ln f_{X^n}(X^n)=\lim_{n\to\infty}\frac{1}{n}\ln f_{X_{k(n)}^{n-k(n)}}(X_{k(n)}^{n-k(n)});\ p-\text{a.e.}$$

In a similar manner we can also choose the sequence so that

$$\lim_{n\to\infty}\frac{1}{n}\ln \bar{f}_{X^n}(X^n)=\lim_{n\to\infty}\frac{1}{n}\ln \bar{f}_{X_{k(n)}^{n-k(n)}}(X_{k(n)}^{n-k(n)});\ \bar{p}-\text{a.e.}$$

From Markov's inequality

$$\bar{p}(\frac{1}{n}\ln f_{X_k^{n-k}}(X_k^{n-k})\geq\frac{1}{n}\ln\bar{f}_{X_k^{n-k}}(X_k^{n-k})+\epsilon)$$

$$=\bar{p}(\frac{f_{X_k^{n-k}}(X_k^{n-k})}{\bar{f}_{X_k^{n-k}}(X_k^{n-k})}\geq e^{n\epsilon})\leq e^{-n\epsilon}\int d\bar{p}\frac{f_{X_k^{n-k}}(X_k^{n-k})}{\bar{f}_{X_k^{n-k}}(X_k^{n-k})}$$

$$=e^{-n\epsilon}\int dm f_{X_k^{n-k}}(X_k^{n-k})=e^{-n\epsilon}.$$

Hence again invoking the Borel-Cantelli lemma we have that

$$\bar{p}(\frac{1}{n}\ln f_{X_k^{n-k}}(X_k^{n-k})\geq\frac{1}{n}\ln\bar{f}_{X_k^{n-k}}(X_k^{n-k})+\epsilon \text{ i.o.})=0$$

and therefore

$$\limsup_{n\to\infty}\frac{1}{n}\ln f_{X_k^{n-k}}(X_k^{n-k})\leq h,\bar{p}-\text{a.e.} \tag{8.4.2}$$

The above event is in the tail σ-field $\bigcap_n \sigma(X_n,X_{n+1},\cdots)$ since h is invariant and $\bar{p}$ dominates p on the tail σ-field. Thus

$$\limsup_{n\to\infty}\frac{1}{n}\ln f_{X_{k(n)}^{n-k(n)}}(X_{k(n)}^{n-k(n)})\leq h;\ p-\text{a.e.}$$

and hence

$$\limsup_{n\to\infty} \frac{1}{n} \ln f_{X^n}(X^n) \leq h;\ p-\text{a.e.}$$

which half proves the lemma.

Since $\bar{p}$ asymptotically dominates p, given $\epsilon > 0$ there is a k such that

$$p(\lim_{n\to\infty} n^{-1}\bar{f}(X_k^{n-k}) = h) \geq 1-\epsilon.$$

Again applying Markov's inequality and the Borel-Cantelli lemma as previously we have that

$$\liminf_{n\to\infty} \frac{1}{n} \ln \frac{f_{X_{k(n)}^{n-k(n)}}(X_{k(n)}^{n-k(n)})}{\bar{f}_{X_{k(n)}^{n-k(n)}}(X_{k(n)}^{n-k(n)})} \geq 0;\ p-\text{a.e.}$$

which implies that

$$p(\liminf_{n\to\infty} \frac{1}{n} f_{X_{k(n)}^{n-k(n)}}(X_k^{n-k}) \geq h) \geq \epsilon$$

and hence also that

$$p(\liminf_{n\to\infty} \frac{1}{n} f_{X^n}(X^n) \geq h) \geq \epsilon.$$

Since ϵ can be made arbitrarily small, this proves that p-a.e. $\liminf n^{-1}h_n \geq h$, which completes the proof of the lemma. □

We can now extend the ergodic theorem for relative entropy densities to the general AMS case.

Corollary 8.4.1: Given the assumptions of Theorem 8.4.1,

$$\lim_{n\to\infty} \frac{1}{n} \ln f_{X^n}(X^n) = \bar{H}_{\bar{p}_\psi}(X),$$

where $\bar{p}_\psi$ is the ergodic component of the stationary mean $\bar{p}$ of p.

Proof: The proof follows immediately from Theorem 8.4.1 and Corollary 8.3.1, the ergodic theorem for the relative entropy density for the stationary mean. □

8.5 Ergodic Theorems for Information Densities.

As an application of the general theorem we prove an ergodic theorem for mutual information densities for stationary and ergodic sources. The result can be extended to AMS sources in the same manner that the results of Section 8.3 were extended to those of Section 8.4. As the stationary and ergodic result suffices for the coding theorems and the AMS conditions are messy, only the stationary case is considered here. The result is due to Barron [9].

Theorem 8.5.1: Let $\{X_n, Y_n\}$ be a stationary ergodic pair random process with standard alphabet. Let $P_{X^nY^n}$, P_{X^n}, and P_{Y^n} denote the induced distributions and assume that for all n $P_{X^n} \times P_{Y^n} >> P_{X^nY^n}$ and hence the information densities

$$i_n(X^n; Y^n) = \frac{dP_{X^nY^n}}{d(P_{X^n} \times P_{Y^n})}$$

are well defined. Assume in addition that both the $\{X_n\}$ and $\{Y_n\}$ processes have the finite-gap information property of (8.3.1) and hence by the comment following Corollary 7.3.1 there is a K such that both processes satisfy the K-gap property

$$I(X_K; X^-|X^K) < \infty, \; I(Y_K; Y^-|Y^K) < \infty.$$

Then

$$\lim_{n\to\infty} \frac{1}{n} i_n(X^n; Y^n) = \bar{I}(X; Y); \; p-\text{a.e.}.$$

Proof: Let $Z_n = (X_n, Y_n)$. Let $M_{X^n} = P^{(K)}_{X^n}$ and $M_{Y^n} = P^{(K)}_{Y^n}$ denote the Kth order Markov approximations of $\{X_n\}$ and $\{Y_n\}$, respectively. The finite-gap approximation implies as in Section 8.3 that the densities

$$f_{X^n} = \frac{dP_{X^n}}{dM_{X^n}}$$

and

$$f_{Y^n} = \frac{dP_{Y^n}}{dM_{Y^n}}$$

are well defined. From Theorem 8.2.1

$$\lim_{n\to\infty} \frac{1}{n} \ln f_{X^n}(X^n) = H_{p_X \| p_X^{(K)}}(X_0|X^-) = I(X_k; X^-|X^k) < \infty,$$

$$\lim_{n\to\infty} \frac{1}{n} \ln f_{Y^n}(Y^n) = I(Y_k; Y^-|Y^k) < \infty.$$

Define the measures M_{Z^n} by $M_{X^n} \times M_{Y^n}$. Then this is a K-step Markov source and since

$$M_{X^n} \times M_{Y^n} >> P_{X^n} \times P_{Y^n}$$

$$>> P_{X^n, Y^n} = P_{Z^n},$$

the density

$$f_{Z^n} = \frac{dP_{Z^n}}{dM_{Z^n}}$$

is well defined and from Theorem 8.2.1 has a limit

$$\lim_{n\to\infty} \frac{1}{n} \ln f_{Z^n}(Z^n) = H_{p||m}(Z_0|Z^-).$$

If the density $i_n(X^n, Y^n)$ is infinite for any n, then it is infinite for all larger n and convergence is trivially to the infinite information rate. If it is finite, the chain rule for densities yields

$$\frac{1}{n} i_n(X^n; Y^n) = \frac{1}{n} \ln f_{Z^n}(Z^n) - \frac{1}{n} \ln f_{X^n}(X^n) - \frac{1}{n} \ln f_{Y^n}(Y^n)$$

$$\underset{n\to\infty}{\to} H_{p||p^{(k)}}(Z_0|Z^-) - H_{p||p^{(k)}}(X_0|X^-) - H_{p||p^{(k)}}(Y_0|Y^-)$$

$$= \bar{H}_{p||p^{(k)}}(X,Y) - \bar{H}_{p||p^{(k)}}(X) - \bar{H}_{p||p^{(k)}}(Y).$$

The limit is not indeterminate (of the form $\infty - \infty$) because the two subtracted terms are finite. Since convergence is to a constant, the constant must also be the limit of the expected values of $n^{-1} i_n(X^n, Y^n)$, that is, $\bar{I}(X;Y)$. □

Chapter 9

Channels and Codes

9.1 Introduction

We have considered a random process or source $\{X_n\}$ as a sequence of random entities, where the object produced at each time could be quite general, e.g., a random variable, vector, or waveform. Hence sequences of pairs of random objects such as $\{X_n, Y_n\}$ are included in the general framework. We now focus on the possible interrelations between the two components of such a pair process. In particular, we consider the situation where we begin with one source, say $\{X_n\}$, called the *input* and use either a random or a deterministic mapping to form a new source $\{Y_n\}$, called the *output*. We generally refer to the mapping as a *channel* if it is random and a *code* if it is deterministic. Hence a code is a special case of a channel and results for channels will immediately imply the corresponding results for codes. The initial point of interest will be conditions on the structure of the channel under which the resulting pair process $\{X_n, Y_n\}$ will inherit stationarity and ergodic properties from the original source $\{X_n\}$. We will also be interested in the behavior resulting when the output of one channel serves as the input to another, that is, when we form a new channel as a cascade of other channels. Such cascades yield models of a communication system which typically has a code mapping (called the *encoder*) followed by a channel followed by another code mapping (called the *decoder*).

A fundamental nuisance in the development is the notion of time. So far we have considered pair processes where at each unit of time, one random object is produced for each coordinate of the pair. In the channel or code example, this corresponds to one output for every input. Interesting communication systems do not always easily fit into this framework, and this can cause serious problems in notation and in the interpretation and development of results. For example, suppose that an input source consists of a sequence of real numbers and let T denote the time shift on the real sequence space. Suppose that the output source consists of a binary sequence and let S denote its shift. Suppose also that the channel is such that for each real number in, three binary symbols are produced. This fits our usual framework if we consider each output variable to consist of a binary three-tuple since then there is one output vector for each input symbol. One must be careful, however, when considering the stationarity of such a system. Do we consider the output process to be physically stationary if it is stationary with respect to S or with respect to S^3? The former might make more sense if we are looking at the output alone, the latter if we are looking at the output in relation to the input. How do we define stationarity for the pair process? Given two sequence spaces, we might first construct a shift on the pair sequence space as simply the cartesian product of the shifts, e.g., given an input sequence x and an output sequence y define a shift T^* by $T^*(x,y) = (Tx, Sy)$. While this might seem natural given simply the pair random process $\{X_n, Y_n\}$, it is not natural in the physical context that one symbol of X yields three symbols of Y. In other words, the two shifts do not correspond to the same amount of *time*. Here the more physically meaningful shift on the pair space would be $T'(x,y) = (Tx, S^3y)$ and the more physically meaningful questions on stationarity and ergodicity relate to T' and not to T^*. The problem becomes even more complicated when channels or codes produce a varying number of output symbols for each input symbol, where the number of symbols depends on the input sequence. Such variable rate codes arise often in practice, especially for noiseless coding applications such as Huffman, Lempel-Ziv, and arithmetic codes. (See [140] for a survey of noiseless coding.) While we will not treat such variable rate systems in any detail, they point out the difficulty that can arise associating the mathematical shift operation with physical time when we are considering cartesian products of spaces, each having their own shift.

There is no easy way to solve this problem notationally. We adopt the following view as a compromise which is usually adequate for fixed-rate systems. We will be most interested in pair processes that are stationary in the physical sense, that is, whose statistics are not changed when both are shifted by an equal amount of *physical* time. This is the same as stationarity with respect to the product shift if the two shifts correspond to equal amounts of physical time. Hence for simplicity we will usually focus on this case. More general cases will be introduced when appropriate to point out their form and how they can be put into the matching shift structure by considering groups of symbols and different shifts. This will necessitate occasional discussions about what is meant by stationarity or ergodicity for a particular system.

The mathematical generalization of Shannon's original notions of sources, codes, and channels are due to Khinchine [72] [73]. Khinchine's results characterizing stationarity and ergodicity of channels were corrected and developed by Adler [2].

9.2 Channels

Say we are given a source $[A, X, \mu]$, that is, a sequence of A-valued random variables $\{X_n; n \in \mathcal{T}\}$ defined on a common probability space $(\Omega, \mathcal{F}, P)$ having a process distribution μ defined on the measurable sequence space $(B^{\mathcal{T}}, \mathcal{B}_A{}^{\mathcal{T}})$. We shall let $X = \{X_n; n \in \mathcal{T}\}$ denote the sequence-valued random variable, that is, the random variable taking values in $A^{\mathcal{T}}$ according to the distribution μ. Let B be another alphabet with a corresponding measurable sequence space $(A^{\mathcal{T}}, \mathcal{B}_B{}^{\mathcal{T}})$. We assume as usual that A and B are standard and hence so are their sequence spaces and cartesian products. A *channel* $[A, \nu, B]$ with input alphabet A and output alphabet B (we denote the channel simply by ν when these alphabets are clear from context) is a family of probability measures $\{\nu_x; x \in A^{\mathcal{T}}\}$ on $(B^{\mathcal{T}}, \mathcal{B}_B{}^{\mathcal{T}})$ (the output sequence space) such that for every output event $F \in \mathcal{B}_B{}^{\mathcal{T}}$ $\nu_x(F)$ is a measurable function of x. This measurability requirement ensures that the set function p specified on the joint input/output space $(A^{\mathcal{T}} \times B^{\mathcal{T}})$, $\mathcal{B}_A{}^{\mathcal{T}} \times \mathcal{B}_B{}^{\mathcal{T}})$ by its values on rectangles as

$$p(G \times F) = \int_G d\mu(x)\nu_x(F); F \in \mathcal{B}_B{}^{\mathcal{T}}, G \in \mathcal{B}_A{}^{\mathcal{T}},$$

is well defined. The set function p is nonnegative, normalized, and countably additive on the field generated by the rectangles $G \times F$, $G \in \mathcal{B}_A{}^{\mathcal{T}}$, $F \in \mathcal{B}_B{}^{\mathcal{T}}$. Thus p extends to a probability measure on the joint input/output space, which is sometimes called the ***hookup*** of the source μ and channel ν. We will often denote this joint measure by $\mu\nu$. The corresponding sequences of random variables are called the ***input/output process.***

Thus a channel is a probability measure on the output sequence space for each input sequence such that a joint input/output probability measure is well-defined. The above equation shows that a channel is simply a regular conditional probability, in particular,

$$\nu_x(F) = p((x, y) : y \in F|x); F \in \mathcal{B}_B{}^{\mathcal{T}}, x \in A^{\mathcal{T}}.$$

We can relate a channel to the notation used previously for conditional distributions by using the sequence-valued random variables $X = \{X_n; n \in \mathcal{T}\}$ and $Y = \{Y_n; n \in \mathcal{T}\}$:

$$\nu_x(F) = P_{Y|X}(F|x). \tag{9.2.1}$$

Eq. (1.6.8) then provides the probability of an arbitrary input/output event:

$$p(F) = \int d\mu(x)\nu_x(F_x),$$

where $F_x = \{y : (x, y) \in F\}$ is the *section* of F at x.

If we start with a hookup p, then we can obtain the input distribution μ as

$$\mu(F) = p(F \times B^{\mathcal{T}}); F \in \mathcal{B}_A{}^{\mathcal{T}}.$$

Similarly we can obtain the output distribution, say η, via

$$\eta(F) = p(A^{\mathcal{T}} \times F); F \in \mathcal{B}_B{}^{\mathcal{T}}.$$

Suppose one now starts with a pair process distribution p and hence also with the induced source distribution μ. Does there exist a channel ν for which $p = \mu\nu$? The answer is yes since the spaces are standard. One can always define the conditional probability $\nu_x(F) = P(F \times A^{\mathcal{T}} | X = x)$ for all input sequences x, but this need not possess a regular version, that is, be a probability measure for all x, in the case of arbitrary alphabets. If the alphabets are standard, however, we have seen that a regular conditional probability measure always exists.

9.3 Stationarity Properties of Channels

We now define a variety of stationarity properties for channels that are related to, but not the same as, those for sources. The motivation behind the various definitions is that stationarity properties of channels coupled with those of sources should imply stationarity properties for the resulting source-channel hookups.

The classical definition of a stationary channel is the following: Suppose that we have a channel $[A, \nu, B]$ and suppose that T_A and T_B are the shifts on the input sequence space and output sequence space, respectively. The channel is ***stationary*** with respect to T_A and T_B or (T_A, T_B)-***stationary*** if

$$\nu_x(T_B^{-1}F) = \nu_{T_A x}(F), x \in \mathcal{A}^{\mathcal{T}}, F \in \mathcal{B}_B{}^{\mathcal{T}}. \tag{9.3.1}$$

If the transformations are clear from context then we simply say that the channel is stationary. Intuitively, a right shift of an output event yields the same probability as the left shift of an input event. The different shifts are required because in general only $T_A x$ and not $T_A^{-1} x$ exists since the shift may not be invertible and in general only $T_B^{-1} F$ and not $T_B F$ exists for the same reason. If the shifts are invertible, e.g., the processes are two-sided, then the definition is equivalent to

$$\nu_{T_A x}(T_B F) = \nu_{T_A^{-1}x}(T_B^{-1}F) = \nu_x(F), \text{ all } x \in A^{\mathcal{T}}, F \in \mathcal{B}_B{}^{\mathcal{T}} \tag{9.3.2}$$

that is, shifting the input sequence and output event in the same direction does not change the probability.

The fundamental importance of the stationarity of a channel is contained in the following lemma.

Lemma 9.3.1: If a source $[A, \mu]$, stationary with respect to T_A, is connected to channel $[A, \nu, B]$, stationary with respect to T_A and T_B, then the resulting hookup $\mu\nu$ is also stationary (with respect to the cartesian product shift $T = T_{A\times B} = T_A \times T_B$ defined by $T(x, y) = (T_A x, T_B y)$).

Proof: We have that

$$\mu\nu(T^{-1}F) = \int d\mu(x)\nu_x((T^{-1}F)_x).$$

Now

$$\begin{aligned}(T^{-1}F)_x &= \{y : T(x, y) \in F\} = \{y : (T_A x, T_B y) \in F\} \\ &= \{y : T_B y \in F_{T_A x}\} = T_B^{-1} F_{T_A x} \end{aligned} \tag{9.3.3}$$

and hence

$$\mu\nu(T^{-1}F) = \int d\mu(x)\nu_x(T_B^{-1}F_{T_A x}).$$

Since the channel is stationary, however, this becomes

$$\mu\nu(T^{-1}F) = \int d\mu(x)\nu_{T_A x}(F_{T_A x}) = \int d\mu T_A^{-1}(x)\nu_x(F_x),$$

where we have used the change of variables formula. Since μ is stationary, however, the right hand side is

$$\int d\mu(x)\nu_x(F),$$

which proves the lemma. □

Suppose next that we are told that a hookup $\mu\nu$ is stationary. Does it then follow that the source μ and channel ν are necessarily stationary? The source must be since

$$\mu(T_A^{-1}F) = \mu\nu((T_A \times T_B)^{-1}(F \times B^{\mathcal{T}})) = \mu\nu(F \times B^{\mathcal{T}}) = \mu(F).$$

The channel need not be stationary, however, since, for example, the stationarity could be violated on a set of μ measure 0 without affecting the proof of the above lemma. This suggests a somewhat weaker notion of stationarity which is more directly related to the stationarity of the hookup. We say that a channel $[A,\nu,B]$ is ***stationary with respect to a source*** $[A,\mu]$ if $\mu\nu$ is stationary. We also state that a channel is stationary μ-a.e. if it satisfies (9.3.1) for all x in a set of μ-probability one. If a channel is stationary μ-a.e. and μ is stationary, then the channel is also stationary with respect to μ. Clearly a stationary channel is stationary with respect to all stationary sources. The reason for this more general view is that we wish to extend the definition of stationary channels to asymptotically mean stationary channels. The general definition extends; the classical definition of stationary channels does not.

Observe that the various definitions of stationarity of channels immediately extend to block shifts since they hold for any shifts defined on the input and output sequence spaces, e.g., a channel stationary with respect to T_A^N and T_B^K could be a reasonable model for a channel or code that puts out K symbols from an alphabet B every time it takes in N symbols from an alphabet A. We shorten the name (T_A^N, T_B^K)-stationary to (N,K)-stationary channel in this case. A stationary channel (without modifiers) is simply a (1,1)-stationary channel in this sense.

The most general notion of stationarity that we are interested in is that of asymptotic mean stationarity We define a channel $[A,\nu,B]$ to be ***asymptotically mean stationary*** or *AMS* for a source $[A,\mu]$ with respect to T_A and T_B if the hookup $\mu\nu$ is AMS with respect to the product shift $T_A \times T_B$. As in the stationary case, an immediate necessary condition is that the input source be AMS with respect to T_A. A channel will be said to be (T_A,T_B)-AMS if the hookup is (T_A,T_B)-AMS for all T_A-AMS sources.

The following lemma shows that an AMS channel is indeed a generalization of the idea of a stationary channel and that the stationary mean of a hookup of an AMS source to a stationary channel is simply the hookup of the stationary mean of the source to the channel.

Lemma 9.3.2: Suppose that ν is (T_A,T_B)-stationary and that μ is AMS with respect to T_A. Let $\bar{\mu}$ denote the stationary mean of μ and observe that $\bar{\mu}\nu$ is stationary. Then the hookup $\mu\nu$ is AMS with stationary mean

$$\overline{\mu\nu} = \bar{\mu}\nu.$$

Thus, in particular, ν is an AMS channel.

Proof: We have that

$$(T^{-i}F)_x = \{y : (x,y) \in T^{-i}F\} = \{y : T^i(x,y) \in F\}$$

$$= \{y : (T_A^i x, T_B^i y) \in F\} = \{y : T_B^i y \in F_{T_A^i x}\} = T_B^{-i} F_{T_A^i x} \qquad (9.3.4)$$

and therefore since ν is stationary

$$\mu\nu(T^{-i}F) = \int d\mu(x)\nu_x(T_B^{-i}F_{T_A^i x})$$

$$= \int d\mu(x)\nu_{T_A^i x}(F_{T_A^i x}) = \int d\mu T_A^{-i}(x)\nu_x(F).$$

Therefore

$$\frac{1}{n}\sum_{i=0}^{n-1}\mu\nu(T^{-i}F)$$

$$= \frac{1}{n}\sum_{i=0}^{n-1}\int d\mu T_A^{-i}(x)\nu_x(F) \underset{n\to\infty}{\to} \int d\bar{\mu}(x)\nu_x(F) = \bar{\mu}\nu(F)$$

from Lemma 6.5.1 of [50]. This proves that $\mu\nu$ is AMS and that the stationary mean is $\bar{\mu}\nu$. □

A final property crucial to quantifying the behavior of random processes is that of ergodicity. Hence we define a (stationary, AMS) channel ν to be ergodic with respect to (T_A, T_B) if it has the property that whenever a (stationary, AMS) ergodic source (with respect to T_A) is connected to the channel, the overall input/output process is (stationary, AMS) ergodic. The following modification of Lemma 6.7.4 of [50] is the principal tool for proving a channel to be ergodic.

Lemma 9.3.3: An AMS (stationary) channel $[A, \nu, B]$ is ergodic if for all AMS (stationary) sources μ and all sets of the form $\bar{F} = F_A \times F_B$, $\bar{G} = G_A \times G_B$ for rectangles $F_A, G_A \in \mathcal{B}_A^\infty$ and $F_B, G_B \in \mathcal{B}_B^\infty$ we have that for $p = \mu\nu$

$$\lim_{n\to\infty}\frac{1}{n}\sum_{i=0}^{n-1}p(T_{A\times B}^{-i}\bar{F}\bigcap\bar{G}) = \bar{p}(\bar{F})p(\bar{G}), \tag{9.3.5}$$

where $\bar{p}$ is the stationary mean of p (p if p is already stationary).

Proof: The proof parallels that of Lemma 6.7.4 of [50]. The result does not follow immediately from that lemma since the collection of given sets does not itself form a field. Arbitrary events $F, G \in \mathcal{B}_{A\times B}^\infty$ can be approximated arbitrarily closely by events in the field generated by the above rectangles and hence given $\epsilon > 0$ we can find finite disjoint rectangles of the given form F_i, G_i, $i = 1, \cdots, L$ such that if $F_0 = \bigcup_{i=1}^L F_i$ and $G_0 = \bigcup_{i=1}^L G_i$, then $p(F\Delta F_0)$, $p(G\Delta G_0)$, $\bar{p}(F\Delta F_0)$, and $\bar{p}(G\Delta G_0)$ are all less than ϵ. Then

$$|\frac{1}{n}\sum_{k=0}^{n-1}p(T^{-k}F\bigcap G) - \bar{p}(F)p(G)|$$

$$\leq |\frac{1}{n}\sum_{k=0}^{n-1} p(T^{-k}F\bigcap G) - \frac{1}{n}\sum_{k=0}^{n-1} p(T^{-k}F_0\bigcap G_0)|$$

$$+|\frac{1}{n}\sum_{k=0}^{n-1} p(T^{-k}F_0\bigcap G_0) - \bar{p}(F_0)p(G_0)| + |\bar{p}(F_0)p(G_0) - \bar{p}(F)p(G)|.$$

Exactly as in Lemma 6.7.4 of [50], the rightmost term is bound above by 2ϵ and the first term on the left goes to zero as $n \to \infty$. The middle term is the absolute magnitude of

$$\frac{1}{n}\sum_{k=0}^{n-1} p(T^{-k}\bigcup_i F_i \bigcap \bigcup_j G_j) - \bar{p}(\bigcup_i F_i)p(\bigcup_j G_j)$$

$$= \sum_{i,j}\left(\frac{1}{n}\sum_{k=0}^{n-1} p(T^{-k}F_i\bigcap G_j) - \bar{p}(F_i)p(G_j)\right).$$

Each term in the finite sum converges to 0 by assumption. Thus p is ergodic from Lemma 6.7.4 of [50]. □

Because of the specific class of sets chosen, the above lemma considered separate sets for shifting and remaining fixed, unlike using the same set for both purposes as in Lemma 6.7.4 of [50]. This was required so that the cross products in the final sum considered would converge accordingly.

9.4 Examples of Channels

In this section a variety of examples of channels are introduced, ranging from the trivially simple to the very complicated. The first two channels are the simplest, the first being perfect and the second being useless (at least for communication purposes).

Example 9.4.1: Noiseless Channel

A channel $[A, \nu, B]$ is said to be *noiseless* if $A = B$ and

$$\nu_x(F) = \begin{cases} 1 & x \in F \\ 0 & x \notin F \end{cases}$$

that is, with probability one the channel puts out what goes in. Such a channel is clearly stationary and ergodic.

Example 9.4.2: Completely Random Channel

Suppose that η is a probability measure on the output space $(B^{\mathcal{T}}, \mathcal{B}_B{}^{\mathcal{T}})$ and define a channel

$$\nu_x(F) = \eta(F), F \in \mathcal{B}_B{}^{\mathcal{T}}, x \in A^{\mathcal{T}}.$$

Then it is easy to see that the input/output measure satisfies

$$p(G \times F) = \eta(F)\mu(G); F \in \mathcal{B}_B{}^{\mathcal{T}}, G \in \mathcal{B}_A{}^{\mathcal{T}},$$

and hence the input/output measure is a product measure and the input and output sequences are therefore independent of each other. This channel is called a *completely random channel* or *product channel* because the output is independent of the input. This channel is quite useless because the output tells us nothing of the input. The completely random channel is stationary (AMS) if the measure η is stationary (AMS). Perhaps surprisingly, such a channel need not be ergodic even if η is ergodic since the product of two stationary and ergodic sources need not be ergodic. (See, e.g., [21].) We shall later see that if η is also assumed to be weakly mixing, then the resulting channel is ergodic.

A generalization of the noiseless channel that is of much greater interest is the deterministic channel. Here the channel is not random, but the output is formed by a general mapping of the input rather than being the input itself.

Example 9.4.3: Deterministic Channel and Sequence Coders

A channel $[A, \nu, B]$ is said to be *deterministic* or *nonrandom* if each input string is mapped into a fixed output string, that is, if there is a mapping $f : A^{\mathcal{T}} \to B^{\mathcal{T}}$ such that

$$\nu_x(G) = \begin{cases} 1 & f(x) \in G \\ 0 & f(x) \notin G \end{cases}.$$

The mapping f must be measurable in order to satisfy the measurability assumption of the channel. Note that such a channel can also be written as

$$\nu_x(G) = 1_{f^{-1}(G)}(x).$$

Define a *sequence coder* as a deterministic channel, that is, a measurable mapping from one sequence space into another. It is easy to see that for a deterministic code we have a hookup specified by

$$p(F \times G) = \mu(F \bigcap f^{-1}(G))$$

and an output process with distribution

$$\eta(G) = \mu(f^{-1}(G)).$$

A sequence coder is said to be (T_A, T_B)-*stationary* (or just *stationary*) or (T_A^N, T_B^K)-*stationary* (or just (N, K)-*stationary*) if the corresponding channel is. Thus a sequence coder f is stationary if and only if $f(T_A x) = T_B f(x)$ and it is (N, K)- stationary if and only if $f(T_A^N x) = T_B^K f(x)$.

Lemma 9.4.1: A stationary deterministic channel is ergodic.

Proof: From Lemma 9.3.3 it suffices to show that

$$\lim_{n\to\infty} \frac{1}{n} \sum_{i=0}^{n-1} p(T_{A\times B}^{-i} F \bigcap G) = p(F)P(G)$$

for all rectangles of the form $F = F_A \times F_B$, $F_A \in \mathcal{B}_B{}^{\mathcal{T}}$, $F_B \in \mathcal{B}_A{}^{\mathcal{T}}$ and $G = G_A \times G_B$. Then

$$p(T_{A\times B}^{-i} F \bigcap G) = p((T_A^{-i} F_A \bigcap G_A) \times (T_B^{-i} F_B \bigcap G_B))$$

$$= \mu((T_A^{-i} F_A \bigcap G_A) \bigcap f^{-1}(T_B^{-i} F_B \bigcap G_B)).$$

Since f is stationary and since inverse images preserve set theoretic operations,

$$f^{-1}(T_B^{-i} F_B \bigcap G_B) = T_A^{-i} f^{-1}(F_B) \bigcap f^{-1}(G_B)$$

and hence

$$\frac{1}{n} \sum_{i=0}^{n-1} p(T_{A\times B}^{-i} F \bigcap G)$$

$$= \frac{1}{n} \sum_{i=0}^{n-1} \mu(T_A^{-i}(F_A \bigcap f^{-1}(F_B)) \bigcap G_A \bigcap f^{-1}(G_B))$$

$$\underset{n\to\infty}{\to} \mu(F_A \bigcap f^{-1}(F_B))\mu(G_A \bigcap f^{-1}(G_B)) = p(F_A \times F_B)p(G_A \times G_B)$$

since μ is ergodic. This means that the rectangles meet the required condition. Some algebra then will show that finite unions of disjoint sets meeting the conditions also meet the conditions and that complements of sets meeting the conditions also meet them. This implies from the good sets principle (see, for example, p. 14 of [50]) that the field generated by the rectangles also meets the condition and hence the lemma is proved. □

A stationary sequence coder has a simple and useful structure. Suppose one has a mapping $f : A^{\mathcal{T}} \to B$, that is, a mapping that maps an input sequence into an output letter. We can define a complete output sequence y corresponding to an input sequence x by

$$y_i = f(T_A^i x); i \in \mathcal{T}, \tag{9.4.1}$$

that is, we produce an output, then shift or slide the input sequence by one time unit, and then we produce another output using the same function, and so on. A mapping of this form is called an infinite length *sliding block code* because it produces outputs by successively sliding an infinite length input sequence and each time using a fixed mapping to produce the output. The sequence-to-letter mapping implies a sequence coder, say $\bar{f}$, defined by $\bar{f}(x) = \{f(T_A^i x); i \in \mathcal{T}\}$. Furthermore, $\bar{f}(T_A x) = T_B \bar{f}(x)$, that is, a sliding block code induces a stationary sequence coder. Conversely, any stationary sequence coder $\bar{f}$ induces a sliding block code f for which (9.4.1) holds by the simple identification $f(x) = (\bar{f}(x))_0$, the output at time 0 of the sequence coder. Thus the ideas of stationary sequence coders mapping sequences into sequences and sliding block codes mapping sequences into letters by sliding the input sequence are equivalent. We can similarly define an (N, K)-sliding block code which is a mapping $f : A^{\mathcal{T}} \to B^K$ which forms an output sequence y from an input sequence x via the construction

$$y_{iK}^K = f(T_A^{Ni} x).$$

By a similar argument, (N, K)-sliding block coders are equivalent to (N, K)-stationary sequence coders. When dealing with sliding block codes we will usually assume for simplicity that K is 1. This involves no loss in generality since it can be made true by redefining the output alphabet.

Example 9.4.4: B-processes

The above construction using sliding block or stationary codes provides an easy description of an important class of random processes that has several nice properties. A process is said to be a *B-process* or *Bernoulli process* if it can be defined as a stationary coding of an independent identically distributed (i.i.d.) process. Let μ denote the original distribution of the i.i.d. process and let η denote the induced output distribution. Then for any output events F and G

$$\eta(F \bigcap T_B^{-n} G) = \mu(\bar{f}^{-1}(F \bigcap T_B^{-n} G)) = \mu(\bar{f}^{-1}(F) \bigcap T_A^{-n} \bar{f}^{-1}(G)),$$

since $\bar{f}$ is stationary. But μ is stationary and mixing since it is i.i.d. (see Section 6.7 of [50]) and hence this probability converges to

$$\mu(f^{-1}(F))\mu(f^{-1}(G)) = \eta(F)\eta(G)$$

and hence η is also mixing. Thus a B-process is mixing of all orders and hence is ergodic with respect to T_B^n for all positive integers n.

While codes that depend on infinite input sequences may not at first glance seem to be a reasonable physical model of a coding system, it is possible for such codes to depend on the infinite sequence only through a finite number of coordinates. In addition, some real codes may indeed depend on an unboundedly large number of past inputs because of feedback.

Suppose that we consider two-sided processes and that we have a measurable mapping

$$\phi : \mathop{\times}_{i=-M}^{D} A_i \to B$$

and we define a sliding block code by

$$f(x) = \phi(x_{i-M}, \cdots, x_0, \cdots, x_{i+D}),$$

then $\bar{f}$ is a stationary sequence coder. The mapping ϕ is also called a sliding block code or a finite-length sliding block code or a finite-window sliding block code. M is called the memory of the code and D is called the delay of the code since M past source symbols and D future symbols are required to produce the current output symbol. The *window length* or *constraint length* of the code is $M + D + 1$, the number of input symbols viewed to produce an output symbol. If $D = 0$ the code is said to be *causal*. If $M = 0$ the code is said to be *memoryless*.

There is a problem with the above model if we wish to code a one-sided source since if we wish to start coding at time 0, there are no input symbols with negative indices. Hence we either must require the code be memoryless ($M = 0$) or we must redefine the code for the first M instances (e.g., by "stuffing" the code register with arbitrary symbols) or we must only define the output for times $i \geq M$. For two-sided sources a finite-length sliding block code is stationary. In the one-sided case it is not even defined precisely unless it is memoryless, in which case it is stationary.

Another case of particular interest is when we have a measurable mapping $\gamma : A^N \to B^K$ and we define a sequence coder $f(x) = y$ by

$$y_{nK}^K = (y_{nK}, y_{nK+1}, \cdots, y_{(n+1)K-1}) = \gamma(x_{nN}^N),$$

that is, the input is parsed into nonoverlapping blocks of length N and each is successively coded into a block of length K outputs without regard to past or previous input or output blocks. Clearly N input time units must correspond to K output time units in physical time if the code is to make sense. A code of this form is called a *block code* and it is a special case of an (N, K) sliding block code. Such a code is trivially (T_A^N, T_A^K)-stationary.

We now return to genuinely random channels. The next example is perhaps the most popular model for a noisy channel because of its simplicity.

Example 9.4.5: Memoryless channels

Suppose that $q_{x_0}(\cdot)$ is a probability measure on $\mathcal{B}_B$ for all $x_0 \in A$ and that for fixed F ,$q_{x_0}(F)$ is a measurable function of x_0. Let ν be a channel specified by its values on output rectangles by

$$\nu_x(\underset{i \in \mathcal{J}}{\times} F_i) = \prod_{i \in \mathcal{J}} q_{x_i}(F_i),$$

for any finite index set $\mathcal{J} \subset \mathcal{T}$. Then ν is said to be a *memoryless channel.* Intuitively,

$$\Pr(Y_i \in F_i; i \in \mathcal{J}|X) = \prod_{i \in \mathcal{J}} \Pr(Y_i \in F_i|X_i).$$

For later use we pause to develop a useful inequality for mutual information between the input and output of a memoryless channel. For contrast we also describe the corresponding result for a memoryless source and an arbitrary channel.

Lemma 9.4.2: Let $\{X_n\}$ be a source with distribution μ and let ν be a channel. Let $\{X_n, Y_n\}$ be the hookup with distribution p. If the channel is memoryless, then for any n

$$I(X^n; Y^n) \leq \sum_{i=0}^{n-1} I(X_i; Y_i)$$

If instead the source is memoryless, then the inequality is reversed:

$$I(X^n; Y^n) \geq \sum_{i=0}^{n-1} I(X_i; Y_i).$$

Thus if both source and channel are memoryless,

$$I(X^n; Y^n) = \sum_{i=0}^{n-1} I(X_i; Y_i).$$

Proof: First suppose that the process is discrete. Then

$$I(X^n; Y^n) = H(Y^n) - H(Y^n|X^n).$$

Since by construction

$$P_{Y^n|X^n}(y^n|x^n) = \prod_{i=0}^{n-1} P_{Y_0|X_0}(y_i|x_i)$$

an easy computation shows that

$$H(Y^n|X^n) = \sum_{i=0}^{n-1} H(Y_i|X_i).$$

This combined with the inequality

$$H(Y^n) \leq \sum_{i=0}^{n-1} H(Y_i)$$

(Lemma 2.3.2 used several times) completes the proof of the memoryless channel result for finite alphabets. If instead the source is memoryless, we have

$$\begin{aligned} I(X^n;Y^n) &= H(X^n) - H(X^n|Y^n) \\ &= \sum_{i=0}^{n-1} H(X_i) - H(X^n|Y^n). \end{aligned}$$

Extending Lemma 2.3.2 to conditional entropy yields

$$H(X^n|Y^n) \leq \sum_{i=0}^{n-1} H(X_i|Y^n)$$

which can be further overbounded by using Lemma 2.5.2 (the fact that reducing conditioning increases conditional entropy) as

$$H(X^n|Y^n) \leq \sum_{i=0}^{n-1} H(X_i|Y_i)$$

which implies that

$$I(X^n;Y^n) \geq \sum_{i=0}^{n-1} H(X_i) - H(X_i|Y_i) = \sum_{i=0}^{n-1} I(X_i;Y_i),$$

which completes the proof for finite alphabets.

To extend the result to standard alphabets, first consider the case where the Y^n are quantized to a finite alphabet. If the Y_k are conditionally independent given X^k, then the same is true for $q(Y_k)$, $k = 0, 1, \cdots, n-1$. Lemma 5.5.6 then implies that as in the discrete case, $I(X^n;Y^n) = H(Y^n) - H(Y^n|X^n)$ and the remainder of the proof follows as in the discrete case. Letting the quantizers become asymptotically accurate then completes the proof. □

In fact two forms of memorylessness are evident in a memoryless channel. The channel is ***input memoryless*** in that the probability of an output event involving $\{Y_i; i \in \{k, k+1, \cdots, m\}\}$ does not involve any inputs before time k, that is, the past inputs. The channel is also ***input nonanticipatory*** since this event does not depend on inputs after time m, that is, the future inputs. The channel is also ***output memoryless*** in the sense that for any given input x, output events involving nonoverlapping times are independent, i.e.,

$$\nu_x(Y_1 \in F_1 \bigcap Y_2 \in F_2) = \nu_x(Y_1 \in F_1)\nu_x(Y_2 \in F_2).$$

We pin down these ideas in the following examples.

Example 9.4.6: Channels with finite input memory and anticipation

A channel ν is said to have finite input memory of order M if for all one-sided events F and all n

$$\nu_x((Y_n, Y_{n+1}, \cdots) \in F) = \nu_{x'}((Y_n, Y_{n+1}, \cdots) \in F)$$

whenever $x_i = x'_i$ for $i \geq n - M$. In other words, for an event involving Y_i's after some time n, knowing only the inputs for the same times and M time units earlier completely determines the output probability. Channels with finite input memory were introduced by Feinstein [40]. Similarly ν is said to have finite anticipation of order L if for all one-sided events F and all n

$$\nu_x((\cdots, Y_n) \in F) = \nu_{x'}((\cdots, Y_n) \in F)$$

provided $x'_i = x_i$ for $i \leq n + L$. That is, at most L future inputs must be known to determine the probability of an event involving current and past outputs.

Example 9.4.7: Channels with finite output memory

A channel ν is said to have *finite output memory of order K* if for all one-sided events F and G and all inputs x, if $k > K$ then

$$\nu_x((\cdots, Y_n) \in F \bigcap (Y_{n+k}, \cdots) \in G) = \nu_x((\cdots, Y_n) \in F)\nu_x((Y_{n+k}, \cdots) \in G);$$

that is, output events involving output samples separated by more than K time units are independent. Channels with finite output memory were introduced by Wolfowitz [150].

Channels with finite memory and anticipation are historically important as the first real generalizations of memoryless channels for which coding theorems could be proved. Furthermore, the assumption of finite anticipation is physically reasonable as a model for real-world communication channels. The finite memory assumptions, however, exclude many important examples, e.g., finite-state or Markov channels and channels with feedback filtering action. Hence we will emphasize more general notions which can be viewed as approximations or asymptotic versions of the finite memory assumption. The generalization of finite input memory channels requires some additional tools and is postponed to the next chapter. The notion of finite output memory can be generalized by using the notion of mixing.

Example 9.4.8: Output mixing channels

A channel is said to be output mixing (or asymptotically output independent or asymptotically output memoryless) if for all output rectangles F and G and all input sequences x

$$\lim_{n\to\infty} |\nu_x(T^{-n}F\bigcap G) - \nu_x(T^{-n}F)\nu_x(G)| = 0.$$

More generally it is said to be *output weakly mixing* if

$$\lim_{n\to\infty} \frac{1}{n}\sum_{i=0}^{n-1} |\nu_x(T^{-i}F\bigcap G) - \nu_x(T^{-i}F)\nu_x(G)| = 0.$$

Unlike mixing systems, the above definitions for channels place conditions only on output rectangles and not on all output events. Output mixing channels were introduced by Adler [2].

The principal property of output mixing channels is provided by the following lemma.

Lemma 9.4.3: If a channel is stationary and output weakly mixing, then it is also ergodic. That is, if ν is stationary and output weakly mixing and if μ is stationary and ergodic, then also $\mu\nu$ is stationary and ergodic.

Proof: The process $\mu\nu$ is stationary by Lemma 9.3.1. To prove that it is ergodic it suffices from Lemma 9.3.3 to prove that for all input/output rectangles of the form $F = F_B \times F_A$, $F_B \in \mathcal{B}_A{}^{\mathcal{T}}$, $F_A \in \mathcal{B}_B{}^{\mathcal{T}}$, and $G = G_B \times G_A$ that

$$\lim_{n\to\infty} \frac{1}{n}\sum_{i=0}^{n-1} \mu\nu(T^{-i}F\bigcap G) = \mu\nu(F)\mu\nu(G).$$

We have that

$$\frac{1}{n}\sum_{i=0}^{n-1} \mu\nu(T^{-i}F\bigcap G) - m(F)m(G)$$

$$= \frac{1}{n}\sum_{i=0}^{n-1} \mu\nu((T_B^{-i}F_B \bigcap G_B) \times (T_A^{-i}F_A \bigcap G_A)) - \mu\nu(F_B \times F_A)\mu\nu(G_B \times G_A)$$

$$= \frac{1}{n}\sum_{i=0}^{n-1} \int_{T_A^{-i}F_A \bigcap G_A} d\mu(x)\nu_x(T_B^{-i}F_B \bigcap G_B) - \mu\nu(F_B \times F_A)\mu(G_B \times G_A)$$

$$= \left(\frac{1}{n}\sum_{i=0}^{n-1} \left(\int_{T_A^{-i}F_A \bigcap G_A} d\mu(x)\nu_x(T_B^{-i}F_B \bigcap G_B)\right.\right.$$

$$\left.\left. - \int_{T_A^{-i}F_A \bigcap G_A} d\mu(x)\nu_x(T_B^{-i}F_B)\nu_x(G_B)\right)\right) + \left(\frac{1}{n}\sum_{i=0}^{n-1}\right.$$

$$\left.\left(\int_{T_A^{-i}F_A \bigcap G_A} d\mu(x)\nu_x(T_B^{-i}F_B)\nu_x(G_B) - \mu\nu(F_B \times F_A)\mu\nu(G_B \times G_A)\right)\right).$$

The first term is bound above by

$$\frac{1}{n}\sum_{i=0}^{n-1} \int_{T_A^{-i}F_A \bigcap G_A} d\mu(x)|\nu_x(T_B^{-i}F_B \bigcap G_B) - \nu_x(T_B^{-i}F_B)\nu_x(G_B)|$$

$$\leq \int d\mu(x)\frac{1}{n}\sum_{i=0}^{n-1} |\nu_x(T_B^{-i}F_B \bigcap G_B) - \nu_x(T^{-i}F_B)\nu_x(G_B)|$$

which goes to zero from the dominated convergence theorem since the integrand converges to zero from the output weakly mixing assumption. The second term can be expressed using the stationarity of the channel as

$$\int_{F_A} d\mu(x)\nu_x(G_B)\frac{1}{n}\sum_{i=0}^{n-1} 1_{F_A}(T_A^i x)\nu_{T_A^i x}(F_B) - \mu\nu(F)\mu\nu(G).$$

The ergodic theorem implies that as $n \to \infty$ the sample average goes to its expectation

$$\int d\mu(x)1_{F_A}(x)\nu_x(F_B) = \mu\nu(F)$$

and hence the above formula converges to 0, proving the lemma. □

The lemma provides an example of a completely random channel that is also ergodic in the following corollary.

Corollary 9.4.1: Suppose that ν is a stationary completely random channel described by an output measure η. If η is weakly mixing, then ν is ergodic. That is, if μ is stationary and ergodic and η is stationary and weakly mixing, then $\mu\nu = \mu \times \eta$ is stationary and ergodic.

Proof: If η is weakly mixing, then the channel ν defined by $\nu_x(F) = \eta(F)$, all $x \in A^{\mathcal{T}}$, $F \in \mathcal{B}_B{}^{\mathcal{T}}$ is output weakly mixing. Thus ergodicity follows from the lemma. □

The idea of a memoryless channel can be extended to a block memoryless or block independent channel, as described next.

Example 9.4.9: Block Memoryless Channels

Suppose now that we have an integers N and K (usually $K = N$) and a probability measure $q_{x^N}(\cdot)$ on $\mathcal{B}_B^K$ for each $x^N \in A^N$ such that $q_{x^N}(F)$ is a measurable function of x^N for each $F \in \mathcal{B}_B^K$. Let ν be specified by its values on output rectangles by

$$\nu_x(y : y_i \in G_i; i = m, \cdots, m+n-1) = \prod_{i=0}^{\lfloor \frac{n}{K} \rfloor} q_{x_{iN}^N}(G_i),$$

where $G_i \in \mathcal{B}_B$, all i, where $\lfloor z \rfloor$ is the largest integer contained in z, and where

$$G_i = \mathop{\times}_{j=m+iK}^{m+(i+1)K-1} F_j$$

with $F_j = B$ if $j \geq m+n$. Such channels are called *block memoryless channels* or *block independent channels*. They are a special case of the following class of channels.

Example 9.4.10: Conditionally Block Independent Channels

A *conditionally block independent* or *CBI* channel resembles the block memoryless channel in that for a given input sequence the outputs are block independent. It is more general, however, in that the conditional probabilities of the output block may depend on the entire input sequence (or at least on parts of the input sequence not in the same time block). Thus a channel is CBI if its values on output rectangles satisfy

$$\nu_x(y : y_i \in F_i; i = m, \cdots, m+n-1) = \prod_{i=0}^{\lfloor \frac{n}{K} \rfloor} \nu_x(y : y_{iN}^N \in G_i).$$

where as before

$$G_i = \mathop{\times}_{j=m+iK}^{m+(i+1)K-1} F_j$$

with $F_j = B$ if $j \geq m+n$. Block memoryless channels are clearly a special case of CBI channels. These channels have only finite output memory, but unlike the block memoryless channels they need not have finite input memory or anticipation.

The primary use of block memoryless channels is in the construction of a channel given finite-dimensional conditional probabilities, that is, one has probabilities for output K-tuples given input N-tuples and one wishes to model a channel consistent with these finite-dimensional distributions. The finite dimensional distributions themselves may be the result of an optimization problem or an estimate based on observed behavior. An immediate problem is that a channel constructed in this manner may not be stationary, although it is clearly (N, K)-stationary. The next example shows how to modify a block memoryless channel so as to produce a stationary channel. The basic idea is to occasionally insert some random spacing between the blocks so as to "stationarize" the channel.

Before turning to the example we first develop the technical details required for producing such random spacing.

Random Punctuation Sequences

We demonstrate that we can obtain a sequence with certain properties by stationary coding of an arbitrary stationary and ergodic process. The lemma is a variant of a theorem of Shields and Neuhoff [133] as simplified by Neuhoff and Gilbert [108] for sliding block codings of finite alphabet processes. One of the uses to which the result will be put is the same as theirs: constructing sliding block codes from block codes.

Lemma 9.4.4: Suppose that $\{X_n\}$ is a stationary and ergodic process. Then given N and $\delta > 0$ there exists a stationary (or sliding block) coding $f : A^{\mathcal{T}} \to \{0, 1, 2\}$ yielding a ternary process $\{Z_n\}$ with the following properties:

(a) $\{Z_n\}$ is stationary and ergodic.

(b) $\{Z_n\}$ has a ternary alphabet $\{0, 1, 2\}$ and it can output only N-cells of the form $011 \cdots 1$ (0 followed by $N - 1$ ones) or individual 2's. In particular, each 0 is always followed by at exactly $N - 1$ 1's.

(c) For all integers k

$$\frac{1-\delta}{N} \leq \Pr(Z_k^N = 011 \cdots 1) \leq \frac{1}{N}$$

and hence for any n

$$\Pr(Z_n \text{ is in an } N-\text{cell}) \geq 1 - \delta.$$

A process $\{Z_n\}$ with these properties is called an (N,δ)-*random blocking process* or *punctuation sequence* $\{Z_n\}$.

Proof: A sliding block coding is stationary and hence coding a stationary and ergodic process will yield a stationary and ergodic process (Lemma 9.4.1) which proves the first part. Pick an $\epsilon > 0$ such that $\epsilon N < \delta$. Given the stationary and ergodic process $\{X_n\}$ (that is also assumed to be aperiodic in the sense that it does not place all of its probability on a finite set of sequences) we can find an event $G \in \mathcal{B}_A{}^T$ having probability less than ϵ. Consider the event $F = G - \bigcup_{i=1}^{N-1} T^{-i}G$, that is, F is the collection of sequences x for which $x \in G$, but $T^i x \notin G$ for $i = 1, \cdots, N-1$. We next develop several properties of this set.

First observe that obviously $\mu(F) \leq \mu(G)$ and hence

$$\mu(F) \leq \epsilon.$$

The sequence of sets $T^{-i}F$ are disjoint since if $y \in T^{-i}F$, then $T^i y \in F \subset G$ and $T^{i+l}y \notin G$ for $l = 1, \cdots, N-1$, which means that $T^j y \notin G$ and hence $T^j y \notin F$ for $N-1 \geq j > i$. Lastly we need to show that although F may have small probability, it is not 0. To see this suppose the contrary, that is, suppose that $\mu(G - \bigcup_{i=1}^{N-1} T^{-i}G) = 0$. Then

$$\mu(G \bigcap (\bigcup_{i=1}^{N-1} T^{-i}G)) = \mu(G) - \mu(G \bigcap (\bigcup_{i=1}^{N-1} T^{-i}G)^c) = \mu(G)$$

and hence $\mu(\bigcup_{i=1}^{N-1} T^{-i}G|G) = 1$. In words, if G occurs, then it is certain to occur again within the next N shifts. This means that with probability 1 the relative frequency of G in a sequence x must be no less than $1/N$ since if it ever occurs (which it must with probability 1), it must thereafter occur at least once every N shifts. This is a contradiction, however, since this means from the ergodic theorem that $\mu(G) \geq 1/N$ when it was assumed that $\mu(G) \leq \epsilon < 1/N$. Thus it must hold that $\mu(F) > 0$.

We now use the rare event F to define a sliding block code. The general idea is simple, but a more complicated detail will be required to handle a special case. Given a sequence x, define $n(x)$ to be the smallest i for which $T^i x \in F$; that is, we look into the future to find the next occurrence of F. Since F has nonzero probability, $n(x)$ will be finite with probability 1. Intuitively, $n(x)$ should usually be large since F has small probability. Once F is found, we code backwards from that point using blocks of a 0 prefix followed by $N-1$ 1's. The appropriate symbol is then the output of the sliding block code. More precisely, if $n(x) = kN + l$, then the sliding block code prints a 0 if $l = 0$ and prints a 1 otherwise. This idea suffices until the event F actually occurs at the present time, that is, when $n(x) = 0$. At this point the sliding block code has just completed printing an N-cell of $0111\cdots 1$. It should not automatically start a new N-cell, because at the next shift it will be looking for a new F in the future to code back from and the new cells may not align with the old cells. Thus the coder looks into the future for the next F,;that is, it again seeks $n(x)$, the smallest i for which $T^ix \in F$. This time $n(x)$ must be greater than or equal to N since x is now in F and $T^{-i}F$ are disjoint for $i = 1, \cdots N-1$. After finding $n(x) = kN + l$, the coder again codes back to the origin of time. If $l = 0$, then the two codes are aligned and the coder prints a 0 and continues as before. If $l \neq 0$, then the two codes are not aligned, that is, the current time is in the middle of a new code word. By construction $l \leq N - 1$. In this case the coder prints l 2's (filler poop) and shifts the input sequence l times. At this point there is an $n(x) = kN$ for such that $T^{n(x)}x \in F$ and the coding can proceed as before. Note that k is at least one, that is, there is at least one complete cell before encountering the new F.

By construction, 2's can occur only following the event F and then no more than N 2's can be produced. Thus from the ergodic theorem the relative frequency of 2's (and hence the probability that Z_n is not in an N-block) is no greater than

$$\lim_{n\to\infty} \frac{1}{n}\sum_{i=0}^{n-1} 1_2(Z_0(T^ix)) \leq \lim_{n\to\infty} \frac{1}{n}\sum_{i=0}^{n-1} 1_F(T^ix)N$$

$$= N\mu(F) \leq N\frac{\delta}{N} = \delta, \tag{9.4.2}$$

that is,

$$\Pr(Z_n \text{ is in an } N - \text{cell}) \geq 1 - \delta.$$

Since Z_n is stationary by construction,

$$\Pr(Z_k^N = 011\cdots 1) = \Pr(Z_0^N = 011\cdots 1) \text{ for all } k.$$

Thus

$$\Pr(Z_0^N = 011\cdots 1) = \frac{1}{N}\sum_{k=0}^{N-1} \Pr(Z_k^N = 011\cdots 1).$$

The events $\{Z_k^N = 011\cdots 1\}$, $k = 0, 1, \cdots, N-1$ are disjoint, however, since there can be at most one 0 in a single block of N symbols. Thus

$$N\Pr(Z^N = 011\cdots 1) = \sum_{k=0}^{N-1} \Pr(Z_k^N = 011\cdots 1)$$

$$= \Pr(\bigcup_{k=0}^{N-1} \{Z_k^N = 011\cdots 1\}). \tag{9.4.3}$$

Thus since the rightmost probability is between $1-\delta$ and 1,

$$\frac{1}{N} \geq \Pr(Z_0^N = 011\cdots 1) \geq \frac{1-\delta}{N}$$

which completes the proof. □

The following corollary shows that a finite length sliding block code can be used in the lemma.

Corollary 9.4.2: Given the assumptions of the lemma, a finite-window sliding block code exists with properties (a)-(c).

Proof: The sets G and hence also F can be chosen in the proof of the lemma to be finite dimensional, that is, to be measurable with respect to $\sigma(X_{-K}, \cdots, X_K)$ for some sufficiently large K. Choose these sets as before with $\delta/2$ replacing δ. Define $n(x)$ as in the proof of the lemma. Since $n(x)$ is finite with probability one, there must be an L such that if

$$B_L = \{x : n(x) > L\},$$

then

$$\mu(B_L) < \frac{\delta}{2}.$$

Modify the construction of the lemma so that if $n(x) > L$, then the sliding block code prints a 2. Thus if there is no occurrence of the desired finite dimensional pattern in a huge bunch of future symbols, a 2 is produced. If $n(x) < L$, then f is chosen as in the proof of the lemma. The proof now proceeds as in the lemma until (9.4.2), which is replaced by

$$\lim_{n\to\infty} \frac{1}{n} \sum_{i=0}^{n-1} 1_2(Z_0(T^i x)) \leq \lim_{n\to\infty} \frac{1}{n} \sum_{i=0}^{n-1} 1_{B_L}(T^i x) + \lim_{n\to\infty} \frac{1}{n} \sum_{i=0}^{n-1} 1_F(T^i x) N$$
$$\leq \delta.$$

The remainder of the proof is the same. □

Application of the lemma to an i.i.d. source and merging the symbols 1 and 2 in the punctuation process immediately yield the following result since coding an i.i.d. process yields a B-process which is therefore mixing.

Corollary 9.4.3: Given an integer N and a $\delta > 0$ there exists an (N, δ)-punctuation sequence $\{Z_n\}$ with the following properties:

(a) $\{Z_n\}$ is stationary and mixing (and hence ergodic).

(b) $\{Z_n\}$ has a binary alphabet $\{0,1\}$ and it can output only N-cells of the form $011\cdots 1$ (0 followed by $N-1$ ones) or individual ones, that is, each zero is always followed by at least $N-1$ ones.

(c) For all integers k

$$\frac{1-\delta}{N} \leq \Pr(Z_k^N = 011\cdots 1) \leq \frac{1}{N}$$

and hence for any n

$$\Pr(Z_n \text{ is in an } N-\text{cell}) \geq 1-\delta.$$

Example 9.4.11: Stationarized Block Memoryless Channel

Intuitively, a stationarized block memoryless (SBM) channel is a block memoryless channel with random spacing inserted between the blocks according to a random punctuation process. That is, when the random blocking process produces N-cells (which is most of the time), the channel uses the N-dimensional conditional distribution. When it is not using an N cell, the channel produces some arbitrary symbol in its output alphabet. We now make this idea precise. Let N, K, and $q_{x^N}(\cdot)$ be as in the previous example. We now assume that $K = N$, that is, one output symbol is produced for every input symbol and hence output blocks have the same number of symbols as input blocks. Given $\delta > 0$ let γ denote the distribution of an (N,δ)-random blocking sequence $\{Z_n\}$. Let $\mu \times \gamma$ denote the product distribution on $(A^{\mathcal{T}} \times \{0,1\}^{\mathcal{T}}, \mathcal{B}_A^{\mathcal{T}} \times \mathcal{B}_{\{0,1\}}^{\mathcal{T}})$; that is, $\mu \times \gamma$ is the distribution of the pair process $\{X_n, Z_n\}$ consisting of the original source $\{X_n\}$ and the random blocking source $\{Z_n\}$ with the two sources being independent of one another. Define a regular conditional probability (and hence a channel) $\pi_{x,z}(F)$, $F \in \{\mathcal{B}_B\}^{\mathcal{T}}$, $x \in A^{\mathcal{T}}$, $z \in \{0,1\}^{\mathcal{T}}$ by its values on rectangles as follows: Given z, let $J_0(z)$ denote the collection of indices i for which z_i is not in an N-cell and let $J_1(z)$ denote those indices i for which $z_i = 0$, that is, those indices where N-cells begin. Let q^* denote a trivial probability mass function on B placing all of its probability on a reference letter b^*. Given an output rectangle

$$F = \{y: y_j \in F_j;\ j \in \mathcal{J}\} = \mathop{\times}_{j\in\mathcal{J}} F_j,$$

define

$$\pi_{x,z}(F) = \prod_{i\in\mathcal{J}\bigcap J_0(z)} q^*(F_i) \prod_{i\in\mathcal{J}\bigcap J_1(z)} q_{x_i^N}(\mathop{\times}_{j=i}^{i+N-1} F_i),$$

where we assume that $F_i = B$ if $i \notin \mathcal{J}$. Connecting the product source $\mu\times\gamma$ to the channel π yields a hookup process $\{X_n, Z_n, Y_n\}$ with distribution, say, r, which in turn induces a distribution p on the pair process $\{X_n, Y_n\}$ having distribution μ on $\{X_n\}$. If the alphabets are standard, p also induces a regular conditional probability for Y given X and hence a channel ν for which $p = \mu\nu$. A channel of this form is said to be an (N, δ)-*stationarized block memoryless* or SBM channel.

Lemma 9.4.5: An SBM channel is stationary and ergodic. Thus if a stationary (and ergodic) source μ is connected to a ν, then the output is stationary (and ergodic).

Proof: The product source $\mu \times \gamma$ is stationary and the channel π is stationary, hence so is the hookup $(\mu \times \gamma)\pi$ or $\{X_n, Z_n, Y_n\}$. Thus the pair process $\{X_n, Y_n\}$ must also be stationary as claimed. The product source $\mu \times \gamma$ is ergodic from Corollary 9.4.1 since it can be considered as the input/output process of a completely random channel described by a mixing (hence also weakly mixing) output measure. The channel π is output strongly mixing by construction and hence is ergodic from Lemma 9.4.1. Thus the hookup $(\mu \times \gamma)\pi$ must be ergodic. This implies that the coordinate process $\{X_n, Y_n\}$ must also be ergodic. This completes the proof. □

The block memoryless and SBM channels are principally useful for proving theorems relating finite-dimensional behavior to sequence behavior and for simulating channels with specified finite dimensional behavior. The SBM channels will also play a key role in deriving sliding block coding theorems from block coding theorems by replacing the block distributions by trivial distributions, i.e., by finite-dimensional deterministic mappings or block codes.

The SMB channel was introduced by Pursley and Davisson [29] for finite alphabet channels and further developed by Gray and Saadat [61], who called it a randomly blocked conditionally independent (RBCI) channel. We opt for the alternative name because these channels resemble block memoryless channels more than CBI channels.

We now consider some examples that provide useful models for real-world channels.

Example 9.4.12: Primitive Channels

Primitive channels were introduced by Neuhoff and Shields [113],[110] as a physically motivated general channel model. The idea is that most physical channels combine the input process with a separate noise process that is independent of the signal and then filter the combination in a stationary fashion. The noise is assumed to be i.i.d. since the filtering can introduce dependence. The construction of such channels strongly resembles that of the SBM channels. Let γ be the distribution of an i.i.d. process $\{Z_n\}$ with alphabet W, let $\mu \times \gamma$ denote the product source formed by an independent joining of the original source distribution μ and the noise process Z_n, let π denote the deterministic channel induced by a stationary sequence coder $f : A^{\mathcal{T}} \times W^{\mathcal{T}} \rightarrow B^{\mathcal{T}}$ mapping an input sequence and a noise sequence into an output sequence. Let $r = (\mu \times \gamma)\pi$ denote the resulting hookup distribution and $\{X_n, Z_n, Y_n\}$ denote the resulting process. Let p denote the induced distribution for the pair process $\{X_n, Y_n\}$. If the alphabets are standard, then p and μ together induce a channel $\nu_x(F)$, $x \in A^{\mathcal{T}}$, $F \in {\mathcal{B}_B}^{\mathcal{T}}$. A channel of this form is called a ***primitive channel.***

Lemma 9.4.6: A primitive channel is stationary with respect to any stationary source and it is ergodic. Thus if μ is stationary and ergodic and ν is primitive, then $\mu\nu$ is stationary and ergodic.

Proof: Since μ is stationary and ergodic and γ is i.i.d. and hence mixing, $\mu \times \nu$ is stationary and ergodic from Corollary 9.4.1. Since the deterministic channel is stationary, it is also ergodic from Lemma 9.4.1 and the resulting triple $\{X_n, Z_n, Y_n\}$ is stationary and ergodic. This implies that the component process $\{X_n, Y_n\}$ must also be stationary and ergodic, completing the proof. □

Example 9.4.13: Additive Noise Channels

Suppose that $\{X_n\}$ is a source with distribution μ and that $\{W_n\}$ is a "noise" process with distribution γ. Let $\{X_n, W_n\}$ denote the induced product source, that is, the source with distribution $\mu \times \gamma$ so that the two processes are independent. Suppose that the two processes take values in a common alphabet A and that A has an addition operation $+$, e.g., it is a semi-group. Define the sliding block code f by $f(x, w) = x_0 + w_0$ and let $\bar{f}$ denote the corresponding sequence coder. Then as in the primitive channels we have an induced distribution r on triples $\{X_n, W_n, Y_n\}$ and hence a distribution on pairs $\{X_n, Y_n\}$ which with μ induces a channel ν if the alphabets are standard. A channel of this form is called a ***additive noise channel*** or a ***signal-independent additive noise channel.*** If the noise process is a B-process, then this is easily seen to be a special case of a primitive channel and hence the channel is stationary with respect to any stationary source and ergodic. If the noise is only known to be stationary, the channel is still stationary with respect to any stationary source. Unless the noise is assumed to be at least weakly mixing, however, it is not known if the channel is ergodic in general.

Example 9.4.14: Markov Channels

We now consider a special case where A and B are finite sets with the same number of symbols. For a fixed positive integer K, let $\mathbf{P}$ denote the space of all $K \times K$ stochastic matrices $P = \{P(i,j); i,j = 1,2,\cdots,K\}$. Using the Euclidean metric on this space we can construct the Borel field $\mathcal{P}$ of subsets of $\mathbf{P}$ generated by the open sets to form a measurable space $(\mathbf{P}, \mathcal{P})$. This, in turn, gives a one-sided or two-sided sequence space $(\mathbf{P}^{\mathcal{T}}, \mathcal{P}^{\mathcal{T}})$.

A map $\phi : A^{\mathcal{T}} \to \mathbf{P}^{\mathcal{T}}$ is said to be *stationary* if $\phi T_A = T_P \phi$. Given a sequence $P \in \mathbf{P}^{\mathcal{T}}$, let $\mathcal{M}(P)$ denote the set of all probability measures on $(B^{\mathcal{T}}, \mathcal{B}^{\mathcal{T}})$ with respect to which $Y_m, Y_{m+1}, Y_{m+2}, \cdots$ forms a Markov chain with transition matrices $P_m, P_{m+1}, \cdots$ for any integer m, that is, $\lambda \in \mathcal{M}(P)$ if and only if for any m

$$\lambda[Y_m = y_m, \cdots, Y_n = y_n]$$

$$= \lambda[Y_m = y_m] \prod_{i=m}^{n-1} P_i(y_i, y_{i+1}),\ n > m, y_m, \cdots, y_n \in B.$$

In the one-sided case only $m = 1$ need be verified. Observe that in general the Markov chain is nonhomogeneous.

A channel $[A, \nu, B]$ is said to be *Markov* if there exists a stationary measurable map $\phi : A^{\mathcal{T}} \to \mathbf{P}^{\mathcal{T}}$ such that $\nu_x \in \mathcal{M}(\phi(x))$, $x \in A^{\mathcal{T}}$.

Markov channels were introduced by Kieffer and Rahe [86] who proved that one-sided and two-sided Markov channels are AMS. Their proof is not included as it is lengthy and involves techniques not otherwise used in this book. The channels are introduced for completeness and to show that several important channels and codes in the literature can be considered as special cases. A variety of conditions for ergodicity for Markov channels are considered in [60]. Most are equivalent to one already considered more generally here: A Markov channel is ergodic if it is output mixing.

The most important special cases of Markov channels are finite state channels and codes. Given a Markov channel with stationary mapping ϕ, the channel is said to be a *finite state channel* (FSC) if we have a collection of stochastic matrices $P_a \in \mathbf{P}$; $a \in A$ and that $\phi(x)_n = P_{x_n}$, that is, the matrix produced by ϕ at time n depends only on the input at that time, x_n. If the matrices P_a; $a \in A$ contain only 0's and 1's, the channel is called a *finite state code.* There are several equivalent models of finite state channels and we pause to consider an alternative form that is more common in information theory. (See Gallager [43], Ch. 4, for a discussion of equivalent models of FSC's and numerous physical examples.) An FSC converts an input sequence x into an output sequence y and a state sequence s according to a conditional probability

$$\Pr(Y_k = y_k, S_k = s_k;\ k = m, \cdots, n | X_i = x_i, S_i = s_i;\ i < m)$$

$$= \prod_{i=m}^{n} P(y_i, s_i | x_i, s_{i-1}),$$

that is, conditioned on X_i, S_{i-1}, the pair Y_i, S_i is independent of all prior inputs, outputs, and states. This specifies a FSC defined as a special case of a Markov channel where the output sequence above is here the joint state-output sequence $\{y_i, s_i\}$. Note that with this setup, saying the Markov channel is AMS implies that the triple process of source, states, and outputs is AMS (and hence obviously so is the Gallager input-output process). We will adapt the Kieffer-Rahe viewpoint and call the outputs $\{Y_n\}$ of the Markov channel states even though they may correspond to state-output pairs for a specific physical model.

In the two-sided case, the Markov channel is significantly more general than the FSC because the choice of matrices $\phi(x)_i$ can depend on the past in a very complicated (but stationary) way. One might think that a Markov channel is not a significant generalization of an FSC in the one-sided case, however, because there stationarity of ϕ does not permit a dependence on past channel inputs, only on future inputs, which might seem physically unrealistic. Many practical communications systems do effectively depend on the future, however, by incorporating delay in the coding. The prime example of such look-ahead coders are trellis and tree codes used in an incremental fashion. Such codes investigate many possible output strings several steps into the future to determine the possible effect on the receiver and select the best path, often by a Viterbi algorithm. (See, e.g., Viterbi and Omura [145].) The encoder then outputs only the first symbol of the selected path. While clearly a finite state machine, this code does not fit the usual model of a finite state channel or code because of the dependence of the transition matrix on future inputs (unless, of course, one greatly expands the state space). It is, however, a Markov channel.

Example 9.4.15: Cascade Channels

We will often wish to connect more than one channel in cascade in order to form a communication system, e.g., the original source is connected to a deterministic channel (encoder) which is connected to a communications channel which is in turn connected to another deterministic channel (decoder). We now make precise this idea. Suppose that we are given two channels $[A, \nu^{(1)}, C]$ and $[C, \nu^{(2)}, B]$. The cascade of $\nu^{(1)}$ and $\nu^{(2)}$ is defined as the channel $[A, \nu, B]$ given by

$$\nu_x(F) = \int_{C^T} \nu_u^{(2)}(F)\, d\nu_x^{(1)}(u).$$

In other words, if the original source sequence is X, the output to the first channel and input to the second is U, and the output of the second channel is Y, then $\nu_x^{(1)}(F) = P_{U|X}(F|x)$, $\nu_u(G) = P_{Y|U}(G|u)$, and $\nu_x(G) = P_{Y|X}(G|x)$. Observe that by construction $X \to U \to Y$ is a Markov chain.

Lemma 9.4.7: A cascade of two stationary channels is stationary.

Proof: Let T denote the shift on all of the spaces. Then

$$\nu_x(T^{-1}F) = \int_{C^{\mathcal{T}}} \nu_u^{(2)}(T^{-1}F) d\nu_x^{(1)}(u).$$

$$= \int_{C^{\mathcal{T}}} \nu_u^{(2)}(F) d\nu_x^{(1)} T^{-1}(u).$$

But $\nu_x^{(1)}(T^{-1}F) = \nu_T x^{(1)}(F)$, that is, the measures $\nu_x^{(1)}T^{-1}$ and $\nu_{Tx}^{(1)}$ are identical and hence the above integral is

$$\int_{C^{\mathcal{T}}} \nu_u^{(2)}(F)\, d\nu_{Tx}^{(1)}(u) = \nu_{Tx}(F),$$

proving the lemma. □

Example 9.4.16: Communication System

A *communication system* consists of a source $[A, \mu]$, a sequence encoder $f : A^{\mathcal{T}} \to B^{\mathcal{T}}$ (a deterministic channel), a channel $[B, \nu, B']$, and a sequence decoder $g : B'^{\mathcal{T}} \to \hat{A}^{\mathcal{T}}$. The overall distribution r is specified by its values on rectangles as

$$r(F_1 \times F_2 \times F_3 \times F_4) = \int_{F_1 \bigcap f^{-1}(F_2)} d\mu(x) \nu_{f(x)}(F_3 \bigcap g^{-1}(F_4)).$$

Denoting the source by $\{X_n\}$, the encoded source or channel input process by $\{U_n\}$, the channel output process by $\{Y_n\}$, and the decoded process by $\{\hat{X}_n\}$, then r is the distribution of the process $\{X_n, U_n, Y_n, \hat{X}_n\}$. If we let X,U,Y, and $\hat{X}$ denote the corresponding sequences, then observe that $X \to U \to Y$ and $U \to Y \to \hat{X}$ are Markov chains. We abbreviate a communication system to $[\mu, f, \nu, g]$.

It is straightforward from Lemma 9.4.7 to show that if the source, channel, and coders are stationary, then so is the overall process.

The following is a basic property of a communication system: If the communication system is stationary, then the mutual information rate between the overall input and output cannot that exceed that over the channel. The result is often called the *data processing theorem*.

Lemma 9.4.8: Suppose that a communication system is stationary in the sense that the process $\{X_n, U_n, Y_n, \hat{X}_n\}$ is stationary. Then

$$\bar{I}(U;Y) \geq \bar{I}(X;Y) \geq \bar{I}(X;\hat{X}). \qquad (9.4.4)$$

If $\{U_n\}$ has a finite alphabet or if it has has the K-gap information property (6.4.11) and $I(U^K, Y) < \infty$, then

$$\bar{I}(X;\hat{X}) \leq \bar{I}(U;Y).$$

Proof: Since $\{\hat{X}_n\}$ is a stationary deterministic encoding of the $\{Y_n\}$

$$\bar{I}(X;\hat{X}) \leq I^*(X;Y).$$

From Theorem 6.4.1 the right hand side is bounded above by $\bar{I}(X;Y)$. For each n

$$I(X^n;Y^n) \leq I((X^n, U);Y^n)$$
$$= I(Y^n;U) + I(X^n;Y^n|U) = I(Y^n;U),$$

where $U = \{U_n, n \in \mathcal{T}\}$ and we have used the fact that $X \to U \to Y$ is a Markov chain and hence so is $X^N \to U \to Y^K$ and hence the conditional mutual information is 0 (Lemma 5.5.2). Thus

$$\bar{I}(X;Y) \leq \lim_{n\to\infty} I(Y^n;U) = \tilde{I}(Y;U).$$

Applying Theorem 6.4.1 then proves that

$$\bar{I}(X;\hat{X}) \leq \tilde{I}(Y;U).$$

If $\{U_n\}$ has finite alphabet or has the K-gap information property and $I(U^K, Y) < \infty$, then from Theorems 6.4.1 or 6.4.3, respectively, $\tilde{I}(Y;U) = \bar{I}((Y;U)$, completing the proof. □

The lemma can be easily extended to block stationary processes.

Corollary 9.4.4: Suppose that the process of the previous lemma is not stationary, but is (N, K)-stationary in the sense that the vector process $\{X^N_{nN}, U^K_{nK}, Y^K_{nK}, \hat{X}^N_{nN}\}$ is stationary. Then

$$\bar{I}(X;\hat{X}) \leq \frac{K}{N}\bar{I}(U;Y).$$

Proof: Apply the previous lemma to the stationary vector sequence to find that

$$\bar{I}(X^N;\hat{X}^N) \leq \bar{I}(U^K;Y^K).$$

But

$$\bar{I}(X^N;\hat{X}^N) = \lim_{n\to\infty} \frac{1}{n} I(X^{nN};\hat{X}^{nN})$$

which is the limit of the expectation of the information densities $n^{-1} i_{X^{nN},\hat{X}^{nN}}$ which is N times a subsequence of the densities $n^{-1} i_{X^n,\hat{X}^n}$, whose expectation converges to $\bar{I}(X;Y)$. Thus

$$\bar{I}(X^N;X^N) = N\bar{I}(X;\hat{X}).$$

A similar manipulation for $\bar{I}(U^K;Y^K)$ completes the proof. □

9.5 The Rohlin-Kakutani Theorem

The punctuation sequences of Section 9.4 provide a means for converting a block code into a sliding block code. Suppose, for example, that $\{X_n\}$ is a source with alphabet A and γ_N is a block code, $\gamma_N : A^N \to B^N$. (The dimensions of the input and output vector are assumed equal to simplify the discussion.) Typically B is binary. As has been argued, block codes are not stationary. One way to stationarize a block code is to use a procedure similar to that used to stationarize a block memoryless channel: Send long sequences of blocks with occasional random spacing to make the overall encoded process stationary. Thus, for example, one could use a sliding block code to produce a punctuation sequence $\{Z_n\}$ as in Corollary 9.4.2 which produces isolated 0's followed by KN 1's and occasionally produces 2's. The sliding block code uses γ_N to encode a sequence of K source blocks $X_n^N, X_{n+N}^N, \cdots, X_{n+(K-1)N}^N$ if and only if $Z_n = 0$. For those rare times l when $Z_l = 2$, the sliding block code produces an arbitrary symbol $b^* \in B$. The resulting sliding block code inherits many of the properties of the original block code, as will be demonstrated when proving theorems for sliding block codes constructed in this manner. In fact this construction suffices for source coding theorems, but an additional property will be needed when treating the channel coding theorems. The shortcoming of the results of Lemma 9.4.4 and Corollary 9.4.2 is that important source events can depend on the punctuation sequence. In other words, probabilities can be changed by conditioning on the occurrence of $Z_n = 0$ or the beginning of a block code word. In this section we modify the simple construction of Lemma 9.4.4 to effectively obtain a new punctuation sequence that is approximately independent of certain prespecified events. The result is a variation of the Rohlin-Kakutani theorem of ergodic theory [127] [71]. The development here is patterned after that in Shields [131].

We begin by recasting the punctuation sequence result in different terms. Given a stationary and ergodic source $\{X_n\}$ with a process distribution μ and a punctuation sequence $\{Z_n\}$ as in Section 9.4, define the set $F = \{x : Z_N(x) = 0\}$, where $x \in A^\infty$ is a two-sided sequence $x = (\cdots, x_{-1}, x_0, x_1, \cdots)$. Let T denote the shift on this sequence space. Restating Corollary 9.4.2 yields the following.

Lemma 9.5.1: Given $\delta > 0$ and an integer N, an L sufficiently large and a set F of sequences that is measurable with respect to $(X_{-L}, \cdots, X_L)$ with the following properties:

(A) The sets $T^i F$, $i = 0, 1, \cdots, N-1$ are disjoint.

(B)

$$\frac{1-\delta}{N} \leq \mu(F) \leq \frac{1}{N}.$$

(C)

$$1-\delta \leq \mu(\bigcup_{i=0}^{N-1} T^i F).$$

So far all that has been done is to rephrase the punctuation result in more ergodic theory oriented terminology. One can think of the lemma as representing sequence space as a "base" S together with its disjoint shifts $T^i S$; $i = 1, 2, \cdots, N-1$, which make up most of the space, together with whatever is left over, a set $G = \bigcup_{i=0}^{N-1} T^i F$, a set which has probability less than δ which will be called the "garbage set." This picture is called a *tower*. The basic construction is pictured in Figure 9.1

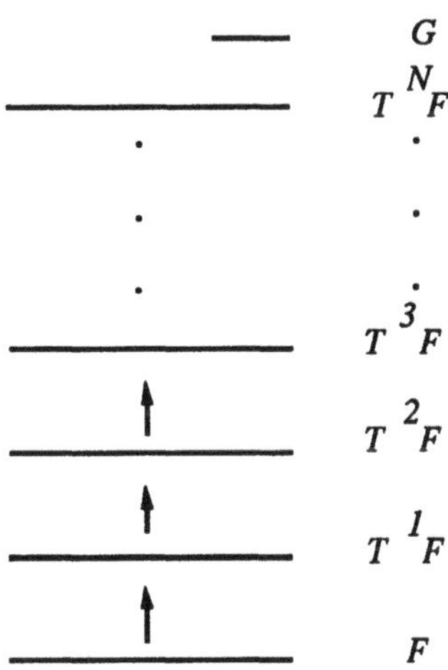

Figure 9.1: Rohlin-Kakutani Tower

Next consider a partition $\mathcal{P} = \{P_i;\ i = 0, 1, \cdots, ||\mathcal{P}|| - 1\}$ of A^∞. One example would be the partition of a finite alphabet sequence space into its possible outputs at time 0, that is, $P_i = \{x : x_0 = a_i\}$ for $i = 0, 1, \cdots, ||A|| - 1$. Another partition would be according to the output of a sliding block coding of x. The most important example, however, will be when there is a finite collection of important events that we wish to force to be approximately independent of the punctuation sequence and $\mathcal{P}$ is chosen so that the important events are unions of atoms of $\mathcal{P}$.

We now can state the main result of this section.

Lemma 9.5.2: Given the assumptions of Lemma 9.5.1, L and F can be chosen so that in addition to properties (A)-(C) it is also true that

(D)

$$\mu(P_i|F) = \mu(P_i|T^l F);\ l = 1, 2, \cdots, N-1, \tag{9.5.1}$$

$$\mu(P_i|F) = \mu(P_i|\bigcup_{k=0}^{N-1} T^k F) \tag{9.5.2}$$

and

$$\mu(P_i \bigcap F) \leq \frac{1}{N}\mu(P_i). \tag{9.5.3}$$

Comment: Eq. (9.5.3) can be interpreted as stating that P_i and F are approximately independent since $1/N$ is approximately the probability of F. Only the upper bound is stated as it is all we need. Eq. (9.5.1) also implies that $\mu(P_i \bigcap F)$ is bound below by $(\mu(P_i) - \delta)\mu(F)$.

Proof: Eq. (9.5.2) follows from (9.5.1) since

$$\mu(P_i|\bigcup_{l=0}^{N-1} T^l F) = \frac{\mu(P_i \bigcap \bigcup_{l=0}^{N-1} T^l F)}{\mu(\bigcup_{l=0}^{N-1} T^l F)} = \frac{\sum_{l=0}^{N-1} \mu(P_i \bigcap T^l F)}{\sum_{l=0}^{N-1} \mu(T^l F)}$$

$$= \frac{\sum_{l=0}^{N-1} \mu(P_i|T^l F)\mu(T^l F)}{N\mu(F)} = \frac{1}{N}\sum_{l=0}^{N-1} \mu(P_i|T^l F)\mu(F).$$

Eq. (9.5.3) follows from (9.5.2) since

$$\mu(P_i \bigcap F) = \mu(P_i|F)\mu(F) = \mu(P_i|\bigcup_{k=0}^{N-1} T^k F)\mu(F)$$

$$= \mu(P_i|\bigcup_{k=0}^{N-1} T^k F)\frac{1}{N}\mu(\bigcup_{k=0}^{N-1} T^k F)),$$

since the $T^k F$ are disjoint and have equal probability,

$$= \frac{1}{N}\mu(P_i \bigcap \bigcup_{k=0}^{N-1} T^k F) \leq \frac{1}{N}\mu(P_i).$$

The remainder of this section is devoted to proving (9.5.1). We begin by reviewing and developing some needed notation.

Given a partition $\mathcal{P}$, we define the *label* function

$$label_{\mathcal{P}}(x) = \sum_{i=0}^{||\mathcal{P}||-1} i 1_{P_i}(x),$$

where as usual 1_P is the indicator function of a set P. Thus the label of a sequence is simply the index of the atom of the partition into which it falls.

As $\mathcal{P}$ partitions the input space into which sequences belong to atoms of $\mathcal{P}$, $T^{-i}\mathcal{P}$ partitions the space according to which shifted sequences $T^i x$ belong to atoms of $\mathcal{P}$, that is, $x \in T^{-i}P_l \in T^{-i}\mathcal{P}$ is equivalent to $T^i x \in P_l$ and hence $label_{\mathcal{P}}(T^i x) = l$. The join

$$\mathcal{P}^N = \bigvee_{i=0}^{N-1} T^{-i}\mathcal{P}$$

partitions the space into sequences sharing N labels in the following sense: Each atom Q of $\mathcal{P}^N$ has the form

$$Q = \{x : label_{\mathcal{P}}(x) = k_0, label_{\mathcal{P}}(Tx) = k_1, \cdots, label_{\mathcal{P}}(T^{N-1}x) = k_N - 1\}$$

for some N tuple of integers $\mathbf{k} = (k_0, \cdots, k_N - 1)$. For this reason we will index the atoms of $\mathcal{P}^N$ as $Q_{\mathbf{k}}$. Thus $\mathcal{P}^N$ breaks up the sequence space into groups of sequences which have the same labels for N shifts.

We first construct using Lemma 9.5.1 a huge tower of size $KN >> N$, the height of the tower to be produced for this lemma. Let S denote the base of this original tower and let ϵ by the probability of the garbage set. This height KN tower with base S will be used to construct a new tower of height N and a base F with the additional desired property. First consider the restriction of the partition $\mathcal{P}^N$ to F defined by $\mathcal{P}^N \bigcap F = \{Q_{\mathbf{k}} \bigcap F$; all KN-tuples $\mathbf{k}$ with coordinates taking values in $\{0, 1, \cdots, ||\mathcal{P}|| - 1\}\}$. $\mathcal{P}^N \bigcap F$ divides up the original base according to the labels of NK shifts of base sequences. For each atom $Q_{\mathbf{k}} \bigcap F$ in this base partition, the sets $\{T^l(Q_{\mathbf{k}} \bigcap F);\ k = 0, 1, \cdots, KN - 1\}$ are disjoint and together form a *column* of the tower $\{T^l F;\ k = 0, 1, \cdots, KN - 1\}$. A set of the form $T^l(Q_{\mathbf{k}} \bigcap F)$ is called the lth *level* of the column containing it. Observe that if $y \in T^l(Q_{\mathbf{k}} \bigcap F)$, then $y = T^l u$ for some $u \in Q_{\mathbf{k}} \bigcap F$ and $T^l u$ has label k_l. Thus we consider k_l to be the label of the column level $T^l(Q_{\mathbf{k}} \bigcap F)$. This complicated structure of columns and levels can be used to recover the original partition by

$$P_j = \bigcup_{l, \mathbf{k}: k_l = j} T^l(Q_{\mathbf{k}} \bigcap F) \bigcap (P_j \bigcap G), \tag{9.5.4}$$

that is, P_j is the union of all column levels with label j together with that part of P_j in the garbage. We will focus on the pieces of P_j in the column levels as the garbage has very small probability.

We wish to construct a new tower with base F so that the probability of P_i for any of N shifts of F is the same. To do this we form F dividing each column of the original tower into N equal parts. We collect a group of these parts to form F so that F will contain only one part at each level, the N shifts of F will be disjoint, and the union of the N shifts will almost contain all of the original tower. By using the equal probability parts the new base will have conditional probabilities for P_j given T^l equal for all l, as will be shown.

Consider the atom $Q = Q_{\mathbf{k}} \bigcap S$ in the partition $\mathcal{P}^N \bigcap S$ of the base of the original tower. If the source is aperiodic in the sense of placing zero probability on individual sequences, then the set Q can be divided into N disjoint sets of equal probability, say $W_0, W_1, \cdots, W_{N-1}$. Define the set F_Q by

$$F_Q = (\bigcup_{i=0}^{(K-2)N} T^{iN} W_0) \bigcup (\bigcup_{i=0}^{(K-2)N} T^{1+iN} W_1) \bigcup$$

$$\cdots (\bigcup_{i=0}^{(K-2)N} T^{N-1+iN} W_{N-1}) = \bigcup_{l=0}^{N-1} \bigcup_{i=0}^{(K-2)N} T^{l+iN} W_l.$$

F_Q contains $(K-2)$ N shifts of W_0, of $TW_1, \cdots$ of $T^l W_l, \cdots$ and of $T^{N-1} W_{N-1}$. Because it only takes N-shifts of each small set and because it does not include the top N levels of the original column, shifting F_Q fewer than N times causes no overlap, that is, $T^l F_Q$ are disjoint for $j = 0, 1, \cdots, N-1$. The union of these sets contains all of the original column of the tower except possibly portions of the top and bottom $N-1$ levels (which the construction may not include). The new base F is now defined to be the union of all of the $F_{Q_{\mathbf{k}} \bigcap S}$. The sets $T^l F$ are then disjoint (since all the pieces are) and contain all of the levels of the original tower except possibly the top and bottom $N-1$ levels. Thus

$$\mu(\bigcup_{l=0}^{N-1} T^l F) \geq \mu(\bigcup_{i=N}^{(K-1)N-1} T^i S) = \sum_{i=N}^{(K-1)N-1} \mu(S)$$

$$\geq K - 2\frac{1-\epsilon}{KN} = \frac{1-\epsilon}{N} - \frac{2}{KN}.$$

by choosing $\epsilon = \delta/2$ and K large this can be made larger than $1-\delta$. Thus the new tower satisfies conditions (A)-(C) and we need only verify the new condition (D), that is, (9.5.1). We have that

$$\mu(P_i | T^l F) = \frac{\mu(P_i \bigcap T^l F)}{\mu(F)}.$$

Since the denominator does not depend on l, we need only show the numerator does not depend on l. From (9.5.4) applied to the original tower we have that

$$\mu(P_i \bigcap T^l F) = \sum_{j,\mathbf{k}:k_j=i} \mu(T^j(Q_{\mathbf{k}} \bigcap S) \bigcap T^l F),$$

that is, the sum over all column levels (old tower) labeled i of the probability of the intersection of the column level and the lth shift of the new base F. The intersection of a column level in the jth level of the original tower with any shift of F must be an intersection of that column level with the jth shift of one of the sets $W_0, \cdots, W_{N-1}$ (which particular set depends on l). Whichever set is chosen, however, the probability within the sum has the form

$$\mu(T^j(Q_{\mathbf{k}} \bigcap S) \bigcap T^l F) = \mu(T^j(Q_{\mathbf{k}} \bigcap S) \bigcap T^j W_m)$$
$$= \mu((Q_{\mathbf{k}} \bigcap S) \bigcap W_m) = \mu(W_m),$$

where the final step follows since W_m was originally chosen as a subset of $Q_{\mathbf{k}} \bigcap S$. Since these subsets were all chosen to have equal probability, this last probability does not depend on m and hence on l and

$$\mu(T^j(Q_{\mathbf{k}} \bigcap S) \bigcap T^l F) = \frac{1}{N}\mu(Q_{\mathbf{k}} \bigcap S)$$

and hence

$$\mu(P_i \bigcap T^l F) = \sum_{j,\mathbf{k}:k_j=i} \frac{1}{N}\mu(Q_{\mathbf{k}} \bigcap S),$$

which proves (9.5.1) since there is no dependence on l. This completes the proof of the lemma. □

Chapter 10

Distortion

10.1 Introduction

We now turn to quantification of various notions of the distortion between random variables, vectors and processes. A distortion measure is not a "measure" in the sense used so far; it is an assignment of a nonnegative real number which indicates how bad an approximation one symbol or random object is of another; the smaller the distortion, the better the approximation. If the two objects correspond to the input and output of a communication system, then the distortion provides a measure of the performance of the system. Distortion measures need not have metric properties such as the triangle inequality and symmetry, but such properties can be exploited when available. We shall encounter several notions of distortion and a diversity of applications, with eventually the most important application being a measure of the performance of a communications system by an average distortion between the input and output. Other applications include extensions of finite memory channels to channels which approximate finite memory channels and different characterizations of the optimal performance of communications systems.

10.2 Distortion and Fidelity Criteria

Given two measurable spaces $(A, \mathcal{B}_A)$ and $(B, \mathcal{B}_B)$, a *distortion measure* on $A \times B$ is a nonnegative measurable mapping $\rho : A \times B \to [0, \infty)$ which assigns a real number $\rho(x, y)$ to each $x \in A$ and $y \in B$ which can be thought of as the cost of reproducing x and y. The principal practical goal is to have a number by which the goodness or badness of communication systems can be compared. For example, if the input to a communication system is a random variable $X \in A$ and the output is $Y \in B$, then one possible measure of the quality of the system is the average distortion $E\rho(X,Y)$. Ideally one would like a distortion measure to have three properties:

- It should be tractable so that one can do useful theory.
- It should be computable so that it can be measured in real systems.
- It should be subjectively meaningful in the sense that small (large) distortion corresponds to good (bad) perceived quality.

Unfortunately these requirements are often inconsistent and one is forced to compromise between tractability and subjective significance in the choice of distortion measures. Among the most popular choices for distortion measures are metrics or distances, but many practically important distortion measures are not metrics, e.g., they are not symmetric in their arguments or they do not satisfy a triangle inequality. An example of a metric distortion measure that will often be emphasized is that given when the input space A is a Polish space, a complete separable metric space under a metric ρ, and B is either A itself or a Borel subset of A. In this case the distortion measure is fundamental to the structure of the alphabet and the alphabets are standard since the space is Polish.

Suppose next that we have a sequence of product spaces A^n and B^n for $n = 1, 2, \cdots$. A *fidelity criterion* ρ_n, $n = 1, 2, \cdots$ is a sequence of distortion measures on $A^n \times B^n$. If one has a pair random process, say $\{X_n, Y_n\}$, then it will be of interest to find conditions under which there is a limiting per symbol distortion in the sense that

$$\rho_\infty(x,y) = \lim_{n\to\infty} \frac{1}{n}\rho_n(x^n, y^n)$$

exists. As one might guess, the distortion measures in the sequence often are interrelated. The simplest and most common example is that of an *additive* or *single-letter* fidelity criterion which has the form

$$\rho_n(x^n, y^n) = \sum_{i=0}^{n-1} \rho_1(x_i, y_i).$$

Here if the pair process is AMS, then the limiting distortion will exist and it is invariant from the ergodic theorem. By far the bulk of the information theory literature considers only single-letter fidelity criteria and we will share this emphasis. We will point out, however, other examples where the basic methods and results apply. For example, if ρ_n is subadditive in the sense that

$$\rho_n(x^n, y^n) \leq \rho_k(x^k, y^k) + \rho_{n-k}(x_k^{n-k}, y_k^{n-k}),$$

then stationarity of the pair process will ensure that $n^{-1}\rho_n$ converges from the subadditive ergodic theorem. For example, if d is a distortion measure on $A \times B$, then

$$\rho_n(x^n, y^n) = \left(\sum_{i=0}^{n-1} d(x_i, y_i)^p\right)^{1/p}$$

for $p > 1$ is subadditive from Minkowski's inequality.

As an even simpler example, if d is a distortion measure on $A \times B$, then the following fidelity criterion converges for AMS pair processes:

$$\frac{1}{n}\rho_n(x^n, y^n) = \frac{1}{n}\sum_{i=0}^{n-1} f(d(x_i, y_i)).$$

This form often arises in the literature with d being a metric and f being a nonnegative nondecreasing function (sometimes assumed convex).

The fidelity criteria introduced here all are *context-free* in that the distortion between n successive input/output samples of a pair process does not depend on samples occurring before or after these n-samples. Some work has been done on context-dependent distortion measures (see, e.g., [93]), but we do not consider their importance sufficient to merit the increased notational and technical difficulties involved. Hence we shall consider only context-free distortion measures.

10.3 Performance

As a first application of the notion of distortion, we define a performance measure of a communication system. Suppose that we have a communication system $[\mu, f, \nu, g]$ such that the overall input/output process is $\{X_n, \hat{X}_n\}$. For the moment let p denote the corresponding distribution. Then one measure of the quality (or rather the lack thereof) of the communication system is the long term time average distortion per symbol between the input and output as determined by the fidelity criterion. Given two sequences x and $\hat{x}$ and a fidelity criterion ρ_n; $n = 1, 2, \cdots$, define the limiting sample average distortion or *sequence distortion* by

$$\rho_\infty(x, y) = \limsup_{n\to\infty} \frac{1}{n}\rho_n(x^n, y^n).$$

Define the performance of a communication system by the expected value of the limiting sample average distortion:

$$\Delta(\mu, f, \nu, g) = E_p\rho_\infty = E_p\left(\limsup_{n\to\infty} \frac{1}{n}\rho_n(X^n, \hat{X}^n)\right). \tag{10.3.1}$$

We will focus on two important special cases. The first is that of AMS systems and additive fidelity criteria. A large majority of the information theory literature is devoted to additive distortion measures and this bias is reflected here. We also consider the case of subadditive distortion measures and systems that are either two-sided and AMS or are one-sided and stationary. Unhappily the overall AMS one-sided case cannot be handled as there is not yet a subadditive ergodic theorem for that case. In all of these cases we have that if ρ_1 is integrable with respect to the stationary mean process $\bar{p}$, then

$$\rho_\infty(x,y) = \lim_{n\to\infty} \frac{1}{n}\rho_n(x^n,y^n);\ p-\text{a.e.}, \tag{10.3.2}$$

and ρ_∞ is an invariant function of its two arguments, i.e.,

$$\rho_\infty(T_A x, T_{\hat{A}} y) = \rho_\infty(x,y);\ p-\text{a.e..} \tag{10.3.3}$$

When a system and fidelity criterion are such that (10.3.2) and (10.3.3) are satisfied we say that we have a *convergent fidelity criterion*. We henceforth make this assumption.

Since ρ_∞ is invariant, we have from Lemma 6.3.1 of [50] that

$$\Delta = E_p\rho_\infty = E_{\bar{p}}\rho_\infty. \tag{10.3.4}$$

If the fidelity criterion is additive, then we have from the stationarity of $\bar{p}$ that the performance is given by

$$\Delta = E_{\bar{p}}\rho_1(X_0,Y_0). \tag{10.3.5}$$

If the fidelity criterion is subadditive, then this is replaced by

$$\Delta = \inf_N \frac{1}{N} E_{\bar{p}}\rho_N(X^N,Y^N). \tag{10.3.6}$$

Assume for the remainder of this section that ρ_n is an additive fidelity criterion. Suppose now that we now that p is N-stationary; that is, if $T = T_A \times T_{\hat{A}}$ denotes the shift on the input/output space $A^{\mathcal{T}} \times \hat{A}^{\mathcal{T}}$, then the overall process is stationary with respect to T^N. In this case

$$\Delta = \frac{1}{N} E\rho_N(X_N,\hat{X}_N). \tag{10.3.7}$$

We will have this N stationarity, for example, if the source and channel are stationary and the coders are N-stationary, e.g., are length N-block codes. More generally, the source could be N-stationary, the first sequence coder (N, K)-stationary, the channel K-stationary (e.g., stationary), and the second sequence coder (K, N)-stationary.

We can also consider the behavior of the N-shift more generally when the system is only AMS This will be useful when considering block codes. Suppose now that p is AMS with stationary mean $\bar{p}$. Then from Theorem 7.3.1 of [50], p is also T^N-AMS with an N-stationary mean, say $\bar{p}_N$. Applying the ergodic theorem to the N shift then implies that if ρ_N is $\bar{p}_N$-integrable, then

$$\lim_{n\to\infty} \frac{1}{n}\sum_{i=0}^{n-1} \rho_N(x_{iN}^N, y_{iN}^N) = \rho_\infty^{(N)} \tag{10.3.8}$$

exists $\bar{p}_N$ (and hence also p) almost everywhere. In addition, $\rho_\infty^{(N)}$ is N-invariant and

$$E_p \rho_\infty^{(N)} = E_{\bar{p}_N} \rho_\infty^{(N)} = E_{\bar{p}_N} \rho_N(X^N, Y^N). \tag{10.3.9}$$

Comparison of (10.3.2) and (10.3.9) shows that $\rho_\infty^{(N)} = N\rho_\infty$ p-a.e. and hence

$$\Delta = \frac{1}{N} E_{\bar{p}_N} \rho_N(X^N, Y^N) = \frac{1}{N} E_p \rho_\infty^{(N)} = E_{\bar{p}} \rho_1(X_0, Y_0). \tag{10.3.10}$$

Given a notion of the performance of a communication system, we can now define the optimal performance achievable for trying to communicate a given source $\{X_n\}$ with distribution μ over a channel ν: Suppose that $\mathcal{E}$ is some class of sequence coders $f : A^{\mathcal{T}} \to B^{\mathcal{T}}$. For example, $\mathcal{E}$ might consist of all sequence coders generated by block codes with some constraint or by finite-length sliding block codes. Similarly let $\mathcal{D}$ denote a class of sequence coders $g : B'^{\mathcal{T}} \to \hat{A}^{\mathcal{T}}$. Define the *optimal performance theoretically achievable* or *OPTA* function for the source μ, channel ν, and code classes $\mathcal{E}$ and $\mathcal{D}$ by

$$\Delta^*(\mu, \nu, \mathcal{E}, \mathcal{D}) = \inf_{f\in\mathcal{E}, g\in\mathcal{D}} \Delta([\mu, f, \nu, g]). \tag{10.3.11}$$

The goal of the coding theorems of information theory is to relate the OPTA function to (hopefully) computable functions of the source and channel.

10.4 The rho-bar distortion

In the previous sections it was pointed out that if one has a distortion measure ρ on two random objects X and Y and a joint distribution on the two random objects (and hence also marginal distributions for each), then a natural notion of the difference between the processes or the poorness of their mutual approximation is the expected distortion $E\rho(X,Y)$. We now consider a different question: What if one does not have a joint probabilistic description of X and Y, but instead knows only their marginal distributions. What then is a natural notion of the distortion or poorness of approximation of the two random objects? In other words, we previously measured the distortion between two random variables whose stochastic connection was determined, possibly by a channel, a code, or a communication system. We now wish to find a similar quantity for the case when the two random objects are only described as individuals. One possible definition is to find the smallest possible distortion in the old sense consistent with the given information, that is, to minimize $E\rho(X,Y)$ over all joint distributions consistent with the given marginal distributions. Note that this will necessarily give a lower bound to the distortion achievable when any specific joint distribution is specified.

To be precise, suppose that we have random variables X and Y with distributions P_X and P_Y and alphabets A and B, respectively. Let ρ be a distortion measure on $A \times B$. Define the ***$\bar{\rho}$-distortion*** (pronounced ρ-bar) between the random variables X and Y by

$$\bar{\rho}(P_X, P_Y) = \inf_{p \in \mathcal{P}} E_p \rho(X,Y),$$

Where $\mathcal{P} = \mathcal{P}(P_X, P_Y)$ is the collection of all measures on $(A \times B, \mathcal{B}_A \times \mathcal{B}_B)$ with P_X and P_Y as marginals; that is,

$$p(A \times F) = P_Y(F); \ F \in \mathcal{B}_B,$$

and

$$p(G \times B) = P_X(G); \ G \in \mathcal{B}_A.$$

Note that $\mathcal{P}$ is not empty since, for example, it contains the product measure $P_X \times P_Y$.

Levenshtein [94] and Vasershtein [144] studied this quantity for the special case where A and B are the real line and ρ is the Euclidean distance. When as in their case the distortion is a metric or distance, the $\bar{\rho}$-distortion is called the $\bar{\rho}$-distance. Ornstein [116] developed the distance and many of its properties for the special case where A and B were common discrete spaces and ρ was the Hamming distance. In this case the $\bar{\rho}$-distance is called the $\bar{d}$-distance. R. L. Dobrushin has suggested that because of the common suffix in the names of its originators, this distance between distributions should be called the shtein or stein distance.

The $\bar{\rho}$-distortion can be extended to processes in a natural way. Suppose now that $\{X_n\}$ is a process with process distribution m_X and that $\{Y_n\}$

is a process with process distribution m_Y. Let P_{X^n} and P_{Y^n} denote the induced finite dimensional distributions. A fidelity criterion provides the distortion ρ_n between these n dimensional alphabets. Let $\bar{\rho}_n$ denote the corresponding $\bar{\rho}$ distortion between the n dimensional distributions. Then

$$\bar{\rho}(m_X, m_Y) = \sup_n \frac{1}{n} \bar{\rho}_n(P_{X^n}, P_{Y^n});$$

that is, the $\bar{\rho}$-distortion between two processes is the maximum of the $\bar{\rho}$-distortions per symbol between n-tuples drawn from the process. The properties of the $\bar{\rho}$ distance are developed in [57] [119] and a detailed development may be found in [50] . The following theorem summarizes the principal properties.

Theorem 10.4.1: Suppose that we are given an additive fidelity criterion ρ_n with a pseudo-metric per-letter distortion ρ_1 and suppose that both distributions m_X and m_Y are stationary and have the same standard alphabet. Then

(a) $\lim_{n\to\infty} n^{-1}\bar{\rho}_n(P_{X^n}, P_{Y^n})$ exists and equals $\sup_n n^{-1}\bar{\rho}_n(P_{X^n}, P_{Y^n})$.

(b) $\bar{\rho}_n$ and $\bar{\rho}$ are pseudo-metrics. If ρ_1 is a metric, then $\bar{\rho}_n$ and $\bar{\rho}$ are metrics.

(c) If m_X and m_Y are both i.i.d., then $\bar{\rho}(m_X, m_Y) = \bar{\rho}_1(P_{X_0}, P_{Y_0})$.

(d) Let $\mathcal{P}_s = \mathcal{P}_s(m_X, m_Y)$ denote the collection of all stationary distributions p_{XY} having m_X and m_Y as marginals, that is, distributions on $\{X_n, Y_n\}$ with coordinate processes $\{X_n\}$ and $\{Y_n\}$ having the given distributions. Define the process distortion measure $\bar{\rho}'$

$$\bar{\rho}'(m_X, m_Y) = \inf_{p_{XY}\in\mathcal{P}_s} E_{p_{XY}}\rho(X_0, Y_0).$$

Then

$$\bar{\rho}(m_X, m_Y) = \bar{\rho}'(m_X, m_Y);$$

that is, the limit of the finite dimensional minimizations is given by a minimization over stationary processes.

(e) Suppose that m_X and m_Y are both stationary and ergodic. Define $\mathcal{P}_e = \mathcal{P}_e(m_X, m_Y)$ as the subset of $\mathcal{P}_s$ containing only ergodic processes, then

$$\bar{\rho}(m_X, m_Y) = \inf_{p_{XY}\in\mathcal{P}_e} E_{p_{XY}}\rho(X_0, Y_0),$$

(f) Suppose that m_X and m_Y are both stationary and ergodic. Let G_X denote a collection of generic sequences for m_X in the sense of Section 8.3 of [50]. Generic sequences are those along which the relative frequencies of a set of generating events all converge and hence by measuring relative frequencies on generic sequences one can deduce the underlying stationary and ergodic measure that produced the sequence. An AMS process produces generic sequences with probability 1. Similarly let G_Y denote a set of generic sequences for m_Y. Define the process distortion measure

$$\bar{\rho}''(m_X, m_Y) = \inf_{x \in G_X, y \in G_Y} \limsup_{n \to \infty} \frac{1}{n} \sum_{i=0}^{n-1} \rho_1(x_0, y_0).$$

Then

$$\bar{\rho}(m_X, m_Y) = \bar{\rho}''(m_X, m_Y);$$

that is, the $\bar{\rho}$ distance gives the minimum long term time average distortion obtainable between generic sequences from the two sources.

(g) The infima defining $\bar{\rho}_n$ and $\bar{\rho}'$ are actually minima.

10.5 d-bar Continuous Channels

We can now generalize some of the notions of channels by using the $\bar{\rho}$-distance to weaken the definitions. The first definition is the most important for channel coding applications. We now confine interest to the $\bar{d}$-bar distance, the $\bar{\rho}$-distance for the special case of the Hamming distance:

$$\rho_1(x, y) = d_1(x, y) = \begin{cases} 0 & \text{if } x = y \\ 1 & \text{if } x \neq y. \end{cases}$$

Suppose that $[A, \nu, B]$ is a discrete alphabet channel and let ν_x^n denote the restriction of the channel to B^n, that is, the output distribution on Y^n given an input sequence x. The channel is said to be *$\bar{d}$-continuous* if for any $\epsilon > 0$ there is an n_0 such that for all $n > n_0$ $\bar{d}_n(\nu_x^n, \nu_{x'}^n) \leq \epsilon$ whenever $x_i = x'_i$ for $i = 0, 1, \cdots, n$. Alternatively, ν is $\bar{d}$-continuous if

$$\limsup_{n \to \infty} \sup_{a^n \in A^n} \sup_{x, x' \in c(a^n)} \bar{d}_n(\nu_x^n, \nu_{x'}^n) = 0,$$

where $c(a^n)$ is the rectangle defined as all x with $x_i = a_i$; $i = 0, 1, \cdots, n-1$. $\bar{d}$-continuity implies the distributions on output n-tuples Y^n given two input sequences are very close provided that the input sequences are identical over the same time period and that n is large. This generalizes the notions of 0 or finite input memory and anticipation since the distributions need only approximate each other and do not have to be exactly the same.

More generally we could consider $\bar{\rho}$-continuous channels in a similar manner, but we will focus on the simpler discrete $\bar{d}$-continuous channel.

$\bar{d}$-continuous channels possess continuity properties that will be useful for proving block and sliding block coding theorems. They are "continuous" in the sense that knowing the input with sufficiently high probability for a sufficiently long time also specifies the output with high probability. The following two lemmas make these ideas precise.

Lemma 10.5.1: Suppose that x, $\bar{x} \in c(a^n)$ and

$$\bar{d}(\nu_x^n, \nu_{\bar{x}}^n) \leq \delta^2.$$

This is the case, for example, if the channel is $\bar{d}$ continuous and n is chosen sufficiently large. Then

$$\nu_x^n(G_\delta) \geq \nu_{\bar{x}}^n(G) - \delta$$

and hence

$$\inf_{x \in c(a^n)} \nu_x^n(G_\delta) \geq \sup_{x \in c(a^n)} \nu_x^n(G) - \delta.$$

Proof: From Theorem 10.4.1 the infima defining the $\bar{d}$ distance are actually minima and hence there is a pmf p on $B^n \times B^n$ such that

$$\sum_{b^n \in B^n} p(y^n, b^n) = \nu_x^n(y^n)$$

and

$$\sum_{b^n \in B^n} p(b^n, y^n) = \nu_{\bar{x}}^n(y^n);$$

that is, p has ν_x^n and $\nu_{\bar{x}}^n$ as marginals, and

$$\frac{1}{n} E_p d_n(Y^n, \bar{Y}^n) = \bar{d}(\nu_x^n, \nu_{\bar{x}}^n).$$

Using the Markov inequality we can write

$$\nu_x^n(G_\delta) = p(Y^n \in G_\delta)$$

$$\geq p(\bar{Y}^n \in G \text{ and } d_n(Y^n, \bar{Y}^n) \leq n\delta) = 1 - p(\bar{Y}^n \notin G \text{ or } d_n(Y^n, \bar{Y}^n) > n\delta)$$

$$\geq 1 - p(\bar{Y}^n \notin G) - p(d_n(Y^n, \bar{Y}^n) > n\delta) \geq \nu_{\bar{x}}^n(G) - \frac{1}{\delta} E(n^{-1} d_n(Y^n, \bar{Y}^n))$$

$$\geq \nu_{\bar{x}}^n(G) - \delta$$

proving the first statement. The second statement follows from the first. □

Next suppose that $[G, \mu, U]$ is a stationary source, f is a stationary encoder which could correspond to a finite length sliding block encoder or to an infinite length one, ν is a stationary channel, and g is a length m sliding block decoder. The probability of error for the resulting hookup is defined by

$$P_e(\mu, \nu, f, g) = \Pr(U_0 \neq \hat{U}_0) = \mu\nu(E) = \int d\mu(u)\nu_{f(u)}(E_u),$$

where E is the error event $\{u, y : u_0 \neq g_m(Y_{-q}{}^m)\}$ and $E_u = \{y : (u, y) \in E\}$ is the section of E at u.

Lemma 10.5.2: Given a stationary channel ν, a stationary source $[G, \mu, U]$, a length m sliding block decoder, and two encoders f and ϕ, then for any positive integer r

$$|P_e(\mu, \nu, f, g) - P_e(\mu, \nu, \phi, g)|$$
$$\leq \frac{m}{r} + r\Pr(f \neq \phi) + m \max_{a^r \in A^r} \sup_{x, x' \in c(a^r)} \bar{d}_r(\nu_x^r, \nu_{x'}^r).$$

Proof: Define $\Lambda = \{u : f(u) = \phi(u)\}$ and

$$\Lambda_r = \{u : f(T^i u) = \phi(T^i u);\ i = 0, 1 \cdots, r-1\} = \bigcap_{i=0}^{r-1} T^i \Lambda.$$

From the union bound

$$\mu(\Lambda_r^c) \leq r\mu(\Lambda^c) = r\Pr(f \neq \phi). \tag{10.5.1}$$

From stationarity, if $g = g_m(Y_{-q}^m)$ then

$$P_e(\mu, \nu, f, g) = \int d\mu(u)\nu_{f(u)}(y : g_m(y_{-q}^m) \neq u_0)$$
$$= \frac{1}{r}\sum_{i=0}^{r-1} \int d\mu(u)\nu_{f(u)}(y : g_m(y_{i-q}^m) \neq u_0)$$
$$\leq \frac{m}{r} + \frac{1}{r}\sum_{i=q}^{r-q} \int_{\Lambda_r} d\mu(u)\nu_{f(u)}^r(y^r : g_m(y_{i-q}^m) \neq u_i) + \mu(\Lambda_r^c). \tag{10.5.2}$$

Fix $u \in \Lambda_r$ and let p_u yield $\bar{d}_r(\nu_{f(u),\phi(u)}^r)$; that is, $\sum_{w^r} p_u(y^r, w^r) = \nu_{f(u)}^r(y^r)$, $\sum_{y^r} p_u(y^r, w^r) = \nu_{\phi(u)}^r(w^r)$, and

$$\frac{1}{r}\sum_{i=0}^{r-1} p_u(y^r, w^r : y_i \neq w_i) = \bar{d}_r(\nu_{f(u),\phi(u)}^r). \tag{10.5.3}$$

We have that

$$\frac{1}{r}\sum_{i=q}^{r-q}\nu^r_{f(u)}(y^r : g_m(y^m_{i-q}) \neq u_i) = \frac{1}{r}\sum_{i=q}^{r-q} p_u(y^r, w^r : g_m(y^m_{i-q}) \neq u_i)$$

$$\leq \frac{1}{r}\sum_{i=q}^{r-q} p_u(y^r, w^r : g_m(y^m_{i-q}) \neq w^m_{i-q}) + \frac{1}{r}\sum_{i=q}^{r-q} p_u(y^r, w^r : g_m(w^m_{i-q}) \neq u_i)$$

$$\leq \frac{1}{r}\sum_{i=q}^{r-q} p_u(y^r, w^r : y^r_{i-q} \neq w^r_{i-q}) + P_e(\mu, \nu, \phi, g)$$

$$\leq \frac{1}{r}\sum_{i=q}^{r-q}\sum_{j=i-q}^{i-q+m} p_u(y^r, w^r : y_j \neq w_j) + P_e(\mu, \nu, \phi, g)$$

$$\leq m\bar{d}_r(\nu^r_{f(u)}, \nu^r_{\phi(u)}) + P_e(\mu, \nu, \phi, g),$$

which with (10.5.1)-(10.5.3) proves the lemma. □

The following corollary states that the probability of error using sliding block codes over a $\bar{d}$-continuous channel is a continuous function of the encoder as measured by the metric on encoders given by the probability of disagreement of the outputs of two encoders.

Corollary 10.5.1: Given a stationary $\bar{d}$-continuous channel ν and a finite length decoder $g_m : B^m \to A$, then given $\epsilon > 0$ there is a $\delta > 0$ so that if f and ϕ are two stationary encoders such that $\Pr(f \neq g) \leq \delta$, then

$$|P_e(\mu, \nu, f, g) - P_e(\mu, \nu, \phi, g)| \leq \epsilon.$$

Proof: Fix $\epsilon > 0$ and choose r so large that

$$\max_{a^r} \sup_{x, x' \in c(a^r)} \bar{d}_r(\nu^r_x, \nu^r_{x'}) \leq \frac{\epsilon}{3m},$$

$$\frac{m}{r} \leq \frac{\epsilon}{3},$$

and choose $\delta = \epsilon/(3r)$. Then Lemma 10.5.2 implies that

$$|P_e(\mu, \nu, f, g) - P_e(\mu, \nu, \phi, g)| \leq \epsilon. \ \square$$

Given an arbitrary channel $[A, \nu, B]$, we can define for any block length N a closely related CBI channel $[A, \tilde{\nu}, B]$ as the CBI channel with the same probabilities on output N-blocks, that is, the same conditional probabilities for Y^N_{kN} given x, but having conditionally independent blocks. We shall call

$\tilde{\nu}$ the *N-CBI* approximation to ν. A channel ν is said to be *conditionally almost block independent* or *CABI* if given ϵ there is an N_0 such that for any $N \geq N_0$ there is an M_0 such that for any x and any N-CBI approximation $\tilde{\nu}$ to ν

$$\bar{d}(\tilde{\nu}_x^M, \nu_x^M) \leq \epsilon, \text{ all } M \geq M_0,$$

where ν_x^M denotes the restriction of ν_x to $\mathcal{B}_B^N$, that is, the output distribution on Y^N given x. A CABI channel is one such that the output distribution is close (in a $\bar{d}$ sense) to that of the N-CBI approximation provided that N is big enough. CABI channels were introduced by Neuhoff and Shields [110] who provided several examples alternative characterizations of the class. In particular they showed that finite memory channels are both $\bar{d}$-continuous and CABI. Their principal result, however, requires the notion of the $\bar{d}$ distance between channels. Given two channels $[A, \nu, B]$ and $[A, \nu', B]$, define the $\bar{d}$ distance between the channels to be

$$\bar{d}(\nu, \nu') = \limsup_{n \to \infty} \sup_x \bar{d}(\nu_x^n, {\nu'}_x^N).$$

Neuhoff and Shields [110] showed that the class of CABI channels is exactly the class of primitive channels together with the $\bar{d}$ limits of such channels.

10.6 The Distortion-Rate Function

We close this chapter on distortion, approximation, and performance with the introduction and discussion of Shannon's distortion-rate function. This function (or functional) of the source and distortion measure will play a fundamental role in evaluating the OPTA functions. In fact, it can be considered as a form of information theoretic OPTA. Suppose now that we are given a source $[A, \mu]$ and a fidelity criterion ρ_n; $n = 1, 2, \cdots$ defined on $A \times \hat{A}$, where $\hat{A}$ is called the *reproduction alphabet*. Then the Shannon distortion rate function (DRF) is defined in terms of a nonnegative parameter called *rate* by

$$D(R, \mu) = \limsup_{N \to \infty} \frac{1}{N} D_N(R, \mu^N)$$

where

$$D_N(R, \mu^N) = \inf_{p^N \in \mathcal{R}_N(R, \mu^N)} E_{p^N} \rho_N(X^N, Y^N),$$

where $\mathcal{R}_N(R, \mu^N)$ is the collection of all distributions p^N for the coordinate random vectors X^N and Y^N on the space $(A^N \times \hat{A}^N, \mathcal{B}_A^N \times \mathcal{B}_{\hat{A}}^N)$ with the properties that

(1) p^N induces the given marginal μ^N; that is, $p^N(\hat{A}^N \times F) = \mu^N(F)$ for all $F \in \mathcal{B}_A^N$, and

(2) the mutual information satisfies

$$\frac{1}{N} I_{p^N}(X^N; \hat{X}^N) \leq R.$$

If $\mathcal{R}_N(R, \mu^N)$ is empty, then $D_N(R, \mu^N)$ is ∞. D_N is called the *Nth order distortion-rate function.*

Lemma 10.6.1: $D_N(R, \mu)$ and $D(R, \mu)$ are nonnegative convex $\bigcup$ functions of R and hence are continuous in R for $R > 0$.

Proof: Nonnegativity is obvious from the nonnegativity of distortion. Suppose that $p_i \in \mathcal{R}_N(R_i, \mu^N)$; $i = 1, 2$ yields

$$E_{p_i} \rho_N(X^N, Y^N) \leq D_N(R_i, \mu) + \epsilon.$$

From Corollary 5.5.5 mutual information is a convex $\bigcup$ function of the conditional distribution and hence if $\bar{p} = \lambda p_1 + (1 - \lambda) p_2$, then

$$I_{\bar{p}} \leq \lambda I_{p_1} + (1 - \lambda) I_{p_2} \leq \lambda R_1 + (1 - \lambda) R_2$$

and hence $\bar{p} \in \mathcal{R}_N(\lambda R_1 + (1 - \lambda) R_2)$ and therefore

$$D_N(\lambda R_1 + (1 - \lambda) R_2) \leq E_{\bar{p}} \rho_N(X^N, Y^N)$$

$$= \lambda E_{p_1} \rho_N(X^N, Y^N) + (1 - \lambda) E_{p_2} \rho_N(X^N, Y^N)$$

$$\leq \lambda D_N(R_1, \mu) + (1 - \lambda) D_N(R_2, \mu).$$

Since $D(R, \mu)$ is the limit of $D_N(R, \mu)$, it too is convex. It is well known from real analysis that convex functions are continuous except possibly at their end points. □

The following lemma shows that when the underlying source is stationary and the fidelity criterion is subadditive (e.g., additive), then the limit defining $D(R, \mu)$ is an infimum.

Lemma 10.6.2: If the source μ is stationary and the fidelity criterion is subadditive, then

$$D(R, \mu) = \lim_{N \to \infty} D_N(R, \mu) = \inf_N \frac{1}{N} D_N(R, \mu).$$

Proof: Fix N and $n < N$ and let $p^n \in \mathcal{R}_n(R, \mu^n)$ yield

$$E_{p^n} \rho_n(X^n, Y^n) \leq D_n(R, \mu^n) + \frac{\epsilon}{2}$$

and let $p^{N-n} \in \mathcal{R}_{N-n}(R, \mu^{N-n})$ yield

$$E_{p^{N-n}} \rho_{N-n}(X^{N-n}, Y^{N-n}) \leq D_{N-n}(R, \mu^{N-n}) + \frac{\epsilon}{2}.$$

p_n together with μ^n implies a regular conditional probability $q(F|x^n)$, $F \in \mathcal{B}^n_{\hat{A}}$. Similarly p_{N-n} and μ^{N-n} imply a regular conditional probability $r(G|x^{N-n})$. Define now a regular conditional probability $t(\cdot|x^N)$ by its values on rectangles as

$$t(F \times G|x^N) = q(F|x^n) r(G|x_n^{N-n});\ F \in \mathcal{B}^n_{\hat{A}}, G \in \mathcal{B}^{N-n}_{\hat{A}}.$$

Note that this is the finite dimensional analog of a block memoryless channel with two blocks. Let $p^N = \mu^N t$ be the distribution induced by μ and t. Then exactly as in Lemma 9.4.2 we have because of the conditional independence that

$$I_{p^N}(X^N; Y^N) \leq I_{p^N}(X^n; Y^n) + I_{p^N}(X_n^{N-n}; Y_n^{N-n})$$

and hence from stationarity

$$I_{p^N}(X^N; Y^N) \leq I_{p^n}(X^n; Y^n) + I_{p^{N-n}}(X^{N-n}; Y^{N-n})$$
$$\leq nR + (N-n)R = NR$$

so that $p^N \in \mathcal{R}_N(R, \mu^N)$. Thus

$$D_N(R, \mu^N) \leq E_{p^N} \rho_N(X^N, Y^N) \leq E_{p^N} \left(\rho_n(X^n, Y^n) + \rho_{N-n}(X_n^{N-n}, Y_n^{N-n})\right)$$
$$= E_{p^n} \rho_n(X^n, Y^n) + E_{p^{N-n}} \rho_{N-n}(X^{N-n}, Y^{N-n})$$
$$\leq D_n(R, \mu^n) + D_{N-n}(R, \mu^{N-n}) + \epsilon.$$

Thus since ϵ is arbitrary we have shown that if $d_n = D_n(R, \mu^n)$, then

$$d_N \leq d_n + d_{N-n};\ n \leq N;$$

that is, the sequence d_n is subadditive. The lemma then follows immediately from Lemma 7.5.1 of [50]. □

As with the $\bar{\rho}$ distance, there are alternative characterizations of the distortion-rate function when the process is stationary. The remainder of this section is devoted to developing these results. The idea of an SBM channel will play an important role in relating nth order distortion-rate functions to the process definitions. We henceforth assume that the input source μ is stationary and we confine interest to additive fidelity criteria based on a per-letter distortion $\rho = \rho_1$.

The basic process DRF is defined by

$$\bar{D}_s(R,\mu) = \inf_{p \in \bar{\mathcal{R}}_s(R,\mu)} E_p \rho(X_0, Y_0),$$

where $\bar{\mathcal{R}}_s(R,\mu)$ is the collection of all stationary processes p having μ as an input distribution and having mutual information rate $\bar{I}_p = \bar{I}_p(X;Y) \leq R$. The original idea of a process rate-distortion function was due to Kolmogorov and his colleagues [87] [45] (see also [23]). The idea was later elaborated by Marton [101] and Gray, Neuhoff, and Omura [55].

Recalling that the L^1 ergodic theorem for information density holds when $\bar{I}_p = I_p^*$; that is, the two principal definitions of mutual information rate yield the same value, we also define the process DRF

$$D_s^*(R,\mu) = \inf_{p \in \mathcal{R}_s^*(R,\mu)} E_p \rho(X_0, Y_0),$$

where $\mathcal{R}_s^*(R,\mu)$ is the collection of all stationary processes p having μ as an input distribution, having mutual information rate $\bar{I}_p \leq R$, and having $\bar{I}_p = I_p^*$. If μ is both stationary and ergodic, define the corresponding ergodic process DRF's by

$$\bar{D}_e(R,\mu) = \inf_{p \in \bar{\mathcal{R}}_e(R,\mu)} E_p \rho(X_0, Y_0),$$

$$D_e^*(R,\mu) = \inf_{p \in \mathcal{R}_e^*(R,\mu)} E_p \rho(X_0, Y_0),$$

where $\bar{\mathcal{R}}_e(R,\mu)$ is the subset of $\bar{\mathcal{R}}_s(R,\mu)$ containing only ergodic measures and $\mathcal{R}_e^*(R,\mu)$ is the subset of $\mathcal{R}_s^*(R,\mu)$ containing only ergodic measures.

Theorem 10.6.1: Given a stationary source which possesses a reference letter in the sense that there exists a letter $a^* \in \hat{A}$ such that

$$E_\mu \rho(X_0, a_*) \leq \rho^* < \infty. \tag{10.6.1}$$

Fix $R > 0$. If $D(R,\mu) < \infty$, then

$$D(R,\mu) = \bar{D}_s(R,\mu) = D_s^*(R,\mu).$$

If in addition μ is ergodic, then also

$$D(R,\mu) = \bar{D}_e(R,\mu) = D_e^*(R,\mu).$$

The proof of the theorem depends strongly on the relations among distortion and mutual information for vectors and for SBM channels. These are stated and proved in the following lemma, the proof of which is straightforward but somewhat tedious. The theorem is proved after the lemma.

Lemma 10.6.3: Let μ be the process distribution of a stationary source $\{X_n\}$. Let ρ_n; $n = 1, 2, \cdots$ be a subadditive (e.g., additive) fidelity criterion. Suppose that there is a reference letter $a^* \in \hat{A}$ for which (10.6.1) holds. Let p^N be a measure on $(A^N \times \hat{A}^N, \mathcal{B}_A^N \times \mathcal{B}_{\hat{A}}^N)$ having μ^N as input marginal; that is, $p^N(F \times \hat{A}^N) = \mu^N(F)$ for $F \in \mathcal{B}_A^N$. Let q denote the induced conditional probability measure; that is, $q_{x^N}(F)$, $x^N \in A^N$, $F \in \mathcal{B}_{\hat{A}}^N$, is a regular conditional probability measure. (This exists because the spaces are standard.) We abbreviate this relationship as $p^N = \mu^N q$. Let X^N, Y^N denote the coordinate functions on $A^N \times \hat{A}^N$ and suppose that

$$E_{p^N} \frac{1}{N} \rho_N(X^N, Y^N) \leq D \tag{10.6.2}$$

and

$$\frac{1}{N} I_{p^N}(X^N; Y^N) \leq R. \tag{10.6.3}$$

If ν is an (N, δ) SBM channel induced by q as in Example 9.4.11 and if $p = \mu\nu$ is the resulting hookup and $\{X_n, Y_n\}$ the input/output pair process, then

$$\frac{1}{N} E_p \rho_N(X^N, Y^N) \leq D + \rho^* \delta \tag{10.6.4}$$

and

$$\bar{I}_p(X; Y) = I_p^*(X; Y) \leq R; \tag{10.6.5}$$

that is, the resulting mutual information rate of the induced stationary process satisfies the same inequality as the vector mutual information and the resulting distortion approximately satisfies the vector inequality provided δ is sufficiently small. Observe that if the fidelity criterion is additive, the (10.6.4) becomes

$$E_p \rho_1(X_0, Y_0) \leq D + \rho^* \delta.$$

Proof: We first consider the distortion as it is easier to handle. Since the SBM channel is stationary and the source is stationary, the hookup p is stationary and

$$\frac{1}{n} E_p \rho_n(X^n, Y^n) = \frac{1}{n} \int dm_Z(z) E_{p_z} \rho_n(X^n, Y^n),$$

where p_z is the conditional distribution of $\{X_n, Y_n\}$ given $\{Z_n\}$. Note that the above formula reduces to $E_p \rho(X_0, Y_0)$ if the fidelity criterion is additive because of the stationarity. Given z, define $J_0^n(z)$ to be the collection of indices of z^n for which z_i is not in an N-cell. (See the discussion in Example 9.4.11.) Let $J_1^n(z)$ be the collection of indices for which z_i begins an N-cell. If we define the event $G = \{z : z_0 \text{ begins an } N-\text{cell}\}$, then $i \in J_1^n(z)$ if

$T^i z \in G$. From Corollary 9.4.3 $m_Z(G) \leq N^{-1}$. Since μ is stationary and $\{X_n\}$ and $\{Z_n\}$ are mutually independent,

$$nE_{p_z}\rho_n(X^n, Y^n) \leq \sum_{i \in J_0^n(z)} E_{p_z}\rho(X_i, a^*) + N \sum_{i \in J_1^n(z)} E_{p_z}\rho(X_i^N, Y_i^N)$$

$$= \sum_{i=0}^{n-1} 1_{G^c}(T^i z)\rho^* + \sum_{i=0}^{n-1} E_{p^N}\rho_N 1_G(T^i z).$$

Since m_Z is stationary, integrating the above we have that

$$E_p\rho_1(X_0, Y_0) = \rho^* m_Z(G^c) + N m_Z(G) E_{p^N}\rho_N$$

$$\leq \rho^*\delta + E_{p^N}\rho_N,$$

proving (10.6.4).

Let r_m and t_m denote asymptotically accurate quantizers on A and $\hat{A}$; that is, as in Corollary 6.2.1 define

$$\hat{X}^n = r_m(X)^n = (r_m(X_0), \cdots, r_m(X_{n-1}))$$

and similarly define $\hat{Y}^n = t_m(Y)^n$. Then

$$I(r_m(X)^n; t_m(Y)^n) \underset{m\to\infty}{\to} I(X^n; Y^n)$$

and

$$\bar{I}(r_m(X); t_m(Y)) \underset{m\to\infty}{\to} I^*(X; Y).$$

We wish to prove that

$$\bar{I}(X;Y) = \lim_{n\to\infty} \lim_{m\to\infty} \frac{1}{n} I(r_m(X)^n; t_m(Y)^n)$$

$$= \lim_{m\to\infty} \lim_{n\to\infty} \frac{1}{n} I(r_m(X)^n; t_m(Y)^n). = I^*(X;Y)$$

Since $\bar{I} \geq I^*$, we must show that

$$\lim_{n\to\infty} \lim_{m\to\infty} \frac{1}{n} I(r_m(X)^n; t_m(Y)^n)$$

$$\leq \lim_{m\to\infty} \lim_{n\to\infty} \frac{1}{n} I(r_m(X)^n; t_m(Y)^n).$$

We have that

$$I(\hat{X}^n; \hat{Y}^n) = I((\hat{X}^n, Z^n); \hat{Y}^n) - I(Z^n, \hat{Y}^n | \hat{X}^n)$$

and

$$I((\hat{X}^n, Z^n); \hat{Y}^n) = I(\hat{X}^n; \hat{Y}^n | Z^n) + I(\hat{X}^n; Z^n) = I(\hat{X}^n; \hat{Y}^n | Z^n)$$

since $\hat{X}^n$ and Z^n are independent. Similarly,

$$\begin{aligned} I(Z^n; \hat{Y}^n | \hat{X}^n) &= H(Z^n | \hat{X}^n) - H(Z^n | \hat{X}^n, \hat{Y}^n) \\ &= H(Z^n) - H(Z^n | \hat{X}^n, \hat{Y}^n) = I(Z^n; (\hat{X}^n, \hat{Y}^n)). \end{aligned}$$

Thus we need to show that

$$\lim_{n\to\infty} \lim_{m\to\infty} \left(\frac{1}{n} I(r_m(X)^n; t_m(Y)^n | Z^n) - \frac{1}{n} I(Z^n, (r_m(X)^n, t_m(Y)^n)) \right)$$

$$\leq \lim_{m\to\infty} \lim_{n\to\infty} \left(\frac{1}{n} I(r_m(X)^n; t_m(Y)^n | Z^n) - \frac{1}{n} I(Z^n, (r_m(X)^n, t_m(Y)^n)) \right).$$

Since Z_n has a finite alphabet, the limits of $n^{-1} I(Z^n, (r_m(X)^n, t_m(Y)^n))$ are the same regardless of the order from Theorem 6.4.1. Thus $\bar{I}$ will equal I^* if we can show that

$$\begin{aligned} \bar{I}(X; Y | Z) &= \lim_{n\to\infty} \lim_{m\to\infty} \frac{1}{n} I(r_m(X)^n; t_m(Y)^n | Z^n) \\ &\leq \lim_{m\to\infty} \lim_{n\to\infty} \frac{1}{n} I(r_m(X)^n; t_m(Y)^n | Z^n) = I^*(X; Y | Z). \end{aligned} \tag{10.6.6}$$

This we now proceed to do. From Lemma 5.5.7 we can write

$$I(r_m(X)^n; t_m(Y)^n | Z^n) = \int I(r_m(X)^n; t_m(Y)^n | Z^n = z^n) \, dP_{Z^n}(z^n).$$

Abbreviate $I(r_m(X)^n; t_m(Y)^n | Z^n = z^n)$ to $I_z(\hat{X}^n; \hat{Y}^n)$. This is simply the mutual information between $\hat{X}^n$ and $\hat{Y}^n$ under the distribution for $(\hat{X}^n, \hat{Y}^n)$ given a particular random blocking sequence z. We have that

$$I_z(\hat{X}^n; \hat{Y}^n) = H_z(\hat{Y}^n) - H_z(\hat{Y}^n | \hat{X}^n).$$

Given z, let $J_0^n(z)$ be as before. Let $J_2^n(z)$ denote the collection of all indices i of z_i for which z_i begins an N cell *except* for the final such index (which may begin an N-cell not completed within z^n). Thus $J_2^n(z)$ is the same as $J_1^n(z)$ except that the largest index in the latter collection may have been removed if the resulting N-cell was not completed within the n-tuple. We have using standard entropy relations that

$$I_z(\hat{X}^n; \hat{Y}^n) \geq \sum_{i \in J_0^n(z)} \left(H_z(\hat{Y}_i | \hat{Y}^i) - H_z(\hat{Y}_i | \hat{Y}^i, \hat{X}^{i+1}) \right)$$

$$+ \sum_{i \in J_2^n(z)} \left(H_z(\hat{Y}_i^N | \hat{Y}^i) - H_z(\hat{Y}_i^N | \hat{Y}^i, \hat{X}^{i+N}) \right). \tag{10.6.7}$$

For $i \in J_0^n(z)$, however, Y_i is a^* with probability one and hence

$$H_z(\hat{Y}_i | \hat{Y}^i) \leq H_z(\hat{Y}_i) \leq H_z(Y_i) = 0$$

and

$$H_z(\hat{Y}_i | \hat{Y}^i, \hat{X}^{i+1}) \leq H_z(\hat{Y}_i) \leq H_z(Y_i) = 0.$$

Thus we have the bound

$$I_z(\hat{X}^n; \hat{Y}^n) \geq \sum_{i \in J_2^n(z)} \left(H_z(\hat{Y}_i^N | \hat{Y}^i) - H_z(\hat{Y}_i^N | \hat{Y}^i, \hat{X}^{i+N}) \right).$$

$$= \sum_{i \in J_2^n(z)} \left(I_z(\hat{Y}_i^N; (\hat{Y}^i, \hat{X}^i + N)) - I_z(\hat{Y}_i^N; \hat{Y}^i) \right)$$

$$\geq \sum_{i \in J_2^n(z)} \left(I_z(\hat{Y}_i^N; \hat{X}_i^N) - I_z(\hat{Y}_i^N; \hat{Y}^i) \right), \tag{10.6.8}$$

where the last inequality follows from the fact that $I(U; (V, W)) \geq I(U; V)$.

For $i \in J_2^n(z)$ we have by construction and the stationarity of μ that

$$I_z(\hat{X}_i^N; \hat{Y}_i^N) = I_{p^N}(\hat{X}^N; \hat{Y}^N). \tag{10.6.9}$$

As before let $G = \{z : z_0 \text{ begins an } N - \text{cell}\}$. Then $i \in J_2^n(z)$ if $T^i z \in G$ and $i < n - N$ and we can write

$$\frac{1}{n} I_z(\hat{X}^n; \hat{Y}^n) \geq \frac{1}{n} I_{p^N}(\hat{X}^N; \hat{Y}^N) \sum_{i=0}^{n-N-1} 1_G(T^i z)$$

$$- \frac{1}{n} \sum_{i=0}^{n-N-1} I_z(\hat{Y}_i^N; \hat{Y}^i) 1_G(T^i z).$$

All of the above terms are measurable functions of z and are nonnegative. Hence they are integrable (although we do not yet know if the integral is finite) and we have that

$$\frac{1}{n} I(\hat{X}^n; \hat{Y}^n) \geq I_{p^n}(\hat{X}^N; \hat{Y}^N) m_Z(G) \frac{n - N}{n}$$

$$- \frac{1}{n} \sum_{i=0}^{n-N-1} \int dm_Z(z) I_z(\hat{Y}_i^N; \hat{Y}^i) 1_G(T^i z).$$

To continue we use the fact that since the processes are stationary, we can consider it to be a two sided process (if it is one sided, we can imbed it in a two sided process with the same probabilities on rectangles). By construction

$$I_z(\hat{Y}_i^N;\hat{Y}^i) = I_{T^i z}(\hat{Y}_0^N;(Y_{-i},\cdots,Y_{-1}))$$

and hence since m_Z is stationary we can change variables to obtain

$$\frac{1}{n}I(\hat{X}^n;\hat{Y}^n) \geq I_{p^n}(\hat{X}^N;\hat{Y}^N)m_Z(G)\frac{n-N}{n}$$

$$-\frac{1}{n}\sum_{i=0}^{n-N-1}\int dm_Z(z)I_z(\hat{Y}_0^N;(\hat{Y}_{-i},\cdots,\hat{Y}_{-1}))1_G(z).$$

We obtain a further bound from the inequalities

$$I_z(\hat{Y}_0^N;(\hat{Y}_{-i},\cdots,\hat{Y}_{-1})) \leq I_z(Y_0^N;(Y_{-i},\cdots,Y_{-1})) \leq I_z(Y_0^N;Y^-)$$

where $Y^- = (\cdots,Y_{-2},Y_{-1})$. Since $I_z(Y_0^N;Y^-)$ is measurable and nonnegative, its integral is defined and hence

$$\lim_{n\to\infty}\frac{1}{n}I(\hat{X}^n;\hat{Y}^n|Z^n) \geq I_{p^n}(\hat{X}^N;\hat{Y}^N)m_Z(G) - \int_G dm_Z(z)I_z(Y_0^N;Y^-).$$

We can now take the limit as $m \to \infty$ to obtain

$$I^*(X;Y|Z) \geq I_{p^n}(X^N;Y^N)m_Z(G) - \int_G dm_Z(z)I_z(Y_0^N;Y^-). \quad (10.6.10)$$

This provides half of what we need.

Analogous to (10.6.7) we have the upper bound

$$I_z(\hat{X}^n;\hat{Y}^n) \leq \sum_{i\in J_1^n(z)}\left(I_z(\hat{Y}_i^N;(\hat{Y}^i,\hat{X}^{i+N})) - I_z(\hat{Y}_i^N;\hat{Y}^i)\right) \quad (10.6.11)$$

We note in passing that the use of J_1 here assumes that we are dealing with a one sided channel and hence there is no contribution to the information from any initial symbols not contained in the first N-cell. In the two sided case time 0 could occur in the middle of an N-cell and one could fix the upper bound by adding the first index less than 0 for which z_i begins an N-cell to the above sum. This term has no affect on the limits. Taking the limits as $m \to \infty$ using Lemma 5.5.1 we have that

$$I_z(X^n;Y^n) \leq \sum_{i\in J_1^n(z)}\left(I_z(Y_i^N;(Y^i,X^{i+N})) - I_z(Y_i^N;Y^i)\right).$$

Given $Z^n = z^n$ and $i \in J_1^n(z)$, $(X^i, Y^i) \to X_i^N \to Y_i^N$ forms a Markov chain because of the conditional independence and hence from Lemma 5.5.2 and Corollary 5.5.3

$$I_z(Y_i^N, (Y^i, X^{i+N})) = I_z(X_i^N; Y_i^N) = I_{p^N}(X^N; Y^N).$$

Thus we have the upper bound

$$\frac{1}{n} I_z(X^n; Y^n) \leq \frac{1}{n} I_{p^N}(X^N; Y^N) \sum_{i=0}^{n-1} 1_G(T^i z) - \frac{1}{n} \sum_{i=0}^{n-1} I_z(Y_i^N; Y^i) 1_G(T^i z).$$

Taking expectations and using stationarity as before we find that

$$\frac{1}{n} I(X^n; Y^n | Z^n) \leq I_{p^N}(X^N; Y^N) m_Z(G)$$

$$- \frac{1}{n} \sum_{i=0}^{n-1} \int_G dm_Z(z) I_z(Y_0^N; (Y_{-i}, \cdots, Y_{-1})).$$

Taking the limit as $n \to \infty$ using Lemma 5.6.1 yields

$$\bar{I}(X; Y|Z) \leq I_{p^N}(X^N; Y^N) m_Z(G) - \int_G dm_Z(z) I_z(Y_0^N; Y^-). \quad (10.6.12)$$

Combining this with (10.6.10) proves that

$$\bar{I}(X; Y|Z) \leq I^*(X; Y|Z)$$

and hence that

$$\bar{I}(X; Y) = I^*(X; Y).$$

It also proves that

$$\bar{I}(X; Y) = \bar{I}(X; Y|Z) - \bar{I}(Z; (X, Y)) \leq \bar{I}(X; Y|Z)$$

$$\leq I_{p^N}(X^N; Y^N) m_Z(G) \leq \frac{1}{N} I_{p^N}(X^N; Y^N)$$

using Corollary 9.4.3 to bound $m_X(G)$. This proves (10.6.5). □

Proof of the theorem: We have immediately that

$$\mathcal{R}_e^*(R, \mu) \subset \mathcal{R}_s^*(R, \mu) \subset \bar{\mathcal{R}}_s(R, \mu)$$

and

$$\mathcal{R}_e^*(R, \mu) \subset \bar{\mathcal{R}}_e(R, \mu) \subset \bar{\mathcal{R}}_s(R, \mu),$$

and hence we have for stationary sources that

$$\bar{D}_s(R,\mu) \leq D_s^*(R,\mu) \tag{10.6.13}$$

and for ergodic sources that

$$\bar{D}_s(R,\mu) \leq D_s^*(R,\mu) \leq D_e^*(R,\mu) \tag{10.6.14}$$

and

$$\bar{D}_s(R,\mu) \leq \bar{D}_e(R,\mu) \leq D_e^*(R,\mu). \tag{10.6.15}$$

We next prove that

$$\bar{D}_s(R,\mu) \geq D(R,\mu). \tag{10.6.16}$$

If $\bar{D}_s(R,\mu)$ is infinite, the inequality is obvious. Otherwise fix $\epsilon > 0$ and choose a $p \in \bar{\mathcal{R}}_s(R,\mu)$ for which $E_p\rho_1(X_0,Y_0) \leq \bar{D}_s(R,\mu)+\epsilon$ and fix $\delta > 0$ and choose m so large that for $n \geq m$ we have that

$$n^{-1}I_p(X^n;Y^n) \leq \bar{I}_p(X;Y)+\delta \leq R+\delta.$$

For $n \geq m$ we therefore have that $p^n \in \mathcal{R}_n(R+\delta,\mu^n)$ and hence

$$\bar{D}_s(R,\mu)+\epsilon = E_{p^n}\rho_n \geq D_n(R+\delta,\mu) \geq D(R+\delta,\mu).$$

From Lemma 10.6.1 $D(R,\mu)$ is continuous in R and hence (10.6.16) is proved.

Lastly, fix $\epsilon > 0$ and choose N so large and $p^N \in \mathcal{R}_N(R,\mu^N)$ so that

$$E_{p^N}\rho_N \leq D_N(R,\mu^N)+\frac{\epsilon}{3} \leq D(R,\mu)+\frac{2\epsilon}{3}.$$

Construct the corresponding (N,δ)-SBM channel as in Example 9.4.11 with δ small enough to ensure that $\delta\rho^* \leq \epsilon/3$. Then from Lemma 10.6.2 we have that the resulting hookup p is stationary and that $\bar{I}_p = I_p^* \leq R$ and hence $p \in \mathcal{R}_s^*(R,\mu) \subset \bar{\mathcal{R}}_s(R,\mu)$. Furthermore, if μ is ergodic then so is p and hence $p \in \mathcal{R}_e^*(R,\mu) \subset \bar{\mathcal{R}}_e(R,\mu)$. From Lemma 10.6.2 the resulting distortion is

$$E_p\rho_1(X_0,Y_0) \leq E_{p^N}\rho_N + \rho^*\delta \leq D(R,\mu)+\epsilon.$$

Since $\epsilon > 0$ this implies the exisitence of a $p \in \mathcal{R}_s^*(R,\mu)$ ($p \in \mathcal{R}_e^*(R,\mu)$ if μ is ergodic) yielding $E_p\rho_1(X_0,Y_0)$ arbitrarily close to $D(R,\mu$. Thus for any stationary source

$$D_s^*(R,\mu) \leq D(R,\mu)$$

and for any ergodic source

$$D_e^*(R,\mu) \leq D(R,\mu).$$

With (10.6.13)–(10.6.16) this completes the proof. □

The previous lemma is technical but important. It permits the construction of a stationary and ergodic pair process having rate and distortion near that of that for a finite dimensional vector described by the original source and a finite-dimensional conditional probability.

Chapter 11

Source Coding Theorems

11.1 Source Coding and Channel Coding

In this chapter and the next we develop the basic coding theorems of information theory. As is traditional, we consider two important special cases first and then later form the overall result by combining these special cases. In the first case we assume that the channel is noiseless, but it is constrained in the sense that it can only pass R bits per input symbol to the receiver. Since this is usually insufficient for the receiver to perfectly recover the source sequence, we attempt to code the source so that the receiver can recover it with as little distortion as possible. This leads to the theory of *source coding* or *source coding subject to a fidelity criterion* or *data compression*, where the latter name reflects the fact that sources with infinite or very large entropy are "compressed" to fit across the given communication link. In the next chapter we ignore the source and focus on a discrete alphabet channel and construct codes that can communicate any of a finite number of messages with small probability of error and we quantify how large the message set can be. This operation is called *channel coding* or *error control coding*. We then develop *joint source and channel codes* which combine source coding and channel coding so as to code a given source for communication over a given channel so as to minimize average distortion. The *ad hoc* division into two forms of coding is convenient and will permit performance near that of the OPTA function for the codes considered.

11.2 Block Source Codes for AMS Sources

We first consider a particular class of codes: block codes. For the time being we also concentrate on additive distortion measures. Extensions to subadditive distortion measures will be considered later. Let $\{X_n\}$ be a source with a standard alphabet A. Recall that an (N, K) block code of a source $\{X_n\}$ maps successive nonoverlapping input vectors $\{X_{nN}^N\}$ into successive channel vectors $U_{nK}^K = \alpha(X_{nN}^N)$, where $\alpha : A^N \rightarrow B^K$ is called the *source encoder*. We assume that the channel is noiseless, but that it is constrained in the sense that N source time units corresponds to the same amount of physical time as K channel time units and that

$$\frac{K \log ||B||}{N} \leq R,$$

where the inequality can be made arbitrarily close to equality by taking N and K large enough subject to the physical stationarity constraint. R is called the *source coding rate* or *resolution* in bits or nats per input symbol. We may wish to change the values of N and K, but the rate is fixed.

A reproduction or approximation of the original source is obtained by a *source decoder*, which we also assume to be a block code. The decoder is a mapping $\beta : B^K \rightarrow \hat{A}^N$ which forms the reproduction process $\{\hat{X}_n\}$ via $\hat{X}_{nN}^N = \beta(U_{nK}^K)$; $n = 1, 2, \cdots$. In general we could have a reproduction dimension different from that of the input vectors provided they corresponded to the same amount of physical time and a suitable distortion measure was defined. We will make the simplifying assumption that they are the same, however.

Because N source symbols are mapped into N reproduction symbols, we will often refer to N alone as the block length of the source code. Observe that the resulting sequence coder is N-stationary. Our immediate goal is now the following: Let $\mathcal{E}$ and $\mathcal{D}$ denote the collection of all block codes with rate no greater than R and let ν be the given channel. What is the OPTA function $\Delta(\mu, \mathcal{E}, \nu, \mathcal{D})$ for this system? Our first step toward evaluating the OPTA is to find a simpler and equivalent expression for the current special case.

Given a source code consisting of encoder α and decoder β, define the *codebook* to be

$$\mathcal{C} = \{ \text{ all } \beta(u^K); u^K \in B^K\},$$

that is, the collection of all possible reproduction vectors available to the receiver. For convenience we can index these words as

$$\mathcal{C} = \{y_i; i = 1, 2, \cdots, M\},$$

where $N^{-1}\log M \leq R$ by construction. Observe that if we are given only a decoder β or, equivalently, a codebook, and if our goal is to minimize the average distortion for the current block, then no encoder can do better than the encoder α^* which maps an input word x^N into the minimum distortion available reproduction word, that is, define $\alpha^*(x^N)$ to be the u^K minimizing $\rho_N(x^N,\beta(u^K))$, an assignment we denote by

$$\alpha^*(x^N) = \min_{u^K}{}^{-1}\rho_N(x^N,\beta(u^K)).$$

Observe that by construction we therefore have that

$$\rho_N(x^N,\beta(\alpha^*(x^N))) = \min_{y\in\mathcal{C}}\rho_N(x^N,y)$$

and the overall mapping of x^N into a reproduction is a minimum distortion or nearest neighbor mapping. Define

$$\rho_N(x^N,\mathcal{C}) = \min_{y\in\mathcal{C}}\rho_N(x^N,y).$$

To formally prove that this is the best decoder, observe that if the source μ is AMS and p is the joint distribution of the source and reproduction, then p is also AMS. This follows since the channel induced by the block code is N-stationary and hence also AMS with respect to T^N. This means that p is AMS with respect to T^N which in turn implies that it is AMS with respect to T (Theorem 7.3.1 of [50]). Letting $\bar{p}$ denote the stationary mean of p and $\bar{p}_N$ denote the N-stationary mean, we then have from (10.3.10) that for any block codes with codebook $\mathcal{C}$

$$\Delta = \frac{1}{N}E_{\bar{p}_N}\rho_N(X^N,Y^N) \geq \frac{1}{N}E_{\bar{p}_N}\rho_N(X^N,\mathcal{C}),$$

with equality if the minimum distortion encoder is used. For this reason we can confine interest to block codes specified by a codebook: the encoder produces the index of the minimum distortion codeword for the observed vector and the decoder is a table lookup producing the codeword being indexed. A code of this type is also called a *vector quantizer* or *block quantizer*. Denote the performance of the block code with codebook $\mathcal{C}$ on the source μ by

$$\rho(\mathcal{C},\mu) = \Delta = E_p\rho_\infty.$$

Lemma 11.2.1: Given an AMS source μ and a block length N code book $\mathcal{C}$, let $\bar{\mu}_N$ denote the N-stationary mean of μ (which exists from Corollary 7.3.1 of [50]), let p denote the induced input/output distribution, and let $\bar{p}$ and $\bar{p}_N$ denote its stationary mean and N-stationary mean, respectively. Then

$$\rho(\mathcal{C},\mu) = E_{\bar{p}}\rho_1(X_0,Y_0) = \frac{1}{N}E_{\bar{p}_N}\rho_N(X^N,Y^N)$$

$$= \frac{1}{N}E_{\bar{\mu}_N}\rho_N(X^N,\mathcal{C}) = \rho(\mathcal{C},\bar{\mu}_N).$$

Proof: The first two equalities follow from (10.3.10), the next from the use of the minimum distortion encoder, the last from the definition of the performance of a block code. □

It need not be true in general that $\rho(\mathcal{C},\mu)$ equal $\rho(\mathcal{C},\bar{\mu})$. For example, if μ produces a single periodic waveform with period N and $\mathcal{C}$ consists of a single period, then $\rho(\mathcal{C},\mu) = 0$ and $\rho(\mathcal{C},\bar{\mu}) > 0$. It is the N-stationary mean and not the stationary mean that is most useful for studying an N-stationary code.

We now define the OPTA for block codes to be

$$\delta(R,\mu) = \Delta^*(\mu,\nu,\mathcal{E},\mathcal{D}) = \inf_N \delta_N(R,\mu),$$

$$\delta_N(R,\mu) = \inf_{\mathcal{C}\in\mathcal{K}(N,R)} \rho(\mathcal{C},\mu),$$

where ν is the noiseless channel as described previously, $\mathcal{E}$ and $\mathcal{D}$ are classes of block codes for the channel, and $\mathcal{K}(N,R)$ is the class of all block length N codebooks $\mathcal{C}$ with

$$\frac{1}{N}\log||\mathcal{C}|| \le R.$$

$\delta(R,\mu)$ is called the ***block source coding OPTA*** or the ***operational block coding distortion-rate function.***

Corollary 11.2.1: Given an AMS source μ, then for any N and $i = 0,1,\cdots,N-1$

$$\delta_N(R,\mu T^{-i}) = \delta_N(R,\bar{\mu}_N T^{-i}).$$

Proof: For $i = 0$ the result is immediate from the lemma. For $i \neq 0$ it follows from the lemma and the fact that the N-stationary mean of μT^{-i} is $\bar{\mu}_N T^{-i}$ (as is easily verified from the definitions). □

Reference Letters

Many of the source coding results will require a technical condition that is a generalization of reference letter condition of Theorem 10.6.1 for stationary sources. An AMS source μ is said to have a *reference letter* $a^* \in \hat{A}$ with respect to a distortion measure $\rho = \rho_1$ on $A \times \hat{A}$ if

$$\sup_n E_{\mu T^{-n}} \rho(X_0, a^*) = \sup_n E_\mu \rho(X_n, a^*) = \rho^* < \infty, \tag{11.2.1}$$

that is, there exists a letter for which $E_\mu \rho(X^n, a^*)$ is uniformly bounded above. If we define for any k the vector $a^{*k} = (a^*, a^*, \cdots, a^*)$ consisting of k a^*'s, then (11.2.1) implies that

$$\sup_n E_{\mu T^{-n}} \frac{1}{k} \rho_k(X^k, a^{*k}) \leq \rho^* < \infty. \tag{11.2.2}$$

We assume for convenience that any block code of length N contains the reference vector a^{*N}. This ensures that $\rho_N(x^N, \mathcal{C}) \leq \rho_N(x^N, a^{*N})$ and hence that $\rho_N(x^N, \mathcal{C})$ is bounded above by a μ-integrable function and hence is itself μ-integrable. This implies that

$$\delta(R, \mu) \leq \delta_N(R, \mu) \leq \rho^*. \tag{11.2.3}$$

The reference letter also works for the stationary mean source $\bar{\mu}$ since

$$\lim_{n \to \infty} \frac{1}{n} \sum_{i=0}^{n-1} \rho(x_i, a^*) = \rho_\infty(x, \mathbf{a}^*),$$

$\bar{\mu}$-a.e. and μ-a.e., where $\mathbf{a}^*$ denotes an infinite sequence of a^*. Since ρ_∞ is invariant we have from Lemma 6.3.1 of [50] and Fatou's lemma that

$$E_{\bar{\mu}} \rho(X_0, a^*) = E_\mu \left(\lim_{n \to \infty} \frac{1}{n} \sum_{i=0}^{n-1} \rho(X_i, a^*) \right)$$

$$\leq \liminf_{n \to \infty} \frac{1}{n} \sum_{i=0}^{n-1} E_\mu \rho(X_i, a^*) \leq \rho^*.$$

Performance and OPTA

We next develop several basic properties of the performance and OPTA functions for block coding AMS sources with additive fidelity criteria.

Lemma 11.2.2: Given two sources μ_1 and μ_2 and $\lambda \in (0, 1)$, then for any block code $\mathcal{C}$

$$\rho(\mathcal{C}, \lambda\mu_1 + (1 - \lambda)\mu_2) = \lambda\rho(\mathcal{C}, \mu_1) + (1 - \lambda)\rho(\mathcal{C}, \mu_2)$$

and for any N

$$\delta_N(R, \lambda\mu_1 + (1-\lambda)\mu_2) \geq \lambda\delta_N(R, \mu_1) + (1-\lambda)\delta_N(R, \mu_2)$$

and

$$\delta(R, \lambda\mu_1 + (1-\lambda)\mu_2) \geq \lambda\delta(R, \mu_1) + (1-\lambda)\delta(R, \mu_2).$$

Thus performance is linear in the source and the OPTA functions are convex $\bigcap$. Lastly,

$$\delta_N(R + \frac{1}{N}, \lambda\mu_1 + (1-\lambda)\mu_2) \leq \lambda\delta_N(R, \mu_1) + (1-\lambda)\delta_N(R, \mu_2).$$

Proof: The equality follows from the linearity of expectation since $\rho(\mathcal{C}, \mu) = E_\mu\rho(X^N, \mathcal{C})$. The first inequality follows from the equality and the fact that the infimum of a sum is bounded below by the sum of the infima. The next inequality follows similarly. To get the final inequality, let $\mathcal{C}_i$ approximately yield $\delta_N(R, \mu_i)$; that is,

$$\rho(\mathcal{C}_i, \mu_i) \leq \delta_N(R, \mu_i) + \epsilon.$$

Form the union code $\mathcal{C} = \mathcal{C}_1 \bigcup \mathcal{C}_2$ containing all of the words in both of the codes. Then the rate of the code is

$$\frac{1}{N}\log||\mathcal{C}|| = \frac{1}{N}\log(||\mathcal{C}_1|| + ||\mathcal{C}_2||)$$

$$\leq \frac{1}{N}\log(2^{NR} + 2^{NR}) = R + \frac{1}{N}.$$

This code yields performance

$$\rho(\mathcal{C}, \lambda\mu_1 + (1-\lambda)\mu_2) = \lambda\rho(\mathcal{C}, \mu_1) + (1-\lambda)\rho(\mathcal{C}, \mu_2)$$

$$\leq \lambda\rho(\mathcal{C}_1, \mu_1) + (1-\lambda)\rho(\mathcal{C}_2, \mu_2) \leq \lambda\delta_N(R, \mu_1) + \lambda\epsilon + (1-\lambda)\delta_N(R, \mu_2) + (1-\lambda)\epsilon.$$

Since the leftmost term in the above equation can be no smaller than $\delta_N(R + 1/N, \lambda\mu_1 + (1-\lambda)\mu_2)$, the lemma is proved. □

The first and last inequalities in the lemma suggest that δ_N is very nearly an affine function of the source and hence perhaps δ is as well. We will later pursue this possibility, but we are not yet equipped to do so.

Before developing the connection between the OPTA functions of AMS sources and those of their stationary mean, we pause to develop some additional properties for OPTA in the special case of stationary sources. These results follow Kieffer [76].

Lemma 11.2.3: Suppose that μ is a stationary source. Then

$$\delta(R, \mu) = \lim_{N\to\infty} \delta_N(R, \mu).$$

Thus the infimum over block lengths is given by the limit so that longer codes can do better.

Proof: Fix an N and an $n < N$ and choose codes $\mathcal{C}_n \subset \hat{A}^n$ and $\mathcal{C}_{N-n} \subset \hat{A}^{N-n}$ for which

$$\rho(\mathcal{C}_n, \mu) \leq \delta_n(R, \mu) + \frac{\epsilon}{2}$$

$$\rho(\mathcal{C}_{N-n}, \mu) \leq \delta_{N-n}(R, \mu) + \frac{\epsilon}{2}.$$

Form the block length N code $\mathcal{C} = \mathcal{C}_n \times \mathcal{C}_{N-n}$. This code has rate no greater than R and has distortion

$$N\rho(\mathcal{C}, \mu) = E \min_{y \in \mathcal{C}} \rho_N(X^N, y)$$

$$= E_{y^n \in \mathcal{C}_n} \rho_n(X^n, y^n) + E_{v^{N-n} \in \mathcal{C}_{N-n}} \rho_{N-n}(X_n^{N-n}, v^{N-n})$$

$$= E_{y^n \in \mathcal{C}_n} \rho_n(X^n, y^n) + E_{v^{N-n} \in \mathcal{C}_{N-n}} \rho_{N-n}(X^{N-n}, v^{N-n})$$

$$= n\rho(\mathcal{C}_n, \mu) + (N-n)\rho(\mathcal{C}_{N-n}, \mu)$$

$$\leq n\delta_n(R, \mu) + (N-n)\delta_{N-n}(R, \mu) + \epsilon, \tag{11.2.4}$$

where we have made essential use of the stationarity of the source. Since ϵ is arbitrary and since the leftmost term in the above equation can be no smaller than $N\delta_N(R, \mu)$, we have shown that

$$N\delta_N(R, \mu) \leq n\delta_n(R, \mu) + (N-n)\delta_{N-n}(R, \mu)$$

and hence that the sequence $N\delta_N$ is subadditive. The result then follows immediately from Lemma 7.5.1 of [50]. □

Corollary 11.2.2: If μ is a stationary source, then $\delta(R, \mu)$ is a convex $\bigcup$ function of R and hence is continuous for $R > 0$.

Proof: Pick $R_1 > R_2$ and $\lambda \in (0, 1)$. Define $R = \lambda R_1 + (1-\lambda)R_2$. For large n define $n_1 = \lfloor \lambda n \rfloor$ be the largest integer less than λn and let $n_2 = n - n_1$. Pick codebooks $\mathcal{C}_i \subset \hat{A}^{n_i}$ with rate R_i with distortion

$$\rho(\mathcal{C}_i, \mu) \leq \delta_{n_i}(R_i, \mu) + \epsilon.$$

Analogous to (11.2.4), for the product code $\mathcal{C} = \mathcal{C}_1 \times \mathcal{C}_2$ we have

$$n\rho(\mathcal{C}, \mu) = n_1\rho(\mathcal{C}_1, \mu) + n_2\rho(\mathcal{C}_2, \mu)$$

$$\leq n_1\delta_{n_1}(R_1, \mu) + n_2\delta_{n_2}(R_2, \mu) + n\epsilon.$$

The rate of the product code is no greater than R and hence the leftmost term above is bounded below by $n\delta_n(R, \mu)$. Dividing by n we have since ϵ is arbitrary that

$$\delta_n(R, \mu) \leq \frac{n_1}{n}\delta_{n_1}(R_1, \mu) + \frac{n_2}{n}\delta_{n_2}(R_2, \mu).$$

Taking $n \to \infty$ we have using the lemma and the choice of n_i that

$$\delta(R,\mu) \leq \lambda\delta(R_1,\mu) + (1-\lambda)\delta(R_2,\mu),$$

proving the claimed convexity. □

Corollary 11.2.3: If μ is stationary, then $\delta(R,\mu)$ is an affine function of μ.

Proof: From Lemma 11.2.2 we need only prove that

$$\delta(R, \lambda\mu_1 + (1-\lambda)\mu_2) \leq \lambda\delta(R,\mu_1) + (1-\lambda)\delta(R,\mu_2).$$

From the same lemma we have that for any N

$$\delta_N(R+\frac{1}{N}, \lambda\mu_1 + (1-\lambda)\mu_2) \leq \lambda\delta_N(R,\mu_1) + (1-\lambda)\delta_N(R,\mu_2)$$

For any $K \leq N$ we have since $\delta_N(R,\mu)$ is nonincreasing in R that

$$\delta_N(R+\frac{1}{K}, \lambda\mu_1 + (1-\lambda)\mu_2) \leq \lambda\delta_N(R,\mu_1) + (1-\lambda)\delta_N(R,\mu_2).$$

Taking the limit as $N \to \infty$ yields from Lemma 11.2.3 that

$$\delta(R+\frac{1}{K}, \mu) \leq \lambda\delta(R,\mu_1) + (1-\lambda)\delta(R,\mu_2).$$

From Corollary 11.2.2, however, δ is continuous in R and the result follows by letting $K \to \infty$. □

The following lemma provides the principal tool necessary for relating the OPTA of an AMS source with that of its stationary mean. It shows that the OPTA of an AMS source is not changed by shifting or, equivalently, by redefining the time origin.

Lemma 11.2.4: Let μ be an AMS source with a reference letter. Then for any integer i $\delta(R,\mu) = \delta(R,\mu T^{-i})$.

Proof: Fix $\epsilon > 0$ and let $\mathcal{C}_N$ be a rate R block length N codebook for which $\rho(\mathcal{C}_N,\mu) \leq \delta(R,\mu) + \epsilon/2$. For $1 \leq i \leq N-1$ choose J large and define the block length $K = JN$ code $\mathcal{C}_K(i)$ by

$$\mathcal{C}_K(i) = a^{*(N-i)} \times \mathop{\times}_{j=0}^{J-2} \mathcal{C}_N \times a^{*i},$$

where a^{*l} is an l-tuple containing all a^*'s. $\mathcal{C}_K(i)$ can be considered to be a code consisting of the original code shifted by i time units and repeated many times, with some filler at the beginning and end. Except for the edges of the long product code, the effect on the source is to use the original code with a delay. The code has at most $(2^{NR})^{J-1} = 2^{KR}2^{-NR}$ words; the rate is no greater than R.

For any K-block x^K the distortion resulting from using $\mathcal{C}_k^{(i)}$ is given by

$$K\rho_K(x^K, \mathcal{C}_K(i)) \leq (N-i)\rho_{N-i}(x^{N-i}, a^{*(N-i)}) + i\rho_i(x_{K-i}^i, a^{*i}). \quad (11.2.5)$$

Let $\{\hat{x}_n\}$ denote the encoded process using the block code $\mathcal{C}_K(i)$. If n is a multiple of K, then

$$n\rho_n(x^n, \hat{x}^n) \leq \sum_{k=0}^{\lfloor \frac{n}{K} \rfloor}((N-i)\rho_{N-i}(x_{kK}^{N-i}, a^{*(N-i)})$$

$$+ i\rho_i(x_{(k+1)K-i}^i, a^{*i})) + \sum_{k=0}^{\lfloor \frac{n}{K} \rfloor J-1} N\rho_N(x_{N-i+kN}^N, \mathcal{C}_N).$$

If n is not a multiple of K we can further overbound the distortion by including the distortion contributed by enough future symbols to complete a K-block, that is,

$$n\rho_n(x^n, \hat{x}^n) \leq n\gamma_n(x, \hat{x})$$

$$= \sum_{k=0}^{\lfloor \frac{n}{K} \rfloor+1} \left((N-i)\rho_{N-i}(x_{kK}^{N-i}, a^{*(N-i}) + i\rho_i(x_{(k+1)K-i}^i, a^{*i})\right)$$

$$+ \sum_{k=0}^{(\lfloor \frac{n}{K} \rfloor+1)J-1} N\rho_N(x_{N-i+kN}^N, \mathcal{C}_N).$$

Thus

$$\rho_n(x^n, \hat{x}^n) \leq \frac{N-i}{K}\frac{1}{n/K} \sum_{k=0}^{\lfloor \frac{n}{K} \rfloor+1} \rho_{N-i}(X^{N-i}(T^{kK}x), a^{*(N-i})$$

$$+ \frac{i}{K}\frac{1}{n/K} \sum_{k=0}^{\lfloor \frac{n}{K} \rfloor+1} \rho_i(X^i(T^{(k+1)K-i}x, a^{*i})$$

$$+ \frac{1}{n/N} \sum_{k=0}^{(\lfloor \frac{n}{K} \rfloor+1)J-1} \rho_N(X^N(T^{(N-i)+kN}x), \mathcal{C}_N).$$

Since μ is AMS these quantities all converge to invariant functions:

$$\lim_{n\to\infty} \rho_n(x^n, \hat{x}^n) \leq \frac{N-i}{K} \lim_{m\to\infty} \frac{1}{m} \sum_{k=0}^{m-1} \rho_{N-i}(X^{N-i}(T^{kK}x), a^{*(N-i})$$

$$+\frac{i}{K}\lim_{m\to\infty}\frac{1}{m}\sum_{k=0}^{m-1}\rho_i(X^i(T^{(k+1)K-i}x,a^{*i})$$

$$+\lim_{m\to\infty}\frac{1}{m}\sum_{k=0}^{m-1}\rho_N(X^N(T^{(N-i)+kN}x),\mathcal{C}_N).$$

We now apply Fatou's lemma, a change of variables, and Lemma 11.2.1 to obtain

$$\delta(R,\mu T^{-i})\leq\rho(\mathcal{C}_K(i),\mu T^{-i})$$

$$\leq\frac{N-i}{K}\limsup_{m\to\infty}\frac{1}{m}\sum_{k=0}^{m}E_{\mu T^{-i}}\rho_{N-i}(X^{N-i}T^{kK},a^{*(N-i)})$$

$$+\frac{i}{K}\lim_{m\to\infty}\frac{1}{m}\sum_{k=0}^{m-1}E_{\mu T^{-i}}\rho_i(X^iT^{(k+1)K-i},a^{*i})$$

$$+E_{\mu T^{-i}}\lim_{m\to\infty}\frac{1}{m}\sum_{k=0}^{m-1}\rho_N(X^NT^{(N-i)+kN}),\mathcal{C}_N).$$

$$\leq\frac{N-i}{K}\rho^*+\frac{i}{K}\rho^*+E_\mu\lim_{m\to\infty}\frac{1}{m}\sum_{k=1}^{m-1}\rho_N(X^NT^{kN}\mathcal{C}_N).\leq\frac{N}{K}\rho^*+\rho(\mathcal{C}_N,\mu).$$

Thus if J and hence K are chosen large enough to ensure that $N/K\leq\epsilon/2$, then

$$\delta(R,\mu T^{-i})\leq\delta(R,\mu),$$

which proves that $\delta(R,\mu T^{-i})\leq\delta(R,\mu)$. The reverse implication is found in a similar manner: Let $\mathcal{C}_N$ be a codebook for μT^{-i} and construct a codebook $\mathcal{C}_K(N-i)$ for use on μ. By arguments nearly identical to those above the reverse inequality is found and the proof completed. □

Corollary 11.2.4: Let μ be an AMS source with a reference letter. Fix N and let $\bar{\mu}$ and $\bar{\mu}_N$ denote the stationary and N-stationary means. Then for $R>0$

$$\delta(R,\bar{\mu})=\delta(R,\bar{\mu}_NT^{-i});\ i=0,1,\cdots,N-1.$$

Proof: It follows from the previous lemma that the $\delta(R,\bar{\mu}_NT^{-i})$ are all equal and hence it follows from Lemma 11.2.2, Theorem 7.3.1 of [50], and Corollary 7.3.1 of [50] that

$$\delta(R,\bar{\mu})\geq\frac{1}{N}\sum_{i=0}^{N-1}\delta(R,\bar{\mu}_NT^{-i})=\delta(R,\bar{\mu}_N).$$

To prove the reverse inequality, take $\mu = \bar{\mu}_N$ in the previous lemma and construct the codes $\mathcal{C}_K(i)$ as in the previous proof. Take the union code $\mathcal{C}_K = \bigcup_{i=0}^{N-1} \mathcal{C}_K(i)$ having block length K and rate at most $R + K^{-1}\log N$. We have from Lemma 11.2.1 and (11.2.5) that

$$\rho(\mathcal{C}_K, \bar{\mu}) = \frac{1}{N}\sum_{i=0}^{N-1} \rho(\mathcal{C}_K, \bar{\mu}_N T^{-i})$$

$$\leq \frac{1}{N}\sum_{i=0}^{N-1} \rho(\mathcal{C}_K(i), \bar{\mu}_N T^{-i}) \leq \frac{N}{K}\rho^* + \rho(\mathcal{C}_N, \bar{\mu}_N)$$

and hence as before

$$\delta(R + \frac{1}{JN}\log N, \bar{\mu}) \leq \delta(R, \bar{\mu}_N).$$

From Corollary 11.2.1 $\delta(R, \bar{\mu})$ is continuous in R for $R > 0$ since $\bar{\mu}$ is stationary. Hence taking J large enough yields $\delta(R, \bar{\mu}) \leq \delta(R, \bar{\mu}_N)$. This completes the proof since from the lemma $\delta(R, \bar{\mu}_N T^{-i}) = \delta(R, \bar{\mu}_N)$. □

We are now prepared to demonstrate the fundamental fact that the block source coding OPTA function for an AMS source with an additive fidelity criterion is the same as that of the stationary mean process. This will allow us to assume stationarity when proving the actual coding theorems.

Theorem 11.2.2: If μ is an AMS source and $\{\rho_n\}$ an additive fidelity criterion with a reference letter, then for $R > 0$

$$\delta(R, \mu) = \delta(R, \bar{\mu}).$$

Proof: We have from Corollaries 11.2.1 and 11.2.4 that

$$\delta(R, \bar{\mu}) \leq \delta(R, \bar{\mu}_N) \leq \delta_N(R, \bar{\mu}_N) = \delta_N(R, \mu).$$

Taking the infimum over N yields

$$\delta(R, \bar{\mu}) \leq \delta(R, \mu).$$

Conversely, fix $\epsilon > 0$ let $\mathcal{C}_N$ be a block length N codebook for which $\rho(\mathcal{C}_N, \bar{\mu}) \leq \delta(R, \bar{\mu}) + \epsilon$. From Lemma 11.2.1, Corollary 11.2.1, and Lemma 11.2.4

$$\delta(R, \bar{\mu}) + \epsilon \leq \rho(\mathcal{C}_N, \bar{\mu}) = \frac{1}{N}\sum_{i=0}^{N-1} \rho(\mathcal{C}_N, \bar{\mu}_N T^{-i})$$

$$\geq \frac{1}{N}\sum_{i=0}^{N-1} \delta_N(R, \bar{\mu}_N T^{-i}) = \frac{1}{N}\sum_{i=0}^{N-1} \delta_N(R, \mu T^{-i})$$

$$\geq \frac{1}{N} \sum_{i=0}^{N-1} \delta(R, \mu T^{-i}) = \delta(R, \mu),$$

which completes the proof since ϵ is arbitrary. □

Since the OPTA functions are the same for an AMS process and its stationary mean, this immediately yields the following corollary from Corollary 11.2.2:

Corollary 11.2.5: If μ is AMS, then $\delta(R, \mu)$ is a convex function of R and hence a continuous function of R for $R > 0$.

11.3 Block Coding Stationary Sources

We showed in the previous section that when proving block source coding theorems for AMS sources, we could confine interest to stationary sources. In this section we show that in an important special case we can further confine interest to only those stationary sources that are ergodic by applying the ergodic decomposition. This will permit us to assume that sources are stationary and ergodic in the next section when the basic Shannon source coding theorem is proved and then extend the result to AMS sources which may not be ergodic.

As previously we assume that we have a stationary source $\{X_n\}$ with distribution μ and we assume that $\{\rho_n\}$ is an additive distortion measure and there exists a reference letter. For this section we now assume in addition that the alphabet A is itself a Polish space and that $\rho_1(r, y)$ is a continuous function of r for every $y \in \hat{A}$. If the underlying alphabet has a metric structure, then it is reasonable to assume that forcing input symbols to be very close in the underlying alphabet should force the distortion between either symbol and a fixed output to be close also. The following theorem is the ergodic decomposition of the block source coding OPTA function.

Theorem 11.3.1: Suppose that μ is the distribution of a stationary source and that $\{\rho_n\}$ is an additive fidelity criterion with a reference letter. Assume also that $\rho_1(\cdot, y)$ is a continuous function for all y. Let $\{\mu_x\}$ denote the ergodic decomposition of μ. Then

$$\delta(R, \mu) = \int d\mu(x) \delta(R, \mu_x),$$

that is, $\delta(R, \mu)$ is the average of the OPTA of its ergodic components.

Proof: Analogous to the ergodic decomposition of entropy rate of Theorem 2.4.1, we need to show that $\delta(R, \mu)$ satisfies the conditions of Theorem 8.9.1 of [50]. We have already seen (Corollary 11.2.3) that it is an affine function. We next see that it is upper semicontinuous. Since the alphabet is Polish, choose a distance $d_{\mathcal{G}}$ on the space of stationary processes having this alphabet with the property that $\mathcal{G}$ is constructed as in Section 8.2 of [50]. Pick an N large enough and a length N codebook $\mathcal{C}$ so that

$$\delta(R,\mu) \geq \delta_N(R,\mu) - \frac{\epsilon}{2} \geq \rho_N(\mathcal{C},\mu) - \epsilon.$$

$\rho_N(x^N, y)$ is by assumption a continuous function of x^N and hence so is $\rho_N(x^N, \mathcal{C}) = \min_{y \in \mathcal{C}} \rho(x^N, y)$. Since it is also nonnegative, we have from Lemma 8.2.4 of [50] that if $\mu_n \rightarrow \mu$ then

$$\limsup_{n\rightarrow\infty} E_{\mu_n} \rho_N(X^N, \mathcal{C}) \leq E_\mu \rho_N(X^N, \mathcal{C}).$$

The left hand side above is bounded below by

$$\limsup_{n\rightarrow\infty} \delta_N(R,\mu_n) \geq \limsup_{n\rightarrow\infty} \delta(R,\mu_n).$$

Thus since ϵ is arbitrary,

$$\limsup_{n\rightarrow\infty} \delta(R,\mu_n) \leq \delta(R,\mu)$$

and hence $\delta(R,\mu)$ upper semicontinuous in μ and hence also measurable. Since the process has a reference letter, $\delta(R,\mu_x)$ is integrable since

$$\delta(R,\mu_X) \leq \delta_N(R,\mu_x) \leq E_{\mu_x} \rho_1(X_0, a^*)$$

which is integrable if $\rho_1(x_0, a^*)$ is from the ergodic decomposition theorem. Thus Theorem 8.9.1 of [50] yields the desired result. □

The theorem was first proved by Kieffer [76] for bounded continuous additive distortion measures. The above extension removes the requirement that ρ_1 be bounded.

11.4 Block Coding AMS Ergodic Sources

We have seen that the block source coding OPTA of an AMS source is given by that of its stationary mean. Hence we will be able to concentrate on stationary sources when proving the coding theorem.

Theorem 11.4.1: Let μ be an AMS ergodic source with a standard alphabet and $\{\rho_n\}$ an additive distortion measure with a reference letter. Then

$$\delta(R,\mu) = D(R,\bar{\mu}),$$

where $\bar{\mu}$ is the stationary mean of μ.

Proof: From Theorem 11.2.2 $\delta(R,\mu) = \delta(R,\bar{\mu})$ and hence we will be done if we can prove that

$$\delta(R,\bar{\mu}) = D(R,\bar{\mu}).$$

This will follow if we can show that $\delta(R,\mu) = D(R,\mu)$ for any stationary ergodic source with a reference letter. Henceforth we assume that μ is stationary and ergodic.

We first prove the negative or converse half of the theorem. First suppose that we have a codebook $\mathcal{C}$ such that

$$\rho_N(C,\mu) = E_\mu \min_{y \in \mathcal{C}} \rho_N(X^N, y) = \delta_N(R,\mu) + \epsilon.$$

If we let $\hat{X}_N$ denote the resulting reproduction random vector and let p^N denote the resulting joint distribution of the input/output pair, then since $\hat{X}^N$ has a finite alphabet, Lemma 5.5.6 implies that

$$I(X^N;\hat{X}^N) \leq H(\hat{X}^N) \leq NR$$

and hence $p^N \in \mathcal{R}_N(R,\mu^N)$ and hence

$$\delta_N(R,\mu) + \epsilon \geq E_{p^N} \rho_N(X^N;\hat{X}^N) \geq D_N(R,\mu).$$

Taking the limits as $N \to \infty$ proves the easy half of the theorem:

$$\delta(R,\mu) \geq D(R,\mu).$$

(Recall that both OPTA and distortion rate functions are given by limits if the source is stationary.)

The fundamental idea of Shannon's positive source coding theorem is this: for a fixed block size N, choose a code at random according to a distribution implied by the distortion-rate function. That is, perform 2^{NR} independent random selections of blocks of length N to form a codebook. This codebook is then used to encode the source using a minimum distortion mapping as above. We compute the average distortion over this double-random experiment (random codebook selection followed by use of the chosen code to encode the random source). We will find that if the code generation distribution is properly chosen, then this average will be no greater than $D(R,\mu)+\epsilon$. If the average over all randomly selected codes is no greater than $D(R,\mu)+\epsilon$, however, than there must be at least one code such that the average distortion over the source distribution for that one code is no greater than $D(R,\mu)+\epsilon$. This means that there exists at least one code with performance not much larger than $D(R,\mu)$. Unfortunately the proof only demonstrates the existence of such codes, it does not show how to construct them.

To find the distribution for generating the random codes we use the ergodic process definition of the distortion-rate function. From Theorem 10.6.1 (or Lemma 10.6.3) we can select a stationary and ergodic pair process with distribution p which has the source distribution μ as one coordinate and which has

$$E_p\rho(X_0,Y_0) = \frac{1}{N}E_{p^N}\rho_N(X^N,Y^N) \leq D(R,\mu)+\epsilon \tag{11.4.1}$$

and which has

$$\bar{I}_p(X;Y) = I^*(X;Y) \leq R \tag{11.4.2}$$

(and hence information densities converge in L^1 from Theorem 6.3.1). Denote the implied vector distributions for (X^N,Y^N), X^N, and Y^N by p^N, μ^N, and η^N, respectively.

For any N we can generate a codebook $\mathcal{C}$ at random according to η^N as described above. To be precise, consider the random codebook as a large random vector $\mathcal{C} = (W_0,W_1,\cdots,W_M)$, where $M = \lfloor e^{N(R+\epsilon)}\rfloor$ (where natural logarithms are used in the definition of R), where W_0 is the fixed reference vector a^{*N} and where the remaining W_n are independent, and where the marginal distributions for the W_n are given by η^N. Thus the distribution for the randomly selected code can be expressed as

$$P_{\mathcal{C}} = \mathop{\times}_{i=1}^{M} \eta^N.$$

This codebook is then used with the optimal encoder and we denote the resulting average distortion (over codebook generation and the source) by

$$\Delta_N = E\rho(\mathcal{C},\mu) = \int dP_{\mathcal{C}}(\mathcal{W})\rho(\mathcal{W},\mu) \tag{11.4.3}$$

where

$$\rho(\mathcal{W},\mu) = \frac{1}{N}E\rho_N(X^N,\mathcal{W}) = \frac{1}{N}\int d\mu^N(x^N)\rho_N(x^N,\mathcal{W}),$$

and where

$$\rho_N(x^N,\mathcal{C}) = \min_{y\in\mathcal{C}}\rho_N(x^N,y).$$

Choose $\delta > 0$ and break up the integral over x into two pieces: one over a set $G_N = \{x : N^{-1}\rho_N(x^N,a^{*N}) \leq \rho^* + \delta\}$ and the other over the complement of this set. Then

$$\Delta_N \leq \int_{G_N^c} \frac{1}{N}\rho_N(x^N,a^{*N})\,d\mu^N(x^N)$$

$$+\frac{1}{N}\int dP_{\mathcal{C}}(\mathcal{W})\int_{G_N} d\mu^N(x^N)\rho_N(x^N,\mathcal{W}), \tag{11.4.4}$$

where we have used the fact that $\rho_N(x^N,m\mathcal{W}) \leq \rho_N(x^N,a^{*N})$. Fubini's theorem implies that because

$$\int d\mu^N(x^N)\rho_N(x^N,a^{*N}) < \infty$$

and

$$\rho_N(x^N,\mathcal{W}) \leq \rho_N(x^N,a^{*N}),$$

the limits of integration in the second integral of (11.4.4) can be interchanged to obtain the bound

$$\Delta_N \leq \frac{1}{N}\int_{G_N^c} \rho_N(x^N,a^{*N})d\mu^N(x^N)$$

$$+\frac{1}{N}\int_{G_N} d\mu^N(x^N)\int dP_{\mathcal{C}}(\mathcal{W})\rho_N(x^N,\mathcal{W}) \tag{11.4.5}$$

The rightmost term in (11.4.5) can be bound above by observing that

$$\frac{1}{N}\int_{G_N} d\mu^N(x^N)[\int dP_{\mathcal{C}}(\mathcal{W})\rho_N(x^N,\mathcal{W})]$$

$$=\frac{1}{N}\int_{G_N} d\mu^N(x^N)[\int_{\mathcal{C}:\rho_N(x^N,\mathcal{C})\leq N(D+\delta)} dP_{\mathcal{C}}(\mathcal{W})\rho_N(x^N,\mathcal{W})$$

$$+\frac{1}{N}\int_{\mathcal{W}:\rho_N(x^N,\mathcal{W})>N(D+\delta)} dP_{\mathcal{C}}(\mathcal{W})\rho_N(x^N,\mathcal{W})]$$

$$\leq \int_{G_N} d\mu^N(x^N)[D+\delta+\frac{1}{N}(\rho^*+\delta)\int_{\mathcal{W}:\rho_N(x^N,\mathcal{W})>N(D+\delta)} dp_{\mathcal{C}}(\mathcal{W})]$$

where we have used the fact that for $x \in G$ the maximum distortion is given by $\rho^* + \delta$. Define the probability

$$P(N^{-1}\rho_N(x^N,\mathcal{C}) > D+\delta|x^N) = \int_{\mathcal{W}:\rho_N(x^N,\mathcal{W})>N(D+\delta)} dp_{\mathcal{C}}(\mathcal{W})$$

and summarize the above bounds by

$$\Delta_N \leq D+\delta+(\rho^*+\delta)\frac{1}{N}\int d\mu^N(x^N)P(N^{-1}\rho_N(x^N,\mathcal{C}) > D+\delta|x^N)$$

$$+\frac{1}{N}\int_{G_N^c} d\mu^N(x^N)\rho_N(x^N,a^{*N}). \tag{11.4.6}$$

The remainder of the proof is devoted to proving that the two integrals above go to 0 as $N \to \infty$ and hence

$$\limsup_{N\to\infty} \Delta_N \leq D + \delta. \tag{11.4.7}$$

Consider first the integral

$$a_N = \frac{1}{N}\int_{G_N^c} d\mu^N(x^N)\rho_N(x^N, a^{*N})$$

$$= \int d\mu^N(x^N)1_{G_N^c}(x^N)\frac{1}{N}\rho_N(x^N, a^{*N}).$$

We shall see that this integral goes to zero as an easy application of the ergodic theorem. The integrand is dominated by $N^{-1}\rho_N(x^N, a^{*N})$ which is uniformly integrable (Lemma 4.7.2 of [50]) and hence the integrand is itself uniformly integrable (Lemma 4.4.4 of [50]). Thus we can invoke the extended Fatou lemma to conclude that

$$\limsup_{N\to\infty} a_N \leq \int d\mu^N(x^N)\limsup_{N\to\infty}\left(1_{G_N^c}(x^N)\frac{1}{N}\rho_N(x^N, a^{*N})\right)$$

$$\leq \int d\mu^N(x^N)(\limsup_{N\to\infty} 1_{G_N^c}(x^N))(\limsup_{N\to\infty}\frac{1}{N}\rho_N(x^N, a^{*N})).$$

We have, however, that $\limsup_{N\to\infty} 1_{G_N^c}(x^N)$ is 0 unless $x^N \in G_N^c$ i.o. But this set has measure 0 since with μ_N probability 1, an x is produced so that

$$\lim_{N\to\infty}\frac{1}{N}\sum_{i=0}^{N-1}\rho(x_i, a^*) = \rho^*$$

exists and hence with probability one one gets an x which can yield

$$N^{-1}\rho_N(x^N, a^{*N}) > \rho^* + \delta$$

at most for a finite number of N. Thus the above integral of the product of a function that is 0 a.e. with a dominated function must itself be 0 and hence

$$\limsup_{N\to\infty} a_N = 0. \tag{11.4.8}$$

We now consider the second integral in (11.4.6):

$$b_N = (\rho^* + \delta)\frac{1}{N}\int d\mu^N(x^N)P(N^{-1}\rho_N(x^N, \mathcal{C}) > D + \delta|x^N).$$

Recall that $P(\rho_N(x^N,\mathcal{C}) > D+\delta|x^N)$ is the probability that for a fixed input block x^N, a randomly selected code will result in a minimum distortion codeword larger than $D+\delta$. This is the probability that none of the M words (excluding the reference code word) selected independently at random according to to the distribution η^N lie within $D+\delta$ of the fixed input word x^N. This probability is bounded above by

$$P(\frac{1}{N}\rho_N(x^N,\mathcal{C}) > D+\delta|x^N) \leq [1-\eta^N(\frac{1}{N}\rho_N(x^N,Y^N) \leq D+\delta)]^M$$

where

$$\eta^N(\frac{1}{N}\rho_N(x^N,Y^N) \leq D+\delta)) = \int_{y^N:\frac{1}{N}\rho_N(x^N,y^N)\leq D+\delta} d\eta^N(y^N).$$

Now mutual information comes into the picture. The above probability can be bounded below by adding a condition:

$$\eta^N(\frac{1}{N}\rho_N(x^N,Y^N) \leq D+\delta)$$

$$\geq \eta^N(\frac{1}{N}\rho_N(x^N,Y^N) \leq D+\delta \text{ and } \frac{1}{N}i_N(x^N,Y^N) \leq R+\delta),$$

where

$$\frac{1}{N}i_N(x^N,y^N) = \frac{1}{N}\ln f_N(x^N,y^N),$$

where

$$f_N(x^N,y^N) = \frac{dp^N(x^N,y^N)}{d(\mu^N\times\eta^N)(x^N,y^N)},$$

the Radon-Nikodym derivative of p^N with respect to the product measure $\mu^N\times\eta^N$. Thus we require both the distortion and the sample information be less than slightly more than their limiting value. Thus we have in the region of integration that

$$\frac{1}{N}i_N(x^N;y^N) = \frac{1}{N}\ln f_N(x^N,y^N) \leq R+\delta$$

and hence

$$\eta_N(\rho_N(x^N,Y^N) \leq D+\delta) \geq \int_{y^N:\rho_N(x^N,y^N)\leq D+\delta, f_N(x^N,y^N)\leq e^{N(R+\delta)}} d\eta^N(y^N)$$

$$\geq e^{-N(R+\delta)} \int_{y^N:\rho_N(x^N,y^N)\leq D+\delta, f_N(x^N,y^N)\leq e^{N(R+\delta)}} d\eta^N(y^N) f_N(x^N,y^N)$$

which yields the bound

$$P(\frac{1}{N}\rho_N(x^N,\mathcal{C}) > D+\delta|x^N) \leq [1-\eta^N(\frac{1}{N}\rho_N(x^N,Y^N) \leq D+\delta)]^M$$

$$\leq [1-e^{-N(R+\delta)}\int_{y^N:\frac{1}{N}\rho_N(x^N,y^N)\leq D+\delta,\frac{1}{N}i_N(x^N,y^N)\leq R+\delta} d\eta^N(y^N)f_N(x^N,y^N)]^M,$$

Applying the inequality

$$(1-\alpha\beta)^M \leq 1-\beta+e^{-M\alpha}$$

for $\alpha,\beta \in [0,1]$ yields

$$P(\frac{1}{N}\rho_N(x^N,\mathcal{C}) > D+\delta|x^N) \leq$$

$$1-\int_{y^N:\frac{1}{N}\rho_N(x^N,y^N)\leq D+\delta,\frac{1}{N}i_N(x^N,y^N)\leq R+\delta} d\eta^N(y^N)$$

$$\times f_N(x^N,y^N)+e^{[-Me^{-N(R+\delta)}]}.$$

Averaging with respect to the distribution μ^N yields

$$\frac{b_N}{\rho^*+\delta} = \int d\mu^N(x^N)P(\rho_N(x^N,\mathcal{C}) > D+\delta|x^N)$$

$$\leq \int d\mu^N(x^N)\left(1-\int_{y^N:\rho_N(x^N,y^N)\leq N(D+\delta),\frac{1}{N}i_N(x^N,y^N)\leq R+\delta} d\eta^N(y^N)\right.$$

$$\left.\times f_N(x^N,y^N)+e^{-Me^{-N(R+\delta)}}\right)$$

$$=1-\int_{y^N:\frac{1}{N}\rho_N(x^N,y^N)\leq D+\delta,\frac{1}{N}i_N(x^N,y^N)\leq R+\delta} d(\mu^N\times\eta^N)(x^N,y^N)$$

$$\times f_N(x^N,y^N)+e^{-Me^{-N(R+\delta)}}$$

$$=1+e^{-Me^{-N(R+\delta)}}-\int_{y^N:\frac{1}{N}\rho_N(x^N,y^N)\leq D+\delta,\frac{1}{N}i_N(x^N,y^N)\leq R+\delta} dp^N(x^N,y^N)$$

$$=1+e^{-Me^{-N(R+\delta)}}$$

$$-p^N(y^N:\frac{1}{N}\rho_N(x^N,y^N)\leq D+\delta,\frac{1}{N}i_N(x^N,y^N)\leq R+\delta). \quad (11.4.9)$$

Since M is bounded below by $e^{N(R+\epsilon)}-1$, the exponential term is bounded above by

$$e^{[-e^{(N(R+\epsilon)}e^{-N(R+\delta)}+e^{-N(R+\delta)}]} = e^{[-e^{N(\epsilon-\delta)}+e^{-N(R+\delta)}]}.$$

If $\epsilon > \delta$, this term goes to 0 as $N \to \infty$.

The probability term in (11.4.9) goes to 1 from the mean ergodic theorem applied to ρ_1 and the mean ergodic theorem for information density since mean convergence (or the almost everywhere convergence proved elsewhere) implies convergence in probability. This implies that

$$\limsup_{n\to\infty} b_N = 0$$

which with (11.4.8) gives (11.4.7). Choosing an N so large that $\Delta_N \leq \delta$, we have proved that there exists a block code $\mathcal{C}$ with average distortion less than $D(R,\mu) + \delta$ and rate less than $R + \epsilon$ and hence

$$\delta(R+\epsilon,\mu) \leq D(R,\mu) + \delta. \tag{11.4.10}$$

Since ϵ and δ can be chosen as small as desired and since $D(R,\mu)$ is a continuous function of R (Lemma 10.6.1), the theorem is proved. □

The source coding theorem is originally due to Shannon [129] [130], who proved it for discrete i.i.d. sources. It was extended to stationary and ergodic discrete alphabet sources and Gaussian sources by Gallager [43] and to stationary and ergodic sources with abstract alphabets by Berger [10] [11], but an error in the information density convergence result of Perez [123] (see Kieffer [74]) left a gap in the proof, which was subsequently repaired by Dunham [35]. The result was extended to nonergodic stationary sources and metric distortion measures and Polish alphabets by Gray and Davisson [53] and to AMS ergodic processes by Gray and Saadat [61]. The method used here of using a stationary and ergodic measure to construct the block codes and thereby avoid the block ergodic decomposition of Nedoma [106] used by Gallager [43] and Berger [11] was suggested by Pursley and Davisson [29] and developed in detail by Gray and Saadat [61].

11.5 Subadditive Fidelity Criteria

In this section we generalize the block source coding theorem for stationary sources to subadditive fidelity criteria. Several of the interim results derived previously are no longer appropriate, but we describe those that are still valid in the course of the proof of the main result. Most importantly, we now consider only stationary and not AMS sources. The result can be extended to AMS sources in the two-sided case, but it is not known for the one-sided case. Source coding theorems for subadditive fidelity criteria were first developed by Mackenthun and Pursley [96].

Theorem 11.5.1: Let μ denote a stationary and ergodic distribution of a source $\{X_n\}$ and let $\{\rho_n\}$ be a subadditive fidelity criterion with a reference letter, i.e., there is an $a^* \in \hat{A}$ such that

$$E\rho_1(X_0, a^*) = \rho^* < \infty.$$

Then the OPTA for the class of block codes of rate less than R is given by the Shannon distortion-rate function $D(R, \mu)$.

Proof: Suppose that we have a block code of length N, e.g., a block encoder $\alpha : A^N \to B^K$ and a block decoder $\beta : B^K \to \hat{A}^N$. Since the source is stationary, the induced input/output distribution is then N-stationary and the performance resulting from using this code on a source μ is

$$\Delta_N = E_p \rho_\infty = \frac{1}{N} E_p \rho_N(X^N, \hat{X}^N),$$

where $\{\hat{X}^N\}$ is the resulting reproduction process. Let $\delta_N(R, \mu)$ denote the infimum over all codes of length N of the performance using such codes and let $\delta(R, \mu)$ denote the infimum of δ_N over all N, that is, the OPTA. We do not assume a codebook/minimum distortion structure because the distortion is now effectively context dependent and it is not obvious that the best codes will have this form. Assume that given an $\epsilon > 0$ we have chosen for each N a length N code such that

$$\delta_N(R, \mu) \geq \Delta_N - \epsilon.$$

As previously we assume that

$$\frac{K \log ||B||}{N} \leq R,$$

where the constraint R is the rate of the code. As in the proof of the converse coding theorem for an additive distortion measure, we have that for the resulting process $I(X^N; \hat{X}^N) \leq RN$ and hence

$$\Delta_N \geq D_N(R, \mu).$$

From Lemma 10.6.2 we can take the infimum over all N to find that

$$\delta(R, \mu) = \inf_N \delta_N(R, \mu) \geq \inf_N D_N(R, \mu) - \epsilon = D(R, \mu) - \epsilon.$$

Since ϵ is arbitrary, $\delta(R, \mu) \leq D(R, \mu)$, proving the converse theorem.

To prove the positive coding theorem we proceed in an analogous manner to the proof for the additive case, except that we use Lemma 10.6.3 instead of Theorem 10.6.1. First pick an N large enough so that

$$D_N(R, \mu) \leq D(R, \mu) + \frac{\delta}{2}$$

and then select a $p^N \in \mathcal{R}_N(R, \mu^N)$ such that

$$E_{p^N} \frac{1}{N} \rho_N(X^N, Y^N) \leq D_N(R, \mu) + \frac{\delta}{2} \leq D(R, \mu) + \delta.$$

Now then construct as in Lemma 10.6.3 a stationary and ergodic process p which will have (10.6.4) and (10.6.5) satisfied (the right Nth order distortion and information). This step taken, the proof proceeds exactly as in the additive case since the reference vector yields the bound

$$\frac{1}{N} \rho_N(x^N, a^{*N}) \leq \frac{1}{N} \sum_{i=0}^{N-1} \rho_1(x_i, a^*),$$

which converges, and since $N^{-1} \rho_N(x^N, y^N)$ converges as $N \to \infty$ with p probability one from the subadditive ergodic theorem. Thus the existence of a code satisfying (11.4.10) can be demonstrated (which uses the minimum distortion encoder) and this implies the result since $D(R, \mu)$ is a continuous function of R (Lemma 10.6.1). □

11.6 Asynchronous Block Codes

The block codes considered so far all assume block synchronous communication, that is, that the decoder knows where the blocks begin and hence can deduce the correct words in the codebook from the index represented by the channel block. In this section we show that we can construct asynchronous block codes with little loss in performance or rate; that is, we can construct a block code so that a decoder can uniquely determine how the channel data are parsed and hence deduce the correct decoding sequence. This result will play an important role in the development in the next section of sliding block coding theorems.

Given a source μ let $\delta_{\text{async}}(R, \mu)$ denote the OPTA function for block codes with the added constraint that the decoder be able to synchronize, that is, correctly parse the channel codewords. Obviously

$$\delta_{\text{async}}(R, \mu) \geq \delta(R, \mu)$$

since we have added a constraint. The goal of this section is to prove the following result:

Theorem 11.6.1: Given an AMS source with an additive fidelity criterion and a reference letter,

$$\delta_{\text{async}}(R, \mu) = \delta(R, \mu),$$

that is, the OPTA for asynchronous codes is the same as that for ordinary codes.

Proof: A simple way of constructing a synchronized block code is to use a prefix code: Every codeword begins with a short prefix or *source synchronization word* or, simply, sync word, that is not allowed to appear anywhere else within a word or as any part of an overlap of the prefix and a piece of the word. The decoder than need only locate the prefix in order to decode the block begun by the prefix. The insertion of the sync word causes a reduction in the available number of codewords and hence a loss in rate, but ideally this loss can be made negligible if properly done. We construct a code in this fashion by finding a good codebook of slightly smaller rate and then indexing it by channel K-tuples with this prefix property.

Suppose that our channel has a rate constraint R, that is, if source N-tuples are mapped into channel K-tuples then

$$\frac{K \log ||B||}{N} \leq R,$$

where B is the channel alphabet. We assume that the constraint is achievable on the channel in the sense that we can choose N and K so that the physical stationarity requirement is met (N source time units corresponds to K channel time units) and such that

$$||B||^K \approx e^{NR}, \tag{11.6.1}$$

at least for large N.

If K is to be the block length of the channel code words, let δ be small and define $k(K) = \lfloor \delta K \rfloor + 1$ and consider channel codewords which have a prefix of $k(K)$ occurrences of a single channel letter, say b, followed by a sequence of $K - k(K)$ channel letters which have the following constraint: no $k(K)$-tuple beginning after the first symbol can be $b^{k(K)}$. We permit b's to occur at the end of a K-tuple so that a $k(K)$-tuple of b's may occur in the overlap of the end of a codeword and the new prefix since this causes no confusion, e.g., if we see an elongated sequence of b's, the actual code information starts at the right edge. Let $M(K)$ denote the number of distinct channel K-tuples of this form. Since $M(K)$ is the number of distinct reproduction codewords that can be indexed by channel codewords, the codebooks will be constrained to have rate

$$R_K = \frac{\ln M(K)}{N}.$$

We now study the behavior of R_K as K gets large. There are a total of $||B||^{K-k(K)}$ K-tuples having the given prefix. Of these, no more than $(K -$

$k(K))||B||^{K-2k(K)}$ have the sync sequence appearing somewhere within the word (there are fewer than $K - k(K)$ possible locations for the sync word and for each location the remaining $K - 2k(K)$ symbols can be anything). Lastly, we must also eliminate those words for which the first i symbols are b for $i = 1, 2, \cdots, k(K) - 1$ since this will cause confusion about the right edge of the sync sequence. These terms contribute

$$\sum_{i=1}^{k(K)-1} ||B||^{K-k(K)-i}$$

bad words. Using the geometric progression formula to sum the above series we have that it is bounded above by

$$\frac{||B||^{K-k(K)-1}}{1 - 1/||B||}.$$

Thus the total number of available channel vectors is at least

$$M(K) \geq ||B||^{K-k(K)} - (K - k(K))||B||^{K-2k(K)} - \frac{||B||^{K-k(K)-1}}{1 - 1/||B||}.$$

Thus

$$\begin{aligned} R_K &= \frac{1}{N} \ln ||B||^{K-k(K)} + \frac{1}{N} \ln \left(1 - (K - k(K))||B||^{-k(K)} - \frac{1}{||B|| - 1}\right) \\ &= \frac{K - k(K)}{N} \ln ||B|| + \frac{1}{N} \ln \left(\frac{||B|| - 2}{||B|| - 1} - (K - k(K))||B||^{-k(K)}\right). \\ &\geq (1 - \delta)R + o(N), \end{aligned}$$

where $o(N)$ is a term that goes to 0 as N (and hence K) goes to infinity. Thus given a channel with rate constraint R and given $\epsilon > 0$, we can construct for N sufficiently large a collection of approximately $e^{N(R-\epsilon)}$ channel K-tuples (where $K \approx NR$) which are synchronizable, that is, satisfy the prefix condition.

We are now ready to construct the desired code. Fix $\delta > 0$ and then choose $\epsilon > 0$ small enough to ensure that

$$\delta(R(1 - \epsilon), \mu) \leq \delta(R, \mu) + \frac{\delta}{3}$$

(which we can do since $\delta(R, \mu)$ is continuous in R). Then choose an N large enough to give a prefix channel code as above and to yield a rate $R - \epsilon$ codebook $\mathcal{C}$ so that

$$\rho_N(\mathcal{C}, \mu) \leq \delta_N(R - \epsilon, \mu) + \frac{\delta}{3}$$

$$\leq \delta(R-\epsilon,\mu)+\frac{2\delta}{3}\leq \delta(R,\mu)+\delta. \tag{11.6.2}$$

The resulting code proves the theorem. □

11.7 Sliding Block Source Codes

We now turn to sliding block codes. For simplicity we consider codes which map blocks into single symbols. For example, a sliding block encoder will be a mapping $f: A^N \to B$ and the decoder will be a mapping $g: B^K \to \hat{A}$. In the case of one-sided processes, for example, the channel sequence would be given by

$$U_n = f(X_n^N)$$

and the reproduction sequence by

$$\hat{X}_n = g(U_n^L).$$

When the processes are two-sided, it is more common to use memory as well as delay. This is often done by having an encoder mapping $f: A^{2N+1} \to B$, a decoder $g: B^{2L+1} \to \hat{A}$, and the channel and reproduction sequences being defined by

$$U_n = f(X_{-N},\cdots,X_0,\cdots,X_N),$$

$$\hat{X}_n = g(U_{-L},\cdots,U_0,\cdots,U_N).$$

We shall emphasize the two-sided case.

The final output can be viewed as a sliding block coding of the input:

$$\hat{X}_n = g(f(X_{n-L-N},\cdots,X_{n-L+N}),\cdots,f(X_{n+L-N},\cdots,X_{n+L+N}))$$

$$= gf(X_{n-(N+L)},\cdots,X_{n+(N+L)}),$$

where we use gf to denote the overall coding, that is, the cascade of g and f. Note that the delay and memory of the overall code are the sums of those for the encoder and decoder. The overall window length is $2(N+L)+1$

Since one channel symbol is sent for every source symbol, the rate of such a code is given simply by $R = \log ||B||$ bits per source symbol. The obvious problem with this restriction is that we are limited to rates which are logarithms of integers, e.g., we cannot get fractional rates. As previously discussed, however, we could get fractional rates by appropriate redefinition of the alphabets (or, equivalently, of the shifts on the corresponding sequence spaces). For example, regardless of the code window lengths involved, if we shift l source symbols to produce a new group of k channel symbols (to yield an (l,k)-stationary encoder) and then shift a group of k

channel symbols to produce a new group of k source symbols, then the rate is

$$R = \frac{k}{l} \log ||B||$$

bits or nats per source symbol and the overall code fg is l-stationary. The added notation to make this explicit is significant and the generalization is straightforward; hence we will stick to the simpler case.

We can define the sliding block OPTA for a source and channel in the natural way. Suppose that we have an encoder f and a decoder g. Define the resulting performance by

$$\rho(fg, \mu) = E_{\mu f g} \rho_\infty,$$

where $\mu f g$ is the input/output hookup of the source μ connected to the deterministic channel fg and where ρ_∞ is the sequence distortion. Define

$$\delta_{\mathrm{SBC}}(R, \mu) = \inf_{f,g} \rho(fg, \mu) = \Delta^*(\mu, \mathcal{E}, \nu, \mathcal{D}),$$

where $\mathcal{E}$ is the class of all finite length sliding block encoders and $\mathcal{D}$ is the collection of all finite length sliding block decoders. The rate constraint R is determined by the channel.

Assume as usual that μ is AMS with stationary mean $\bar{\mu}$. Since the cascade of stationary channels fg is itself stationary (Lemma 9.4.7), we have from Lemma 9.3.2 that $\mu f g$ is AMS with stationary mean $\bar{\mu} f g$. This implies from (10.3.10) that for any sliding block codes f and g

$$E_{\mu f g} \rho_\infty = E_{\bar{\mu} f g} \rho_\infty$$

and hence

$$\delta_{\mathrm{SBC}}(R, \mu) = \delta_{\mathrm{SBC}}(R, \bar{\mu}).$$

A fact we now formalize as a lemma.

Lemma 11.7.1: Suppose that μ is an AMS source with stationary mean $\bar{\mu}$ and let $\{\rho_n\}$ be an additive fidelity criterion. Let $\delta_{\mathrm{SBC}}(R, \mu)$ denote the sliding block coding OPTA function for the source and a channel with rate constraint R. Then

$$\delta_{\mathrm{SBC}}(R, \mu) = \delta_{\mathrm{SBC}}(R, \bar{\mu}).$$

The lemma permits us to concentrate on stationary sources when quantifying the optimal performance of sliding block codes.

The principal result of this section is the following:

Theorem 11.7.1: Given an AMS and ergodic source μ and an additive fidelity criterion with a reference letter,

$$\delta_{\mathrm{SBC}}(R, \mu) = \delta(R, \mu),$$

that is, the class of sliding block codes is capable of exactly the same performance as the class of block codes. If the source is only AMS and not ergodic, then at least

$$\delta_{\text{SBC}}(R,\mu) \geq \delta(R,\mu), \tag{11.7.1}$$

Proof: The proof of (11.7.1) follows that of Shields and Neuhoff [133] for the finite alphabet case, except that their proof was for ergodic sources and coded only typical input sequences. Their goal was different because they measured the rate of a sliding block code by the entropy rate of its output, effectively assuming that further almost-noiseless coding was to be used. Because we consider a fixed channel and measure the rate in the usual way as a coding rate, this problem does not arise here. From the previous lemma we need only prove the result for stationary sources and hence we henceforth assume that μ is stationary. We first prove that sliding block codes can perform no better than block codes, that is, (11.7.1) holds. Fix $\delta > 0$ and suppose that $f : A^{2N+1} \to B$ and $g : B^{2L+1} \to \hat{A}$ are finite-length sliding block codes for which

$$\rho(fg,\mu) \leq \delta_{\text{SBC}}(R,\mu) + \delta.$$

This yields a cascade sliding block code $fg : A^{2(N+L)+1} \to \hat{A}$ which we use to construct a block codebook. Choose K large (to be specified later). Observe an input sequence x^n of length $n = 2(N+L)+1+K$ and map it into a reproduction sequence $\hat{x}^n$ as follows: Set the first and last $(N+L)$ symbols to the reference letter a^*, that is, $x_0^{N+L} = x_{n-N-L}^{N+L} = a^{*(N+L)}$. Complete the remaining reproduction symbols by sliding block coding the source word using the given codes, that is,

$$\hat{x}_i = fg(x_{i-(N+L)}^{2(N+L)+1});\ i = N+L+1,\cdots,K+N+L.$$

Thus the long block code is obtained by sliding block coding, except at the edges where the sliding block code is not permitted to look at previous or future source symbols and hence are filled with a reference symbol. Call the resulting codebook $\mathcal{C}$. The rate of the block code is less than $R = \log ||B||$ because n channel symbols are used to produce a reproduction word of length n and hence the codebook can have no more that $||B||^n$ possible vectors. Thus the rate is $\log ||B||$ since the codebook is used to encode a source n-tuple. Using this codebook with a minimum distortion rule can do no worse (except at the edges) than if the original sliding block code had been used and therefore if $\hat{X}_i$ is the reproduction process produced by the block code and Y_i that produced by the sliding block code, we have (invoking stationarity) that

$$n\rho(\mathcal{C},\mu) \leq E(\sum_{i=0}^{N+L-1} \rho(X_i, a^*))+$$

$$E(\sum_{i=N+L}^{K+N+L} \rho(X_i, Y_i)) + E(\sum_{i=K+N+L+1}^{K+2(L+N)} \rho(X_i, a^*))$$

$$\leq 2(N+L)\rho^* + K(\delta_{\mathrm{SBC}}(R,\mu)+\delta)$$

and hence

$$\delta(R,\mu) \leq \frac{2(N+L)}{2(N+L)+K}\rho^* + \frac{K}{2(N+L)+K}(\delta_{\mathrm{SBC}}(R,\mu)+\delta).$$

By choosing δ small enough and K large enough we can make make the right hand side arbitrarily close to $\delta_{\mathrm{SBC}}(R,\mu)$, which proves (11.7.1).

We now proceed to prove the converse inequality,

$$\delta(R,\mu) \geq \delta_{\mathrm{SBC}}(R,\mu), \tag{11.7.2}$$

which involves a bit more work.

Before carefully tackling the proof, we note the general idea and an "almost proof" that unfortunately does not quite work, but which may provide some insight. Suppose that we take a very good block code, e.g., a block code $\mathcal{C}$ of block length N such that

$$\rho(\mathcal{C},\mu) \leq \delta(R,\mu) + \delta$$

for a fixed $\delta > 0$. We now wish to form a sliding block code for the same channel with approximately the same performance. Since a sliding block code is just a stationary code (at least if we permit an infinite window length), the goal can be viewed as "stationarizing" the nonstationary block code. One approach would be the analogy of the SBM channel: Since a block code can be viewed as a deterministic block memoryless channel, we could make it stationary by inserting occasional random spacing between long sequences of blocks. Ideally this would then imply the existence of a sliding block code from the properties of SBM channels. The problem is that the SBM channel so constructed would no longer be a deterministic coding of the input since it would require the additional input of a random punctuation sequence. Nor could one use a random coding argument to claim that there must be a specific (nonrandom) punctuation sequence which could be used to construct a code since the deterministic encoder thus constructed would not be a stationary function of the input sequence, that is, it is only stationary if both the source and punctuation sequences are shifted together. Thus we are forced to obtain the punctuation sequence from the source input itself in order to get a stationary mapping. The original proofs that this could be done used a strong form of the Rohlin-Kakutani theorem of Section 9.5 given by Shields [131]. [56] [58]. The Rohlin-Kakutani theorem demonstrates the existence of a punctuation sequence with the property that the punctuation sequence is very nearly independent of the source. Lemma 9.5.2 is a slightly weaker result than the strong form considered by Shields.

The code construction described above can therefore be approximated by using a coding of the source instead of an independent process. Shields and Neuhoff [133] provided a simpler proof of a result equivalent to the Rohlin-Kakutani theorem and provided such a construction for finite alphabet sources. Davisson and Gray [27] provided an alternative heuristic development of a similar construction. We here adopt a somewhat different tack in order to avoid some of the problems arising in extending these approaches to general alphabet sources and to nonergodic sources. The principal difference is that we do not try to prove or use any approximate independence between source and the punctuation process derived from the source (which is code dependent in the case of continuous alphabets). Instead we take a good block code and first produce a much longer block code that is insensitive to shifts or starting positions using the same construction used to relate block coding performance of AMS processes and that of their stationary mean. This modified block code is then made into a sliding block code using a punctuation sequence derived from the source. Because the resulting block code is little affected by starting time, the only important property is that most of the time the block code is actually in use. Independence of the punctuation sequence and the source is no longer required. The approach is most similar to that of Davisson and Gray [27], but the actual construction differs in the details. An alternative construction may be found in Kieffer [79].

Given $\delta > 0$ and $\epsilon > 0$, choose for large enough N an asynchronous block code $\mathcal{C}$ of block length N such that

$$\frac{1}{N}\log ||\mathcal{C}|| \leq R - 2\epsilon$$

and

$$\rho(\mathcal{C},\mu) \leq \delta(R,\mu) + \delta. \tag{11.7.3}$$

The continuity of the block OPTA function and the theorem for asynchronous block source coding ensure that we can do this. Next we construct a longer block code that is more robust against shifts. For $i = 0, 1, \cdots, N-1$ construct the codes $\mathcal{C}_K(i)$ having length $K = JN$ as in the proof of Lemma 11.2.4. These codebooks look like $J-1$ repetitions of the codebook $\mathcal{C}$ starting from time i with the leftover symbols at the beginning and end being filled by the reference letter. We then form the union code $\mathcal{C}_K = \bigcup_i \mathcal{C}_K(i)$ as in the proof of Corollary 11.2.4 which has all the shifted versions. This code has rate no greater than $R - 2\epsilon + (JN)^{-1}\log N$. We assume that J is large enough to ensure that

$$\frac{1}{JN}\log N \leq \epsilon \tag{11.7.4}$$

so that the rate is no greater than $R - \epsilon$ and that

$$\frac{3}{J}\rho^* \leq \delta. \tag{11.7.5}$$

We now construct a sliding block encoder f and decoder g from the given block code. From Corollary 9.4.2 we can construct a finite length sliding block code of $\{X_n\}$ to produce a two-sided (NJ, γ)-random punctuation sequence $\{Z_n\}$. From the lemma $P(Z_0 = 2) \leq \gamma$ and hence by the continuity of integration (Corollary 4.4.2 of [50]) we can choose γ small enough to ensure that

$$\int_{x:Z_0(x)=2} \rho(X_0, a^*) \leq \delta. \tag{11.7.6}$$

Recall that the punctuation sequence usually produces 0's followed by $NJ-1$ 1's with occasional 2's interspersed to make things stationary. The sliding block encoder f begins with time 0 and scans backward NJ time units to find the first 0 in the punctuation sequence. If there is no such 0, then put out an arbitrary channel symbol b. If there is such a 0, then the block codebook $\mathcal{C}_K$ is applied to the input K-tuple x^K_{-n} to produce the minimum distortion codeword

$$u^K = \min_{y\in\mathcal{C}_K}{}^{-1}\rho_K(x^K_{-n}, y)$$

and the appropriate channel symbol, u_n, produced by the channel. The sliding block encoder thus has length at most $2NJ+1$.

The decoder sliding block code g scans left N symbols to see if it finds a codebook sync sequence (remember the codebook is asynchronous and begins with a unique prefix or sync sequence). If it does not find one, it produces a reference letter. (In this case it is not in the middle of a code word.) If it does find one starting in position $-n$, then it produces the corresponding length N codeword from $\mathcal{C}$ and then puts out the reproduction symbol in position n. Note that the decoder sliding block code has a finite window length of at most $2N+1$.

We now evaluate the average distortion resulting from use of this sliding block code. As a first step we mimic the proof of Lemma 10.6.3 up to the assumption of mutual independence of the source and the punctuation process (which is not the case here) to get that for a long source sequence of length n if the punctuation sequence is z, then

$$\rho_n(x^n, \hat{x}^n) = \sum_{i \in J_0^n(z)} \rho(x_i, a^*) + \sum_{i \in J_1^n(z)} \rho_{NJ}(x_i^{NJ}, \hat{x}_i^{NJ}),$$

where $J_0^n(z)$ is the collection of all i for which z_i is not in an NJ-cell (and hence filler is being sent) and $J_1^n(z)$ is the collection of all i for which z_i is 0 and hence begins an NJ-cell and hence an NJ length codeword. Each one of these length NJ codewords contains at most N reference letters at the beginning and N references letters at the end the end and in the middle it contains all shifts of sequences of length N codewords from $\mathcal{C}$. Thus for any $i \in J_1^n(z)$, we can write that

$$\rho_{NJ}(x_i^{NJ}, \hat{x}_i^{NJ}) \le \rho_N(x_i^N, a^{*N}) + \rho_N(x_{i+NJ-N}^N, a^{*N}) + \sum_{j=\lfloor \frac{i}{N} \rfloor}^{\lfloor \frac{i}{N} \rfloor + JN - 1} \rho_N(x_j^N, \mathcal{C}).$$

This yields the bound

$$\frac{1}{n}\rho_n(x^n, \hat{x}^n) \le \frac{1}{n} \sum_{i \in J_0^n(z)} \rho(x_i, a^*)$$

$$+\frac{1}{n} \sum_{i \in J_1^n(z)} \left(\rho_N(x_i^N, a^{*N}) + \rho_N(x_{i+NJ-N}^N, a^{*N})\right)$$

$$+\frac{1}{n} \sum_{j=0}^{\lfloor \frac{n}{N} \rfloor} \rho_N(x_{jN}^N, \mathcal{C}) = \frac{1}{n} \sum_{i=0}^{n-1} 1_2(z_i) \rho(x_i, a^*)$$

$$+\frac{1}{n} \sum_{i=0}^{n-1} 1_0(z_i) \left(\rho_N(x_i^N, a^{*N}) + \rho_N(x_{i+NJ-N}^N, a^{*N})\right) + \frac{1}{n} \sum_{j=0}^{\lfloor \frac{n}{N} \rfloor} \rho_N(x_{jN}^N, \mathcal{C}),$$

where $1_a(z_i)$ is 1 if $z_i = a$ and 0 otherwise. Taking expectations above we have that

$$E(\frac{1}{n}\rho_n(X^n, \hat{X}^n)) \le \frac{1}{n} \sum_{i=0}^{n-1} E(1_2(Z_i) \rho(X_i, a^*))$$

$$+\frac{1}{n} \sum_{i=0}^{n-1} (1_0(Z_i) \left(\rho_N(X_i^N, a^{*N}) + \rho_N(X_{i+NJ-N}^N, a^{*N})\right)) + \frac{1}{n} \sum_{j=0}^{\lfloor \frac{n}{N} \rfloor} \rho_N(X_{jN}^N, \mathcal{C}).$$

Invoke stationarity to write

$$E(\frac{1}{n}\rho_n(X^n,\hat{X}^n)) \leq E(1_2(Z_0)\rho(X_0,a^*))$$

$$+\frac{1}{NJ}E(1_0(Z_0)\rho_2N+1(X^{2N+1},a^{*(2N+1)}))+\frac{1}{N}\rho_N(X^N,\mathcal{C}).$$

The first term is bounded above by δ from (11.7.6). The middle term can be bounded above using (11.7.5) by

$$\frac{1}{JN}E(1_0(Z_0)\rho_{2N+1}(X^{2N+1},a^{*(2N+1)}) \leq \frac{1}{JN}E\rho_{2N+1}(X^{2N+1},a^{*(2N+1)})$$

$$=\frac{1}{JN}(2N+1)\rho^* \leq (\frac{2}{J}+1)\rho^* \leq \delta.$$

Thus we have from the above and (11.7.3) that

$$E\rho(X_0,Y_0) \leq \rho(\mathcal{C},\mu)+3\delta.$$

This proves the existence of a finite window sliding block encoder and a finite window length decoder with performance arbitrarily close to that achievable by block codes. □

The only use of ergodicity in the proof of the theorem was in the selection of the source sync sequence used to imbed the block code in a sliding block code. The result would extend immediately to nonergodic stationary sources (and hence to nonergodic AMS sources) if we could somehow find a single source sync sequence that would work for all ergodic components in the ergodic decomposition of the source. Note that the source synch sequence affects only the encoder and is irrelevant to the decoder which looks for asynchronous codewords prefixed by channel synch sequences (which consisted of a single channel letter repeated several times). Unfortunately, one cannot guarantee the existence of a single source sequence with small but nonzero probability under all of the ergodic components. Since the components are ergodic, however, an infinite length sliding block encoder could select such a source sequence in a simple (if impractical) way: Proceed as in the proof of the theorem up to the use of Corollary 9.4.2. Instead of using this result, we construct by brute force a punctuation sequence for the ergodic component in effect. Suppose that $\mathcal{G}=\{G_i;\ i=1,2,\cdots\}$ is a countable generating field for the input sequence space. Given δ, the infinite length sliding block encoder first finds the smallest value of i for which

$$0<\lim_{n\to\infty}\frac{1}{n}\sum_{k=0}^{n-1}1_{G_i}(T^kx),$$

and

$$\lim_{n\to\infty}\frac{1}{n}\sum_{k=0}^{n-1}1_{G_i}(T^k x)\rho(x_k,a^*)\leq\delta,$$

that is, we find a set with strictly positive relative frequency (and hence strictly positive probability with respect to the ergodic component in effect) which occurs rarely enough to ensure that the sample average distortion between the symbols produced when G_i occurs and the reference letter is smaller than δ. Given N and δ there must exist an i for which these relations hold (apply the proof of Lemma 9.4.4 to the ergodic component in effect with γ chosen to satisfy (11.7.6) for that component and then replace the arbitrary set G by a set in the generating field having very close probability). Analogous to the proof of Lemma 9.4.4 we construct a punctuation sequence $\{Z_n\}$ using the event G_i in place of G. The proof then follows in a like manner except that now from the dominated convergence theorem we have that

$$E(1_2(Z_0)\rho(X_0,a^*))=\lim_{n\to\infty}\frac{1}{n}\sum_{i=0}^{n-1}E(1_2(Z_i)\rho(X_i,a^*)$$

$$=E(\lim_{n\to\infty}\frac{1}{n}\sum_{i=0}^{n-1}1_2(Z_i)\rho(X_i,a^*))\leq\delta$$

by construction.

The above argument is patterned after that of Davisson and Gray [27] and extends the theorem to stationary nonergodic sources if infinite window sliding block encoders are allowed. We can then approximate this encoder by a finite-window encoder, but we must make additional assumptions to ensure that the resulting encoder yields a good approximation in the sense of overall distortion. Suppose that f is the infinite window length encoder and g is the finite window-length (say $2L+1$) encoder. Let $\mathcal{G}$ denote a countable generating field of rectangles for the input sequence space. Then from Corollary 4.2.2 applied to $\mathcal{G}$ given $\epsilon>0$ we can find for sufficiently large N a finite window sliding block code $r: A^{2N+1}\to B$ such that $\Pr(r\neq f')\leq\epsilon/(2L+1)$, that is, the two encoders produce the same channel symbol with high probability. The issue is when does this imply that $\rho(fg,\mu)$ and $\rho(rg,\mu)$ are therefore also close, which would complete the proof. Let $\bar{r}: A^{\mathcal{T}}\to B$ denote the infinite-window sliding block encoder induced by r, i.e., $\bar{r}(x)=r(x_{-N}^{2N+1})$. Then

$$\rho(fg,\mu)=E(\rho(X_0,\hat{X}_0))=\sum_{b\in B^{2L+1}}\int_{x\in V_f(b)}d\mu(x)\rho(x_0,g(b)),$$

where

$$V_f(b) = \{x : f(x)^{2L+1} = b\},$$

where $f(x)^{2L+1}$ is shorthand for $f(x_i)$, $i = -L, \cdots, L$, that is, the channel $(2L+1)$-tuple produced by the source using encoder x. We therefore have that

$$\rho(\bar{r}g, \mu) \leq \sum_{b \in B^{2L+1}} \int_{x \in V_f(b)} d\mu(x) \rho(x_0, g(b))$$

$$+ \sum_{b \in B^{2L+1}} \int_{x \in V_r(b) - V_f(b)} d\mu(x) \rho(x_0, g(b))$$

$$= \rho(f, \mu) + \sum_{b \in B^{2L+1}} \int_{x \in V_r(b) - V_f(b)} d\mu(x) \rho(x_0, g(b))$$

$$\leq \rho(f, \mu) + \sum_{b \in B^{2L+1}} \int_{x \in V_r(b) \Delta V_f(b)} d\mu(x) \rho(x_0, g(b)).$$

By making N large enough, however, we can make

$$\mu(V_{\bar{r}}(f) \Delta V_f(b))$$

arbitrarily small simultaneously for all $b \in \hat{A}^{2L} + 1$ and hence force all of the integrals above to be arbitrarily small by the continuity of integration. With Lemma 11.7.1 and Theorem 11.7.1 this completes the proof of the following theorem.

Theorem 11.7.2: Given an AMS source μ and an additive fidelity criterion with a reference letter,

$$\delta_{\text{SBC}}(R, \mu) = \delta(R, \mu),$$

that is, the class of sliding block codes is capable of exactly the same performance as the class of block codes.

The sliding block source coding theorem immediately yields an alternative coding theorem for a code structure known as *trellis encoding* source codes wherein the sliding block decoder is kept but the encoder is replaced by a tree or trellis search algorithm such as the Viterbi algorithm [41]. The details of inferring the trellis encoding source coding theorem from the sliding-block source coding theorem can be found in [52].

11.8 A Geometric Interpretation of OPTA's

We close this chapter on source coding theorems with a geometric interpretation of the OPTA functions in terms of the $\bar{\rho}$ distortion between sources. Suppose that μ is a stationary and ergodic source and that $\{\rho_n\}$ is an additive fidelity criterion with a fidelity criterion. Suppose that we have a nearly optimal sliding block encoder and decoder for μ and a channel with rate R, that is, if the overall process is $\{X_n, \hat{X}_n\}$ and

$$E\rho(X_0, \hat{X}_0) \leq \delta(R, \mu) + \delta.$$

If the overall hookup (source/encoder/channel/decoder) yields a distribution p on $\{X_n, \hat{X}_n\}$ and distribution η on the reproduction process $\{\hat{X}_n\}$, then clearly

$$\bar{\rho}(\mu, \eta) \leq \delta(R, \mu) + \delta.$$

Furthermore, since the channel alphabet is B the channel process must have entropy rate less than $R = \log ||B||$ and hence the reproduction process must also have entropy rate less than B from Corollary 4.2.5. Since δ is arbitrary,

$$\delta(R, \mu) \geq \inf_{\eta: \bar{H}(\eta) \leq R} \bar{\rho}(\mu, \eta).$$

Suppose next that p, μ and η are stationary and ergodic and that $\bar{H}(\eta) \leq R$. Choose a stationary p having μ and η as coordinate processes such that

$$E_p \rho(X_0, Y_0) \leq \bar{\rho}(\mu, \nu) + \delta.$$

We have easily that $\bar{I}(X; Y) \leq \bar{H}(\eta) \leq R$ and hence the left hand side is bounded below by the process distortion rate function $\bar{D}_s(R, \mu)$. From Theorem 10.6.1 and the block source coding theorem, however, this is just the OPTA function. We have therefore proved the following:

Theorem 11.8.1: Let μ be a stationary and ergodic source and let $\{\rho_n\}$ be an additive fidelity criterion with a reference letter. Then

$$\delta(R, \mu) = \inf_{\eta: \bar{H}(\eta) \leq R} \bar{\rho}(\mu, \eta),$$

that is, the OPTA function (and hence the distortion-rate function) of a stationary ergodic source is just the "distance" in the $\bar{\rho}$ sense to the nearest stationary and ergodic process with the specified reproduction alphabet and with entropy rate less than R.

This result originated in [55].

Chapter 12

Coding for noisy channels

12.1 Noisy Channels

In the treatment of source coding the communication channel was assumed to be noiseless. If the channel is noisy, then the coding strategy must be different. Now some form of error control is required to undo the damage caused by the channel. The overall communication problem is usually broken into two pieces: A source coder is designed for a noiseless channel with a given resolution or rate and an error correction code is designed for the actual noisy channel in order to make it appear almost noiseless. The combination of the two codes then provides the desired overall code or joint source and channel code. This division is natural in the sense that optimizing a code for a particular source may suggest quite different structure than optimizing it for a channel. The structures must be compatible at some point, however, so that they can be used together.

This division of source and channel coding is apparent in the subdivision of this chapter. We shall begin with a basic lemma due to Feinstein [38] which is at the basis of traditional proofs of coding theorems for channels. It does not consider a source at all, but finds for a given conditional distribution the maximum number of inputs which lead to outputs which can be distinguished with high probability. Feinstein's lemma can be thought of as a channel coding theorem for a channel which is used only once and which has no past or future. The lemma immediately provides a coding theorem for the special case of a channel which has no input memory or anticipation. The difficulties enter when the conditional distributions of output blocks given input blocks depend on previous or future inputs. This difficulty is handled by imposing some form of continuity on the channel with respect to its input, that is, by assuming that if the channel input is known for a big enough block, then the conditional probability of outputs during the same block is known nearly exactly regardless of previous or future inputs. The continuity condition which we shall consider is that of $\bar{d}$-continuous channels. Joint source and channel codes have been obtained for more general channels called *weakly continuous channels* (see, e.g., Kieffer [80] [81]), but these results require a variety of techniques not yet considered here and do not follow as a direct descendent of Feinstein's lemma.

Block codes are extended to sliding-block codes in a manner similar to that for source codes: First it is shown that asynchronous block codes can be synchronized and then that the block codes can be "stationarized" by the insertion of random punctuation. The approach to synchronizing channel codes is based on a technique of Dobrushin [33].

We consider stationary channels almost exclusively, thereby not including interesting nonstationary channels such as finite state channels with an arbitrary starting state. We will discuss such generalizations and we point out that they are straightforward for two-sided processes, but the general theory of AMS channels for one-sided processes is not in a satisfactory state. Lastly, we emphasize ergodic channels. In fact, for the sliding block codes the channels are also required to be totally ergodic, that is, ergodic with respect to all block shifts.

As previously discussed, we emphasize digital, i.e., discrete, channels. A few of the results, however, are as easily proved under somewhat more general conditions and hence we shall do so. For example, given the background of this book it is actually easier to write things in terms of measures and integrals than in terms of sums over probability mass functions. This additional generality will also permit at least a description of how the results extend to continuous alphabet channels.

12.2 Feinstein's Lemma

Let $(A, \mathcal{B}_A)$ and $(B, \mathcal{B}_B)$ be measurable spaces called the *input space* and the *output space*, respectively. Let P_X denote a probability distribution on $(A, \mathcal{B}_A)$ and let $\nu(F|x)$, $F \in \mathcal{B}_B$, $x \in B$ denote a regular conditional probability distribution on the output space. ν can be thought of as a "channel" with random variables as input and output instead of sequences. Define the hookup $P_X\nu = P_{XY}$ by

$$P_{XY}(F) = \int dP_X(x)\nu(F_x|x).$$

Let P_Y denote the induced output distribution and let $P_X \times P_Y$ denote the resulting product distribution. Assume that $P_{XY} << (P_X \times P_Y)$ and define the Radon-Nikodym derivative

$$f = \frac{dP_{XY}}{d(P_X \times P_Y)} \tag{12.2.1}$$

and the information density

$$i(x, y) = \ln f(x, y).$$

We use abbreviated notation for densities when the meanings should be clear from context, e.g., f instead of f_{XY}. Observe that for any set F

$$\int_F dP_X(x) \left(\int dP_Y(y) f(x,y) \right) = \int_{F \times B} d(P_X \times P_Y)(x,y) f(x,y)$$

$$= \int_{F \times B} dP_{XY}(x,y) = P_X(B) \le 1$$

and hence

$$\int dP_Y(y) f(x,y) \le 1; \ P_X - \text{a.e.} \tag{12.2.2}$$

Feinstein's lemma shows that we can pick M inputs $\{x_i \in A;\ i = 1, 2, \cdots, M\}$, and a corresponding collection of M disjoint output events $\{\Gamma_i \in \mathcal{B}_B;\ i = 1, 2, \cdots, M\}$, with the property that given an input x_i with high probability the output will be in Γ_i. We call the collection $\mathcal{C} = \{x_i, \Gamma_i;\ i = 1, 2, \cdots, M\}$ a code with codewords x_i and decoding regions Γ_i. We do not require that the Γ_i exhaust B.

The generalization of Feinstein's original proof for finite alphabets to general measurable spaces is due to Kadota [70] and the following proof is based on his.

Lemma 12.2 Feinstein's Lemma: Given an integer M and $a > 0$ there exist $x_i \in A$; $i = 1, \cdots, M$ and a measurable partition $\mathcal{F} = \{\Gamma_i;\ i = 1, \cdots, M\}$ of B such that

$$\nu(\Gamma_i^c | x_i) \le M e^{-a} + P_{XY}(i \le a).$$

Proof: Define $G = \{x, y : i(x, y) > a\}$ Set $\epsilon = Me^{-a} + P_{XY}(i \le a) = Me^{-a} + P_{XY}(G^c)$. The result is obvious if $\epsilon \ge 1$ and hence we assume that $\epsilon < 1$ and hence also that

$$P_{XY}(G^c) \le \epsilon < 1$$

and therefore that

$$P_{XY}(i > a) = P_{XY}(G) = \int dP_X(x)\nu(G_x|x) > 1 - \epsilon > 0.$$

This implies that the set $\tilde{A} = \{x : \nu(G_x|x) > 1 - \epsilon$ and (12.2.2) holds$\}$ must have positive measure under P_X We now construct a code consisting of input points x_i and output sets Γ_{x_i}. Choose an $x_1 \in \tilde{A}$ and define $\Gamma_{x_1} = G_{x_1}$. Next choose if possible a point $x_2 \in \tilde{A}$ for which $\nu(G_{x_2} - \Gamma_{x_1}|x_2) > 1 - \epsilon$. Continue in this way until either M points have been selected or all the points in $\tilde{A}$ have been exhausted. In particular, given the pairs $\{x_j, \Gamma_j\}$; $j = 1, 2, \cdots, i-1$, satisfying the condition, find an x_i for which

$$\nu(G_{x_i} - \bigcup_{j<i} \Gamma_{x_j}|x_i) > 1 - \epsilon. \tag{12.2.3}$$

If the procedure terminates before M points have been collected, denote the final point's index by n. Observe that

$$\nu(\Gamma_{x_i}{}^c|x_i) \le \nu(G_{x_i}{}^c|x_i) \le \epsilon;\ i = 1, 2, \cdots, n$$

and hence the lemma will be proved if we can show that necessarily n cannot be strictly less than M. We do this by assuming the contrary and finding a contradiction.

Suppose that the selection has terminated at $n < M$ and define the set $F = \bigcup_{i=1}^{n} \Gamma_{x_i} \in \mathcal{B}_B$. Consider the probability

$$P_{XY}(G) = P_{XY}(G\bigcap(A \times F)) + P_{XY}(G\bigcap(A \times F^c)). \tag{12.2.4}$$

The first term can be bounded above as

$$P_{XY}(G\bigcap(A \times F)) \le P_{XY}(A \times F) = P_Y(F)$$

$$= \sum_{i=1}^{n} P_Y(\Gamma_{x_i}).$$

We also have from the definitions and from (12.2.2) that

$$P_Y(\Gamma_{x_i}) = \int_{\Gamma_{x_i}} dP_Y(y) \le \int_{G_{x_i}} dP_Y(y)$$

$$\le \int_{G_{x_i}} \frac{f(x_i, y)}{e^a} dP_Y(y) \le e^{-a} \int dP_Y(y) f(x_i, y) \le e^{-a}$$

and hence

$$P_{XY}(G\bigcap(A\times F))\leq ne^{-a}. \tag{12.2.5}$$

Consider the second term of (12.2.3):

$$P_{XY}(G\bigcap(A\times F^c))=\int dP_X(x)\nu((G\bigcap(A\times F^c))_x|x)$$

$$=\int dP_X(x)\nu(G_x\bigcap F^c|x)=\int dP_X(x)\nu(G_x-\bigcup_{i=1}^{n}\Gamma_i|x). \tag{12.2.6}$$

We must have, however, that

$$\nu(G_x-\bigcup_{i=1}^{n}\Gamma_i|x)\leq 1-\epsilon$$

with P_X probability 1 or there would be a point x_{n+1} for which

$$\nu(G_{x_{n+1}}-\bigcup_{i=1}^{n+1}\Gamma_i|x_{n+1})>1-\epsilon,$$

that is, (12.2.3) would hold for $i=n+1$, contradicting the definition of n as the largest integer for which (12.2.3) holds. Applying this observation to (12.2.6) yields

$$P_{XY}(G\bigcap(A\times F^c))\leq 1-\epsilon$$

which with (12.2.4) and (12.2.5) implies that

$$P_{XY}(G)\leq ne^{-a}+1-\epsilon. \tag{12.2.7}$$

From the definition of ϵ, however, we have also that

$$P_{XY}(G)=1-P_{XY}(G^c)=1-\epsilon+Me^{-a}$$

which with (12.2.7) implies that $M\leq n$, completing the proof. □

12.3 Feinstein's Theorem

Given a channel $[A,\nu,B]$ an (M,n,ϵ) block channel code for ν is a collection $\{w_i,\Gamma_i\}$; $i=1,2,\cdots,M$, where $w_i\in A^n$, $\Gamma_i\in\mathcal{B}_B^n$, all i, with the property that

$$\sup_{x\in c(w_i)}\max_{i=1,\cdots,M}\nu_x^n(\Gamma_i)\leq\epsilon, \tag{12.3.1}$$

where $c(a^n) = \{x : x^n = a^n\}$ and where ν_x^n is the restriction of ν_x to $\mathcal{B}_B^n$. The rate of the code is defined as $n^{-1} \log M$. Thus an (n, M, ϵ) channel code is a collection of M input n-tuples and corresponding output cells such that regardless of the past or future inputs, if the input during time 1 to n is a channel codeword, then the output during time 1 to n is very likely to lie in the corresponding output cell. Channel codes will be useful in a communication system because they permit nearly error free communication of a select group of messages or codewords. A communication system can then be constructed for communicating a source over the channel reliably by mapping source blocks into channel codewords. If there are enough channel codewords to assign to all of the source blocks (at least the most probable ones), then that source can be reliably reproduced by the receiver. Hence a fundamental issue for such an application will be the number of messages M or, equivalently, the rate R of a channel code.

Feinstein's lemma can be applied fairly easily to obtain something that resembles a coding theorem for a noisy channel. Suppose that $[A, \nu, B]$ is a channel and $[A, \mu]$ is a source and that $[A \times B, p = \mu\nu]$ is the resulting hookup. Denote the resulting pair process by $\{X_n, Y_n\}$ For any integer K let p^K denote the restriction of p to $(A^K \times B^K, \mathcal{B}_A^K \times \mathcal{B}_B^K)$, that is, the distribution on input/output K-tuples (X^K, Y^K). The joint distribution p^K together with the input distribution μ^K induce a regular conditional probability $\hat{\nu}^K$ defined by $\hat{\nu}^K(F|x^K) = \Pr(Y^K \in F|X^K = x^K)$. In particular,

$$\hat{\nu}^K(G|a^K) = \Pr(Y^K \in G|X^K = a^K) = \frac{1}{\mu^K(a^K)} \int_{c(a^K)} \nu_x^K(G) d\mu(x). \tag{12.3.2}$$

where $c(a^K) = \{x : x^K = a^K\}$ is the rectangle of all sequences with a common K-dimensional output. We call $\hat{\nu}^K$ the *induced K-dimensional channel* of the channel ν and the source μ. It is important to note that the induced channel depends on the source as well as on the channel, a fact that will cause some difficulty in applying Feinstein's lemma. An exception to this case which proves to be an easy application is that of a channel without input memory and anticipation, in which case we have from the definitions that

$$\hat{\nu}^K(F|a^K) = \nu_x(Y^K \in F);\ x \in c(a^K),$$

Application of Feinstein's lemma to the induced channel yields the following result, which was proved by Feinstein for stationary finite alphabet channels and is known as Feinstein's theorem:

Lemma 12.3.1: Suppose that $[A \times B, \mu\nu]$ is an AMS and ergodic hookup of a source μ and channel ν. Let $\bar{I}_{\mu\nu} = \bar{I}_{\mu\nu}(X;Y)$ denote the average mutual information rate and assume that $\bar{I}_{\mu\nu} = I^*_{\mu\nu}$ is finite (as is the case if the alphabets are finite (Theorem 6.4.1) or have the finite-gap information property (Theorem 6.4.3)). Then for any $R < \bar{I}_{\mu\nu}$ and any $\epsilon > 0$ there exists for sufficiently large n a code $\{w_i^n; \Gamma_i;\ i = 1, 2, \cdots, M\}$, where $M = \lfloor e^{nR} \rfloor$, $w_i^n \in A^n$, and $\Gamma_i \in \mathcal{B}_B^n$, with the property that

$$\hat{\nu}^n(\Gamma_i^c | w_i^n) \leq \epsilon, i = 1, 2, \cdots, M. \tag{12.3.3}$$

Comment: We shall call a code $\{w_i, \Gamma_i;\ i = 1, 2, \cdots, M\}$ which satisfies (12.3.3) for a channel input process μ a (μ, M, n, ϵ)-Feinstein code. The quantity $n^{-1} \log M$ is called the *rate* of the Feinstein code.

Proof: Let η denote the output distribution induced by μ and ν. Define the information density

$$i_n = \frac{dp^n}{(d\mu^n \times \eta^n)}$$

and define

$$\delta = \frac{\bar{I}_{\mu\nu} - R}{2} > 0.$$

Apply Feinstein's lemma to the n-dimensional hookup $(\mu\nu)^n$ with $M = \lfloor e^{nR} \rfloor$ and $a = n(R + \delta)$ to obtain a code $\{w_i, \Gamma_i\}$; $i = 1, 2, \cdots, M$ with

$$\max_i \hat{\nu}^n(\Gamma_i^c | w_i^n) \leq M e^{-n(R+\delta)} + p^n(i_n \leq n(R + \delta))$$

$$= \lfloor e^{nR} \rfloor e^{-n(R+\delta)} + p(\frac{1}{n} i_n(X^n; Y^n) \leq R + \delta) \tag{12.3.4}$$

and hence

$$\max_i \hat{\nu}^n(\Gamma_i^c | w_i^n) \leq e^{-n\delta} + p(\frac{1}{n} i_n(X^n; Y^n) \leq \bar{I}_{\mu\nu} - \delta). \tag{12.3.5}$$

From Theorem 6.3.1 $n^{-1} i_n$ converges in L^1 to $\bar{I}_{\mu\nu}$ and hence it also converges in probability. Thus given ϵ we can choose an n large enough to ensure that the right hand side of (12.3.4) is smaller than ϵ, which completes the proof of the theorem. □

We said that the lemma "resembled" a coding theorem because a real coding theorem would prove the existence of an (M, n, ϵ) channel code, that is, it would concern the channel ν itself and not the induced channel $\hat{\nu}$, which depends on a channel input process distribution μ. The difference between a Feinstein code and a channel code is that the Feinstein code has a similar property for an induced channel which in general depends on a source distribution, while the channel code has this property independent of any source distribution and for any past or future inputs.

Feinstein codes will be used to construct block codes for noisy channels. The simplest such construction is presented next.

Corollary 12.3.1: Suppose that a channel $[A, \nu, B]$ is input memoryless and input nonanticipatory (see Section 9.4). Then a (μ, M, n, ϵ)-Feinstein code for some channel input process μ is also an (M, n, ϵ)-code.

Proof: Immediate since for a channel without input memory and anticipation we have that $\nu_x^n(F) = \nu_u^n(F)$ if $x^n = u^n$. □

The principal idea of constructing channel codes from Feinstein codes for more general channels will be to place assumptions on the channel which ensure that for sufficiently large n the channel distribution ν_x^n and the induced finite dimensional channel $\hat{\nu}^n(\cdot|x^n)$ are close. This general idea was proposed by McMillan [103] who suggested that coding theorems would follow for channels that were sufficiently continuous in a suitable sense.

The previous results did not require stationarity of the channel, but in a sense stationarity is implicit if the channel codes are to be used repeatedly (as they will be in a communication system). Thus the immediate applications of the Feinstein results. will be to stationary channels.

The following is a rephrasing of Feinstein's theorem that will be useful.

Corollary 12.3.2: Suppose that $[A \times B, \mu\nu]$ is an AMS and ergodic hookup of a source μ and channel ν. Let $\bar{I}_{\mu\nu} = \bar{I}_{\mu\nu}(X;Y)$ denote the average mutual information rate and assume that $\bar{I}_{\mu\nu} = I^*_{\mu\nu}$ is finite. Then for any $R < \bar{I}_{\mu\nu}$ and any $\epsilon > 0$ there exists an n_0 such that for all $n \geq n_0$ there are $(\mu, \lfloor e^{nR} \rfloor, n, \epsilon)$-Feinstein codes.

As a final result of the Feinstein variety, we point out a variation that applies to nonergodic channels.

Corollary 12.3.3: Suppose that $[A \times B, \mu\nu]$ is an AMS hookup of a source μ and channel ν. Suppose also that the information density converges a.e. to a limiting density

$$i_\infty = \lim_{n \to \infty} \frac{1}{n} i_n(X^n; Y^n).$$

(Conditions for this to hold are given in Theorem 8.5.1.) Then given $\epsilon > 0$ and $\delta > 0$ there exists for sufficiently large n a $[\mu, M, n, \epsilon + \mu\nu(i_\infty \leq R + \delta)]$ Feinstein code with $M = \lfloor e^{nR} \rfloor$.

Proof: Follows from the lemma and from Fatou's lemma which implies that

$$\limsup_{n \to \infty} p(\frac{1}{n} i_n(X^n; Y^n) \leq a) \leq p(i_\infty \leq a). \ \square$$

12.4 Channel Capacity

The form of the Feinstein lemma and its corollaries invites the question of how large R (and hence M) can be made while still getting a code of the desired form. From Feinstein's theorem it is seen that for an ergodic channel R can be any number less than $\bar{I}(\mu\nu)$ which suggests that if we define the quantity

$$C_{\text{AMS, e}} = \sup_{\text{AMS and ergodic } \mu} \bar{I}_{\mu\nu}, \tag{12.4.1}$$

then if $\bar{I}_{\mu\nu} = I^*_{\mu\nu}$ (e.g., the channel has finite alphabet), then we can construct for some μ a Feinstein code for μ with rate R arbitrarily near $C_{\text{AMS, e}}$. $C_{\text{AMS, e}}$ is an example of a quantity called an *information rate capacity* or, simply, *capacity* of a channel. We shall encounter a few variations on this definition just as there were various ways of defining distortion-rate functions for sources by considering either vectors or processes with different constraints. In this section a few of these definitions are introduced and compared.

A few possible definitions of information rate capacity are

$$C_{\text{AMS}} = \sup_{\text{AMS } \mu} \bar{I}_{\mu\nu}, \tag{12.4.2}$$

$$C_{\text{s}} = \sup_{\text{stationary } \mu} \bar{I}_{\mu\nu}, \tag{12.4.3}$$

$$C_{\text{s, e}} = \sup_{\text{stationary and ergodic } \mu} \bar{I}_{\mu\nu}, \tag{12.4.4}$$

$$C_{\text{ns}} = \sup_{n-\text{stationary } \mu} \bar{I}_{\mu\nu}, \tag{12.4.5}$$

$$C_{\text{bs}} = \sup_{\text{block stationary } \mu} \bar{I}_{\mu\nu} = \sup_{n} \sup_{n-\text{stationary } \mu} \bar{I}_{\mu\nu}. \tag{12.4.6}$$

Several inequalities are obvious from the definitions:

$$C_{\text{AMS}} \geq C_{\text{bs}} \geq C_{\text{ns}} \geq C_{\text{s}} \geq C_{\text{s, e}} \tag{12.4.7}$$

$$C_{\text{AMS}} \geq C_{\text{AMS,e}} \geq C_{\text{s, e}}. \tag{12.4.8}$$

In order to relate these definitions we need a variation on Lemma 12.3.1 described in the following lemma.

Lemma 12.4.1: Given a stationary finite-alphabet channel $[A, \nu, B]$, let μ be the distribution of a stationary channel input process and let $\{\mu_x\}$ be its ergodic decomposition. Then

$$\bar{I}_{\mu\nu} = \int d\mu(x) \bar{I}_{\mu_x\nu}. \tag{12.4.9}$$

Proof: We can write

$$\bar{I}_{\mu\nu} = h_1(\mu) - h_2(\mu)$$

where

$$h_1(\mu) = \bar{H}_\eta(Y) = \inf_n \frac{1}{n} H_\eta(Y^n)$$

is the entropy rate of the output, where η is the output measure induced by μ and ν, and where

$$h_2(\mu) = \bar{H}_{\mu\nu}(Y|X) = \lim_{n\to\infty} \frac{1}{n} H_{\mu\nu}(Y^n|X^n)$$

is the conditional entropy rate of the output given the input. If $\mu_k \to \mu$ on any finite dimensional rectangle, then also $\eta_k \to \eta$ and hence

$$H_{\eta_k}(Y^n) \to H_\eta(Y^n)$$

and hence it follows as in the proof of Corollary 2.4.1 that $h_1(\mu)$ is an upper semicontinuous function of μ. It is also affine because $\bar{H}_\eta(Y)$ is an affine function of η (Lemma 2.4.2) which is in turn a linear function of μ. Thus from Theorem 8.9.1 of [50]

$$h_1(\mu) = \int d\mu(x) h_1(\mu_x).$$

$h_2(\mu)$ is also affine in μ since $h_1(\mu)$ is affine in μ and $\bar{I}_{\mu\nu}$ is affine in μ (since it is affine in $\mu\nu$ from Lemma 6.2.2). Hence we will be done if we can show that $h_2(\mu)$ is upper semicontinuous in μ since then Theorem 8.9.1 of [50] will imply that

$$h_2(\mu) = \int d\mu(x) h_2(\mu_x)$$

which with the corresponding result for h_1 proves the lemma. To see this observe that if $\mu_k \to \mu$ on finite dimensional rectangles, then

$$H_{\mu_k\nu}(Y^n|X^n) \to H_{\mu\nu}(Y^n|X^n). \tag{12.4.10}$$

Next observe that for stationary processes

$$H(Y^n|X^n) \le H(Y^m|X^n) + H(Y_m^{n-m}|X^n)$$

$$\le H(Y^m|X^m) + H(Y_m^{n-m}|X_m^{n-m}) = H(Y^m|X^m) + H(Y^{n-m}|X^{n-m})$$

which as in Section 2.4 implies that $H(Y^n|X^n)$ is a subadditive sequence and hence

$$\lim_{n\to\infty} \frac{1}{n} H(Y^n|X^n) = \inf_n \frac{1}{n} H(Y^n|X^n).$$

Coupling this with (12.4.10) proves upper semicontinuity exactly as in the proof of Corollary 2.4.1, which completes the proof of the lemma. □

Lemma 12.4.2: If a channel ν has a finite alphabet and is stationary, then all of the above information rate capacities are equal.

Proof: From Theorem 6.4.1 $\bar{I} = I^*$ for finite alphabet processes and hence from Lemma 6.6.2 and Lemma 9.3.2 we have that if μ is AMS with stationary mean $\bar{\mu}$, then

$$\bar{I}_{\mu\nu} = \bar{I}_{\overline{\mu}\nu} = \bar{I}_{\bar{\mu}\nu}$$

and thus the supremum over AMS sources must be the same as that over stationary sources. The fact that $C_s \leq C_{s,\,e}$ follows immediately from the previous lemma since the best stationary source can do no better than to put all of its measure on the ergodic component yielding the maximum information rate. Combining these facts with (12.4.7)–(12.4.8) proves the lemma. □

Because of the equivalence of the various forms of information rate capacity for stationary channels, we shall use the symbol C to represent the information rate capacity of a stationary channel and observe that it can be considered as the solution to any of the above maximization problems.

Shannon's original definition of channel capacity applied to channels without input memory or anticipation. We pause to relate this definition to the process definitions. Suppose that a channel $[A, \nu, B]$ has no input memory or anticipation and hence for each n there are regular conditional probability measures $\hat{\nu}^n(G|x^n)$; $x \in A^n$, $G \in \mathcal{B}_B^n$, such that

$$\nu_x^n(G) = \hat{\nu}^n(G|x^n).$$

Define the finite-dimensional capacity of the $\hat{\nu}^n$ by

$$C_n(\hat{\nu}^n) = \sup_{\mu^n} I_{\mu^n\hat{\nu}^n}(X^n; Y^n),$$

where the supremum is over all vector distributions μ^n on A^n. Define the Shannon capacity of the channel μ by

$$C_{\text{Shannon}} = \lim_{n\to\infty} \frac{1}{n} C^n(\hat{\nu}^n)$$

if the limit exists. Suppose that the Shannon capacity exists for a channel ν without memory or anticipation. Choose N large enough so that C_N is very close to C_{Shannon} and let μ^N approximately yield C_N. Then construct a block memoryless source using μ^N. A block memoryless source is AMS and hence if the channel is AMS we must have an information rate

$$\bar{I}_{\mu\nu}(X; Y) = \lim_{n\to\infty} \frac{1}{n} I_{\mu\nu}(X^n; Y^n) = \lim_{k\to\infty} \frac{1}{kN} I_{\mu\nu}(X^{kN}; Y^{kN}).$$

Since the input process is block memoryless, we have from Lemma 9.4.2 that

$$I(X^{kN};Y^{kN}) \geq \sum_{i=0}^{k} I(X_{iN}^N;Y_{iN}^N).$$

If the channel is stationary then $\{X_n, Y_n\}$ is N-stationary and hence if

$$\frac{1}{N} I_{\mu^N \hat{\nu}^N}(X^N;Y^N) \geq C_{\text{Shannon}} - \epsilon,$$

then

$$\frac{1}{kN} I(X^{kN};Y^{kN}) \geq C_{\text{Shannon}} - \epsilon.$$

Taking the limit as $k \to \infty$ we have that

$$C_{\text{AMS}} = C \geq \bar{I}(X;Y) = \lim_{k\to\infty} \frac{1}{kN} I(X^{kN};Y^{kN}) \geq C_{\text{Shannon}} - \epsilon$$

and hence

$$C \geq C_{\text{Shannon}}.$$

Conversely, pick a stationary source μ which nearly yields $C = C_s$, that is,

$$\bar{I}_{\mu\nu} \geq C_s - \epsilon.$$

Choose n_0 sufficiently large to ensure that

$$\frac{1}{n} I_{\mu\nu}(X^n;Y^n) \geq \bar{I}_{\mu\nu} - \epsilon \geq C_s - 2\epsilon.$$

This implies, however, that for $n \geq n_0$

$$C_n \geq C_s - 2\epsilon,$$

and hence application of the previous lemma proves the following lemma.

Lemma 12.4.3: Given a finite alphabet stationary channel ν with no input memory or anticipation,

$$C = C_{\text{AMS}} = C_s = C_{s,\,e} = C_{\text{Shannon}}.$$

The Shannon capacity is of interest because it can be numerically computed while the process definitions are not always amenable to such computation.

With Corollary 12.3.2 and the definition of channel capacity we have the following result.

Lemma 12.4.4: If ν is an AMS and ergodic channel and $R < C$, then there is an n_0 sufficiently large to ensure that for all $n \geq n_0$ there exist $(\mu, \lfloor e^{nR} \rfloor, n, \epsilon)$ Feinstein codes for some channel input process μ.

Corollary 12.4.1: Suppose that $[A, \nu, B]$ is an AMS and ergodic channel with no input memory or anticipation. Then if $R < C$, the information rate capacity or Shannon capacity, then for $\epsilon > 0$ there exists for sufficiently large n a $(\lfloor e^{nR} \rfloor, n, \epsilon)$ channel code.

Proof: Follows immediately from Corollary 12.3.3 by choosing a stationary and ergodic source μ with $\bar{I}_{\mu\nu} \in (R, C)$. □

There is another, quite different, notion of channel capacity that we introduce for comparison and to aid the discussion of nonergodic stationary channels. Define for an AMS channel ν and any $\lambda \in (0, 1)$ the quantile

$$C^*(\lambda) = \sup_{\text{AMS } \mu} \sup\{r : \ \mu\nu(i_\infty \leq r) < \lambda)\},$$

where the supremum is over all AMS channel input processes and i_∞ is the limiting information density (which exists because $\mu\nu$ is AMS and has finite alphabet). Define the *information quantile capacity* C^* by

$$C^* = \lim_{\lambda \to 0} C^*(\lambda).$$

The limit is well defined since the $C^*(\lambda)$ are bounded and nonincreasing. The information quantile capacity was introduced by Winkelbauer [149] and its properties were developed by him and by Kieffer [75]. Fix an $R < C^*$ and define $\delta = (C^* - R)/2$. Given $\epsilon > 0$ we can find from the definition of C^* an AMS channel input process μ for which $\mu\nu(i_\infty \leq R + \delta) \leq \epsilon$. Applying Corollary 12.3.3 with this δ and $\epsilon/2$ then yields the following result for nonergodic channels.

Lemma 12.4.5: If ν is an AMS channel and $R < C^*$, then there is an n_0 sufficiently large to ensure that for all $n \geq n_0$ there exist $(\mu, fe^{nR}f, n, \epsilon)$ Feinstein codes for some channel input process μ.

We close this section by relating C and C^* for AMS channels.

Lemma 12.4.6: Given an AMS channel ν,

$$C \geq C^*.$$

Proof: Fix $\lambda > 0$. If $r < C^*(\lambda)$ there is a μ such that $\lambda > \mu\nu(i_\infty \leq r) = 1 - \mu\nu(i_\infty > r) \geq 1\bar{I}_{\mu\nu}/r$, where we have used the Markov inequality. Thus for all $r < C^*$ we have that $\bar{I}_{\mu\nu} \geq r(1 - \mu\nu(i_\infty \leq r))$ and hence

$$C \geq \bar{I}_{\mu\nu} \geq C^*(\lambda)(1 - \lambda) \underset{\lambda \to 0}{\to} C^*. \ \square$$

It can be shown that if a stationary channel is also ergodic, then $C = C^*$ by using the ergodic decomposition to show that the supremum defining $C(\lambda)$ can be taken over ergodic sources and then using the fact that for ergodic μ and ν, i_∞ equals $\bar{I}_{\mu\nu}$ with probability one. (See Kieffer [75].)

12.5 Robust Block Codes

Feinstein codes immediately yield channel codes when the channel has no input memory or anticipation because the induced vector channel is the same with respect to vectors as the original channel. When extending this technique to channels with memory and anticipation we will try to ensure that the induced channels are still reasonable approximations to the original channel, but the approximations will not be exact and hence the conditional distributions considered in the Feinstein construction will not be the same as the channel conditional distributions. In other words, the Feinstein construction guarantees a code that works well for a conditional distribution formed by averaging the channel over its past and future using a channel input distribution that approximately yields channel capacity. This does not in general imply that the code will also work well when used on the unaveraged channel with a particular past and future input sequence. We solve this problem by considering channels for which the two distributions are close if the block length is long enough.

In order to use the Feinstein construction for one distribution on an actual channel, we will modify the block codes slightly so as to make them *robust* in the sense that if they are used on channels with slightly different conditional distributions, their performance as measured by probability of error does not change much. In this section we prove that this can be done. The basic technique is due to Dobrushin [33] and a similar technique was studied by Ahlswede and Gács [4]. (See also Ahlswede and Wolfowitz [5].) The results of this section are due to Gray, Ornstein, and Dobrushin [59].

A channel block length n code $\{w_i, \Gamma_i;\ i = 1, 2, \cdots, M$ will be called *δ-robust* (in the Hamming distance sense) if the decoding sets Γ_i are such that the expanded sets

$$(\Gamma_i)_\delta \equiv \{y^n : \frac{1}{n} d_n(y^n, \Gamma_i) \leq \delta\}$$

are disjoint, where

$$d_n(y^n, \Gamma_i) = \min_{u^n \in \Gamma_i} d_n(y^n, u^n)$$

and

$$d_n(y^n, u^n) = \sum_{i=0}^{n-1} d_H(y_i, u_i)$$

and $d_H(a,b)$ is the Hamming distance (1 if $a \neq b$ and 0 if $a = b$). Thus the code is δ robust if received n-tuples in a decoding set can be changed by an average Hamming distance of up to δ without falling in a different decoding set. We show that by reducing the rate of a code slightly we can always make a Feinstein code robust.

Lemma 12.5.1: Let $\{w_i', \Gamma_i'; \ i = 1, 2, \cdots, M'\}$ be a $(\mu, e^{nR'}, n, \epsilon)$-Feinstein code for a channel ν. Given $\delta \in (0, 1/4)$ and

$$R < R' - h_2(2\delta) - 2\delta \log(||B|| - 1),$$

where as before $h_2(a)$ is the binary entropy function $-a \log a - (1-a) \log(1-a)$, there exists a δ-robust $(\mu, \lfloor e^{nR} \rfloor, n, \epsilon_n)$ Feinstein code for ν with

$$\epsilon_n \leq \epsilon + e^{-n(R' - R - h_2(2\delta) - 2\delta \log(||B|| - 1) - 3/n)}.$$

Proof: For $i = 1, 2, \cdots, M'$ let $r_i(y^n)$ denote the indicator function for $(\Gamma_i)_{2\delta}$. For a fixed y^n there can be at most

$$\sum_{i=0}^{2\delta n} \binom{n}{i} (||B|| - 1)^i = ||B||^n \sum_{i=0}^{2\delta n} \binom{n}{i} (1 - \frac{1}{||B||})^i \frac{(1}{||B||})^{n-i}$$

n-tuples $b^n \in B^n$ such that $n^{-1} d_n(y^n, b^n) \leq 2\delta$. Set $p = 1 - 1/||B||$ and apply Lemma 2.3.5 to the sum to obtain the bound

$$||B||^n \sum_{i=0}^{2\delta n} \binom{n}{k} (1 - \frac{1}{||B||})^i (\frac{1}{||B||})^{n-i} \leq ||B||^n e^{-n h_2(2\delta || p)}$$

$$= e^{-n h_2(2\delta || p) + n \log ||B||},$$

where

$$h_2(2\delta || p) = 2\delta \ln \frac{2\delta}{p} + (1 - 2\delta) \ln \frac{1 - 2\delta}{1 - p}$$

$$= -h_2(\delta) + 2\delta \ln \frac{||B||}{||B|| - 1} + (1 - 2\delta) \ln ||B|| = -h_2(\delta) + \ln ||B|| - 2\delta \ln(||B|| - 1).$$

Combining this bound with the fact that the Γ_i are disjoint we have that

$$\sum_{i=1}^{M'} r_i(y^n) \leq \sum_{i=0}^{2\delta n} \binom{n}{i} (||B|| - 1)^i \leq e^{-n(h_2(2\delta) + 2\delta \ln(||B|| - 1)}.$$

Set $M = \lfloor e^{nR} \rfloor$ and select $2M$ subscripts $k_1, \cdots, k_{2M}$ from $\{1, \cdots, M'\}$ by random equally likely independent selection without replacement so that each index pair (k_j, k_m); $j, m = 1, \cdots, 2M$; $j \neq m$, assumes any unequal pair with probability $(M'(M'-1))^{-1}$. We then have that

$$E\left(\frac{1}{2M}\sum_{j=1}^{2M}\sum_{m=1,m\neq j}^{2M} \hat{\nu}(\Gamma'_{k_j}\bigcap(\Gamma'_{k_m})_{2\delta}|w'_{k_j})\right)$$

$$= \frac{1}{2M}\sum_{j=1}^{2M}\sum_{m=1,m\neq j}^{2M}\sum_{k=1}^{M'}\sum_{i=1,i\neq k}^{M'}\frac{1}{M'(M'-1)}\sum_{y^n\in\Gamma'_k}\hat{\nu}(y^n|w'_k)r_i(y^n)$$

$$\leq \frac{1}{2M}\sum_{j=1}^{2M}\sum_{m=1,m\neq j}^{2M}\sum_{k=1}^{M'}\frac{1}{M'(M'-1)}\sum_{y^n\in\Gamma'_k}\hat{\nu}(y^n|w'_k)\sum_{i=1,i\neq k}^{M'} r_i(y^n)$$

$$\leq \frac{2M}{M-1}e^{n(h_2(2\delta)+2\delta\log(||B||-1)} \leq 4e^{-n(R'-R-h_2(2\delta)-2\delta\log(||B||-1)} \equiv \lambda_n,$$

where we have assumed that $M' \geq 2$ so that $M'-1 \geq M'/2$. Analogous to a random coding argument, since the above expectation is less than λ_n, there must exist a fixed collection of subscripts $i_1, \cdots, i_{2M'}$ such that

$$\frac{1}{2M}\sum_{j=1}^{2M}\sum_{m=1,m\neq j}^{2M} \hat{\nu}(\Gamma'_{i_j}\bigcap(\Gamma'_{i_m})_{2\delta}|w_{i_j}') \leq \lambda_n.$$

Since no more than half of the above indices can exceed twice the expected value, there must exist indices $k_1, \cdots, k_M \in \{j_1, \cdots, j_{2M}\}$ for which

$$\sum_{m=1,m\neq j}^{M} \hat{\nu}(\Gamma'_{k_j}\bigcap(\Gamma'_{k_m})_{2\delta}|w'_{k_j}) \leq 2\lambda_n;\ i = 1, 2, \cdots, M.$$

Define the code $\{w_i, \Gamma_i;\ i = 1, \cdots, M\}$ by $w_i = w'_{k_i}$ and

$$\Gamma_i = \Gamma'_{k_i} - \bigcup_{m=1,m\neq i}^{M'} (\Gamma'_{k_m})_{2\delta}.$$

The $(\Gamma_i)_\delta$ are obviously disjoint since we have removed from Γ'_{k_i} all words within 2δ of a word in any other decoding set. Furthermore, we have for all $i = 1, 2, \cdots, M$ that

$$1 - \epsilon \leq \hat{\nu}(\Gamma'_{k_i}|w'_{k_i})$$

$$= \hat{\nu}(\Gamma'_{k_i} \bigcap \left(\bigcup_{m \neq i} (\Gamma'_{k_m})_{2\delta} \right) |w'_{k_i}) + \hat{\nu}(\Gamma'_{k_i} \bigcap \left(\bigcup_{m \neq i} (\Gamma'_{k_m})_{2\delta} \right)^c |w'_{k_i})$$

$$\leq \sum_{m \neq i} \hat{\nu}(\Gamma'_{k_i} \bigcap (\Gamma'_{k_m})_{2\delta} |w'_{k_i}) + \hat{\nu}(\Gamma_i|w_i)$$

$$< 2\lambda_n + \hat{\nu}(\Gamma_i|w_i)$$

and hence

$$\hat{\nu}(\Gamma_i|w_i) \geq 1 - \epsilon - 8e^{-n(R'-R-h_2(2\delta)-2\delta\log(||B||-1))},$$

which proves the lemma. □

Corollary 12.5.1: Let ν be a stationary channel and let $\mathcal{C}_n$ be a sequence of $(\mu_n, \lfloor e^{nR'} \rfloor, n, \epsilon/2)$ Feinstein codes for $n \geq n_0$. Given an $R > 0$ and $\delta > 0$ such that $R < R' - h_2(2\delta) - 2\delta\log(||B|| - 1)$, there exists for n_1 sufficiently large a sequence $\mathcal{C}'_n$; $n \geq n_1$, of δ-robust $(\mu_n, \lfloor e^{nR} \rfloor, n, \epsilon)$ Feinstein codes.

Proof: The corollary follows from the lemma by choosing n_1 so that

$$e^{-n_1(R'-R-h_2(2\delta)-2\delta\ln(||B||-1)-3/n_1)} \leq \frac{\epsilon}{2}. \quad \square$$

Note that the sources may be different for each n and that n_1 does not depend on the channel input measure.

12.6 Block Coding Theorems for Noisy Channels

Suppose now that ν is a stationary finite alphabet $\bar{d}$-continuous channel. Suppose also that for $n \geq n_1$ we have a sequence of δ-robust $(\mu_n, \lfloor e^{nR} \rfloor, n, \epsilon)$ Feinstein codes $\{w_i, \Gamma_i\}$ as in the previous section. We now quantify the performance of these codes when used as channel block codes, that is, used on the actual channel ν instead of on an induced channel. As previously let $\hat{\nu}^n$ be the n-dimensional channel induced by μ_n and the channel ν, that is, for $\mu_n^n(a^n) > 0$

$$\hat{\nu}^n(G|a^n) = \Pr(Y^n \in G|X^n = a^n) = \frac{1}{\mu_n^n(a^n)} \int_{c(a^n)} \nu_x^n(G)\, d\mu(x), \quad (12.6.1)$$

where $c(a^n)$ is the rectangle $\{x : x \in A^T;\ x^n = a^n\}$, $a^n \in A^n$, and where $G \in \mathcal{B}_B^n$. We have for the Feinstein codes that

$$\max_i \hat{\nu}^n(\Gamma_i^c|w_i) \leq \epsilon.$$

We use the same codewords w_i for the channel code, but we now use the expanded regions $(\Gamma_i)_\delta$ for the decoding regions. Since the Feinstein codes were δ-robust, these sets are disjoint and the code well defined. Since the channel is $\bar{d}$-continuous we can choose an n large enough to ensure that if $x^n = \bar{x}^n$, then

$$\bar{d}_n(\nu_x^n, \nu_{\bar{x}}^n) \leq \delta^2.$$

Suppose that we have a Feinstein code such that for the induced channel

$$\hat{\nu}(\Gamma_i|w_i) \geq 1 - \epsilon.$$

Then if the conditions of Lemma 10.5.1 are met and μ_n is the channel input source of the Feinstein code, then

$$\hat{\nu}^n(\Gamma_i|w_i) = \frac{1}{\mu_n^n(w_i)} \int_{c(w_i)} \nu_x^n(\Gamma_i)\, d\mu(x)$$

$$\leq \sup_{x \in c(w_i)} \nu_x^n(\Gamma_i) \leq \inf_{x \in c(w_i)} \nu_x^n((\Gamma_i)_\delta) + \delta$$

and hence

$$\inf_{x \in c(w_i)} \nu_x^n((\Gamma_i)_\delta) \geq \hat{\nu}^n(\Gamma_i|w_i) - \delta \geq 1 - \epsilon - \delta.$$

Thus if the channel block code is constructed using the expanded decoding sets, we have that

$$\max_i \sup_{x \in c(w_i)} \nu_x((\Gamma_i)_\delta^c) \leq \epsilon + \delta;$$

that is, the code $\{w_i, (\Gamma_i)_\delta\}$ is a $(\lfloor e^{nR} \rfloor, n, \epsilon + \delta)$ channel code. We have now proved the following result.

Lemma 12.6.1: Let ν be a stationary $\bar{d}$-continuous channel and $\mathcal{C}_n$; $n \geq n_0$, a sequence of δ-robust $(\mu_n, \lfloor e^{nR} \rfloor, n, \epsilon)$ Feinstein codes. Then for n_1 sufficiently large and each $n \geq n_1$ there exists a $(\lfloor e^{nR} \rfloor, n, \epsilon + \delta)$ block channel code.

Combining the lemma with Lemma 12.4.4 and Lemma 12.4.5 yields the following theorem.

Theorem 12.6.1: Let ν be an AMS ergodic $\bar{d}$-continuous channel. If $R < C$ then given $\epsilon > 0$ there is an n_0 such that for all $n \geq n_0$ there exist $(\lfloor e^{nR} \rfloor, n, \epsilon)$ channel codes. If the channel is not ergodic, then the same holds true if C is replaced by C^*.

Up to this point the channel coding theorems have been "one shot" theorems in that they consider only a single use of the channel. In a communication system, however, a channel will be used repeatedly in order to communicate a sequence of outputs from a source.

12.7 Joint Source and Channel Block Codes

We can now combine a source block code and a channel block code of comparable rates to obtain a block code for communicating a source over a noisy channel. Suppose that we wish to communicate a source $\{X_n\}$ with a distribution μ over a stationary and ergodic $\bar{d}$-continuous channel $[B, \nu, \hat{B}]$. The channel coding theorem states that if K is chosen to be sufficiently large, then we can reliably communicate length K messages from a collection of $\lfloor e^{KR} \rfloor$ messages if $R < C$. Suppose that $R = C - \epsilon/2$. If we wish to send the given source across this channel, then instead of having a source coding rate of $(K/N)\log ||B||$ bits or nats per source symbol for a source (N, K) block code, we reduce the source coding rate to slightly less than the channel coding rate R, say $R_{\text{source}} = (K/N)(R - \epsilon/2) = (K/N)(C - \epsilon)$. We then construct a block source codebook $\mathcal{C}$ of this rate with performance near $\delta(R_{\text{source}}, \mu)$. Every codeword in the source codebook is assigned a channel codeword as index. The source is encoded by selecting the minimum distortion word in the codebook and then inserting the resulting channel codeword into the channel. The decoder then uses its decoding sets to decide which channel codeword was sent and then puts out the corresponding reproduction vector. Since the indices of the source code words are accurately decoded by the receiver with high probability, the reproduction vector should yield performance near that of $\delta((K/N)(C - \epsilon), \mu)$. Since ϵ is arbitrary and $\delta(R, \mu)$ is a continuous function of R, this implies that the OPTA for block coding μ for ν is given by $\delta((K/N)C, \mu)$, that is, by the OPTA for block coding a source evaluated at the channel capacity normalized to bits or nats per source symbol. Making this argument precise yields the block joint source and channel coding theorem.

A *joint source and channel* (N, K) *block code* consists of an encoder $\alpha : A^N \to B^K$ and decoder $\beta : \hat{B}^K \to \hat{A}^N$. It is assumed that N source time units correspond to K channel time units. The block code yields sequence coders $\bar{\alpha} : A^T \to B^T$ and $\bar{\beta} : \hat{B}^T \to \hat{A}^T$ defined by

$$\bar{\alpha}(x) = \{\alpha(x_{iN}^N);\ \text{all } i\}$$

$$\bar{\beta}(x) = \{\beta(x_{iN}^N);\ \text{all } i\}.$$

Let $\mathcal{E}$ denote the class of all such codes (all N and K consistent with the physical stationarity requirement). Let $\Delta^*(\mu, \nu, \mathcal{E})$ denote the block coding OPTA function and $D(R, \mu)$ the distortion-rate function of the source with respect to an additive fidelity criterion $\{\rho_n\}$. We assume also that ρ_n is bounded, that is, there is a finite value $\rho_{\max}$ such that

$$\frac{1}{n}\rho_n(x^n, \hat{x}^n) \leq \rho_{\max}$$

for all n. This assumption is an unfortunate restriction, but it yields a simple proof of the basic result.

Theorem 12.7: Let $\{X_n\}$ be a stationary source with distribution μ and let ν be a stationary and ergodic $\bar{d}$-continuous channel with channel capacity C. Let $\{\rho_n\}$ be a bounded additive fidelity criterion. Given $\epsilon > 0$ there exists for sufficiently large N and K (where K channel time units correspond to N source time units) an encoder $\alpha : A^N \to B^K$ and decoder $\beta : \hat{B}^K \to \hat{A}^N$ such that if $\bar{\alpha} : A^{\mathcal{T}} \to B^{\mathcal{T}}$ and $\bar{\beta} : \hat{B}^{\mathcal{T}} \to \hat{A}^{\mathcal{T}}$ are the induced sequence coders, then the resulting performance is bounded above as

$$\Delta(\mu, \bar{\alpha}, \nu, \bar{\beta}) = E\rho_N(X^N, \hat{X}^N) \leq \delta(\frac{K}{N}C, \mu) + \epsilon.$$

Proof: Given ϵ, choose $\gamma > 0$ so that

$$\delta(\frac{K}{N}(C - \gamma), \mu) \leq \delta(\frac{K}{N}C, \mu) + \frac{\epsilon}{3}$$

and choose N large enough to ensure the existence of a source codebook $\mathcal{C}$ of length N and rate $R_{\text{source}} = (K/N)(C - \gamma)$ with performance

$$\rho(\mathcal{C}, \mu) \leq \delta(R_{\text{source}}, \mu) + \frac{\epsilon}{3}.$$

We also assume that N and hence K is chosen large enough so that for a suitably small δ (to be specified later) there exists a channel $(\lfloor e^{KR} \rfloor, K, \delta)$ code, with $R = C - \gamma/2$. Index the $\lfloor e^{NR_{\text{source}}} \rfloor$ words in the source codebook by the $\lfloor e^{K(C-\gamma/2}\rfloor$ channel codewords. By construction there are more indices than source codewords so that this is possible. We now evaluate the performance of this code.

Suppose that there are M words in the source codebook and hence M of the channel words are used. Let $\hat{x}_i$ and w_i denote corresponding source and channel codewords, that is, if $\hat{x}_i$ is the minimum distortion word in the source codebook for an observed vector, then w_i is transmitted over the channel. Let Γ_i denote the corresponding decoding region. Then

$$E\rho_N(X^N, \hat{X}^N) = \sum_{i=1}^{M}\sum_{j=1}^{M} \int_{x:\alpha(x^N)=w_i} d\mu(x)\nu_x^K(\Gamma_j)\rho_N(x^N, \hat{x}_j)$$

$$= \sum_{i=1}^{M} \int_{x:\alpha(x^N)=w_i} d\mu(x)\nu_x^K(\Gamma_i)\rho_N(x^N, \hat{x}_i)$$

$$+\sum_{i=1}^{M}\sum_{j=1,j\neq i}^{M}\int_{x:\alpha(x^N)=w_i} d\mu(x)\nu_x^K(\Gamma_j)\rho_N(x^N,\hat{x}_j)$$

$$\leq \sum_{i=1}^{M}\int_{x:\alpha(x^N)=w_i} d\mu(x)\rho_N(x^N,\hat{x}_i)$$

$$+\sum_{i=1}^{M}\sum_{j=1,J\neq i}^{M}\int_{x:\alpha(x^N)=w_i} d\mu(x)\nu_x^K(\Gamma_j)\rho_N(x^N,\hat{x}_j)$$

The first term is bounded above by $\delta(R_{\text{source}},\mu)+\epsilon/3$ by construction. The second is bounded above by $\rho_{\max}$ times the channel error probability, which is less than δ by assumption. If δ is chosen so that $\rho_{\max}\delta$ is less than $\epsilon/2$, the theorem is proved. □

Theorem 12.7.2: Let $\{X_n\}$ be a stationary source source with distribution μ and let ν be a stationary channel with channel capacity C. Let $\{\rho_n\}$ be a bounded additive fidelity criterion. For any block stationary communication system (μ,f,ν,g), the average performance satisfies

$$\Delta(\mu,f,\nu,g)\leq\int_x d\bar{\mu}(x)D(C,\bar{\mu}_x),$$

where $\bar{\mu}$ is the stationary mean of μ and $\{\bar{\mu}_x\}$ is the ergodic decomposition of $\bar{\mu}$, C is the capacity of the channel, and $D(R,\mu)$ the distortion-rate function.

Proof: Suppose that the process $\{X_{nN}^N, U_{nK}^K, Y_{nK}^K, \hat{X}_{nN}^N\}$ is stationary and consider the overall mutual information rate $\bar{I}(X;\hat{X})$. From the data processing theorem (Lemma 9.4.8)

$$\bar{I}(X;\hat{X})\leq\frac{K}{N}\bar{I}(U;Y)\leq\frac{K}{N}C.$$

Choose L sufficiently large so that

$$\frac{1}{n}I(X^n;\hat{X}^n)\leq\frac{K}{N}C+\epsilon$$

and

$$D_n(\frac{K}{N}C+\epsilon,\mu)\geq D(\frac{K}{N}C+\epsilon,\mu)-\delta$$

for $n\geq L$. Then if the ergodic component μ_x is in effect, the performance can be no better than

$$E_{\mu_x}\rho_N(X^n,\hat{X}^N)\geq\inf_{p^N\in\mathcal{R}_N(\frac{K}{N}C+\epsilon,\mu_x^N)}\rho_N(X^N,\hat{X}^N)\geq D_N(\frac{K}{N}C+\epsilon,\mu_x)$$

which when integrated yields a lower bound of

$$\int d\mu(x)D(\frac{K}{N}C+\epsilon,\mu_x)-\delta.$$

Since δ and ϵ are arbitrary, the lemma follows from the continuity of the distortion rate function. □

Combining the previous results yields the block coding OPTA for stationary sources and stationary and ergodic $\bar{d}$-continuous channels.

Corollary 12.7.1: Let $\{X_n\}$ be a stationary source with distribution μ and let ν be a stationary and ergodic $\bar{d}$-continuous process with channel capacity C. Let $\{\rho_n\}$ be a bounded additive fidelity criterion. The block coding OPTA function is given by

$$\Delta^*(\mu,\nu,\mathcal{E},\mathcal{D})=\int d\bar{\mu}(x)D(C,\bar{\mu}_x).$$

12.8 Synchronizing Block Channel Codes

As in the source coding case, the first step towards proving a sliding block coding theorem is to show that a block code can be synchronized, that is, that the decoder can determine (at least with high probability) where the block code words begin and end. Unlike the source coding case, this cannot be accomplished by the use of a simple synchronization sequence which is prohibited from appearing within a block code word since channel errors can cause the appearance of the sync word at the receiver by accident. The basic idea still holds, however, if the codes are designed so that it is very unlikely that a non-sync word can be converted into a valid sync word. If the channel is $\bar{d}$-continuous, then good robust Feinstein codes as in Corollary 12.5.1 can be used to obtain good codebooks . The basic result of this section is Lemma 12.8.1 which states that given a sequence of good robust Feinstein codes, the code length can be chosen large enough to ensure that there is a sync word for a slightly modified codebook; that is, the synch word has length a specified fraction of the codeword length and the sync decoding words never appear as a segment of codeword decoding words. The technique is due to Dobrushin [33] and is an application of Shannon's random coding technique. The lemma originated in [59].

The basic idea of the lemma is this: In addition to a good long code, one selects a short good robust Feinstein code (from which the sync word will be chosen) and then performs the following experiment. A word from the short code and a word from the long code are selected independently and at random. The probability that the short decoding word appears in the long decoding word is shown to be small. Since this average is small, there must be at least one short word such that the probability of its decoding word appearing in the decoding word of a randomly selected long code word is small. This in turn implies that if all long decoding words containing the short decoding word are removed from the long code decoding sets, the decoding sets of most of the original long code words will not be changed by much. In fact, one must remove a bit more from the long word decoding sets in order to ensure the desired properties are preserved when passing from a Feinstein code to a channel codebook.

Lemma 12.8.1: Assume that $\epsilon \leq 1/4$ and $\{\mathcal{C}_n; n \geq n_0\}$ is a sequence of ϵ-robust $\{\tau, M(n), n, \epsilon/2\}$ Feinstein codes for a $\bar{d}$-continuous channel ν having capacity $C > 0$. Assume also that $h(2\epsilon) + 2\epsilon \log(||B|| - 1) < C$, where B is the channel output alphabet. Let $\delta \in (0, 1/4)$. Then there exists an n_1 such that for all $n \geq n_1$ the following statements are true.

(A) If $\mathcal{C}_n = \{v_i, \Gamma_i; i = 1, \cdots, M(n)\}$, then there is a modified codebook $\mathcal{W}_n = \{w_i; W_i; i = 1, \cdots, K(n)\}$ and a set of $K(n)$ indices $\mathcal{K}_n = \{k_1, \cdots, k_{K(n)} \subset \{1, \cdots, M(n)\}$ such that $w_i = v_{k_i}$, $W_i \subset (\Gamma_i)_{\epsilon^2}$; $i = 1, \cdots, K(n)$, and

$$\max_{1 \leq j \leq K(n)} \sup_{x \in c(w_j)} \nu_x^n(W_j^c) \leq \epsilon. \tag{12.8.1}$$

(B) There is a sync word $\sigma \in A^r$, $r = r(n) = \lceil \delta n \rceil$ = smallest integer larger than δn, and a sync decoding set $S \in \mathcal{B}_B^r$ such that

$$\sup_{x \in c(\sigma)} \nu_x^r(S^c) \leq \epsilon. \tag{12.8.2}$$

and such that no r-tuple in S appears in any n-tuple in W_i; that is, if $G(b^r) = \{y^n : y_i^r = b^r \text{ some } i = 0, \cdots, n-r\}$ and $G(S) = \bigcup_{b^r \in S} G(b^r)$, then

$$G(S) \bigcap W_i = \emptyset, i = 1, \cdots, K(n). \tag{12.8.3}$$

(C) We have that

$$||\{k : k \notin \mathcal{K}_n\}|| \leq \epsilon \delta M(n). \tag{12.8.4}$$

The modified code $\mathcal{W}_n$ has fewer words than the original code $\mathcal{C}_n$, but (12.8.4) ensures that $\mathcal{W}_n$ cannot be much smaller since

$$K(n) \geq (1 - \epsilon\delta) M(n). \tag{12.8.5}$$

Given a codebook $\mathcal{W}_n = \{w_i, W_i; i = 1, \cdots, K(n)\}$, a sync word $\sigma \in A^r$, and a sync decoding set S, we call the length $n+r$ codebook $\{\sigma \times w_i, S \times W_i; i = 1, \cdots, K(n)\}$ a *prefixed* or *punctuated* codebook.

Proof: Since ν is $\bar{d}$-continuous, n_2 can be chosen so large that for $n \geq n_2$

$$\max_{a^n \in A^n} \sup_{x,x' \in c(a^n)} \bar{d}_n(\nu_x^n, \nu_{x'}^n) \leq (\frac{\delta\epsilon}{2})^2. \tag{12.8.6}$$

From Corollary 12.5.1 there is an n_3 so large that for each $r \geq n_3$ there exists an $\epsilon/2$-robust $(\tau, J, r, \epsilon/2)$-Feinstein code $\mathcal{C}_s = \{s_j, S_j : j = 1, \cdots, J\}$; $J \geq 2^{rR_s}$, where $R_s \in (0, C - h(2\epsilon) - 2\epsilon\log(||B|| - 1))$. Assume that n_1 is large enough to ensure that $\delta n_1 \geq n_2$; $\delta n_1 \geq n_3$, and $n_1 \geq n_0$. Let 1_F denote the indicator function of the set F and define λ_n by

$$\lambda_n = J^{-1}\sum_{j=1}^{J}\frac{1}{M(n)}\sum_{i=1}^{M(n)}\hat{\nu}^n(G((S_j)_\epsilon)\bigcap\Gamma_i|v_i)$$

$$= J^{-1}\sum_{j=1}^{J}\frac{1}{M(n)}\sum_{i=1}^{M(n)}\sum_{b' \in (S_j)_\epsilon}\sum_{y^n \in \Gamma_i}\hat{\nu}^n(y^n|v_i)1_{G(b')}(y^n)$$

$$= J^{-1}\frac{1}{M(n)}\sum_{i=1}^{M(n)}\sum_{y^n \in \Gamma_i}\hat{\nu}^n(y^n|v_i)\left[\sum_{j=1}^{J}\sum_{b' \in (S_j)_\epsilon}1_{G(b')}(y^n)\right]. \tag{12.8.7}$$

Since the $(S_j)_\epsilon$ are disjoint and a fixed y^n can belong to at most $n - r \leq n$ sets $G(b^r)$, the bracket term above is bound above by n and hence

$$\lambda_n \leq \frac{n}{J}\frac{1}{M(n)}\sum_{i=1}^{M(n)}\hat{\nu}^n(y^n|v_i) \leq \frac{n}{J} \leq n2^{-rR_s} \leq n2^{-\delta n R_s} \underset{n\to\infty}{\to} 0$$

so that choosing n_1 also so that $n_1 2^{-\delta n R_s} \leq (\delta\epsilon)^2$h we have that $\lambda_n \leq (\delta\epsilon)^2$ if $n \geq n_1$. From (12.8.7) this implies that for $n \geq n_1$ there must exist at least one j for which

$$\sum_{i=1}^{M(n)}\hat{\nu}^n(G((S_j)_\epsilon)\bigcap\Gamma_i|v_i) \leq (\delta\epsilon)^2$$

which in turn implies that for $n \geq n_1$ there must exist a set of indices $\mathcal{K}_n \subset \{1, \cdots, M(n)\}$ such that

$$\hat{\nu}^n(G((S_j)_\epsilon)\bigcap\Gamma_i|v_i) \leq \delta\epsilon, i \in \mathcal{K}_n,$$

$$||\{i : i \notin \mathcal{K}_n\}|| \leq \delta\epsilon.$$

Define $\sigma = s_j$; $S = (S_j)_{\epsilon/2}$, $w_i = v_{k_i}$, and $W_i = (\Gamma_{k_i} \bigcap G((S_j)_\epsilon)^c)_{\epsilon\delta}$; $i = 1, \cdots, K(n)$. We then have from Lemma 12.6.1 and (12.8.6) that if $x \in c(\sigma)$, then since $\epsilon\delta \leq \epsilon/2$

$$\nu_x^r(S) = \nu_x^r((S_j)_{\epsilon/2}) \geq \hat{\nu}^r(S_j|\sigma) - \frac{\epsilon}{2} \geq 1 - \epsilon,$$

proving (12.8.2). Next observe that if $y^n \in (G((S_j)_\epsilon)^c)_{\epsilon\delta}$, then there is a $b^n \in G((S_j)_\epsilon)^c$ such that $d_n(y^n, b^n) \leq \epsilon\delta$ and thus for $i = 0, 1, \cdots, n - r$ we have that

$$d_r(y_i^r, b_i^r) \leq \frac{n}{r}\frac{\epsilon\delta}{2} \leq \frac{\epsilon}{2}.$$

Since $b^n \in G((S_j)_\epsilon)^c$, it has no r-tuple within ϵ of an r-tuple in S_j and hence the r-tuples y_i^r are at least $\epsilon/2$ distant from S_j and hence $y^n \in H((S)_{\epsilon/2})^c)$. We have therefore that $(G((S_j)_\epsilon)^c)_{\epsilon\delta} \subset G((S_j)_\epsilon)^c$ and hence

$$G(S) \bigcap W_i = G((S_j)_\epsilon) \bigcap (\Gamma_{k_i} \bigcap G((S_j)_\epsilon)^c)_{\delta\epsilon}$$

$$\subset G((S_j)_{\epsilon/2}) \bigcap (G((S_j)_\epsilon)^c)_{\delta\epsilon} = \emptyset,$$

completing the proof. □

Combining the preceding lemma with the existence of robust Feinstein codes at rates less than capacity (Lemma 12.6.1) we have proved the following synchronized block coding theorem.

Corollary 12.8.1: Le ν be a stationary ergodic $\bar{d}$-continuous channel and fix $\epsilon > 0$ and $R \in (0, C)$. Then there exists for sufficiently large blocklength N, a length N codebook $\{\sigma \times w_i, S \times W_i;\ i = 1, \cdots, M\}$, $M \geq 2^{NR}$, $\sigma \in A^r$, $w_i \in A^n$, $r + n = N$, such that

$$\sup_{x \in c(\sigma)} \nu_x^r(S^c) \leq \epsilon,$$

$$\max_{i \leq j \leq M} \nu_x^n(W_j^c) \leq \epsilon,$$

$$W_j \bigcap G(S) = \emptyset.$$

Proof: Choose $\delta \in (0, \epsilon/2)$ so small that $C - h(2\delta) - 2\delta\log(||B|| - 1) > (1 + \delta)R(1 - \log(1 - \delta^2))$ and choose $R' \in ((1 + \delta)R(1 - \log(1 - \delta^2)), C - h(2\delta) - 2\delta\log(||B|| - 1)$. From Lemma 12.6.1 there exists an n_0 such that for $n \geq n_0$ there exist δ-robust (τ, μ, n, δ) Feinstein codes with $M(n) \geq 2^{nR'}$. From Lemma 12.8.1 there exists a codebook $\{w_i, W_i;\ i = 1, \cdots, K(n)\}$, a sync word $\sigma \in A^r$, and a sync decoding set $S \in \mathcal{B}_B^r$, $r = \lceil\delta n\rceil$ such that

$$\max_j \sup_{x \in c(w_j)} \nu_x^n(W_j^c) \leq 2\delta \leq \epsilon,$$

$$\sup_{x\in c(\sigma)} \nu_x^r(S) \leq 2\delta \leq \epsilon,$$

$G(S)\bigcap W_j = \emptyset$; $j = 1, \cdots, K(n)$, and from (12.8.5)

$$M = K(n) \geq (1-\delta^2)M(n).$$

Therefore for $N = n + r$

$$N^{-1}\log M \geq (n\lceil n\delta \rceil)^{-1}\log((1-\delta^2)2^{nR'})$$

$$= \frac{nR' + \log(1-\delta^2)}{n + n\delta} = \frac{R' + n^{-1}\log(1-\delta^2)}{1+\delta}$$

$$\geq \frac{R' + \log(1-\delta^2)}{1+\delta} \geq R,$$

completing the proof. □

12.9 Sliding Block Source and Channel Coding

Analogous to the conversion of block source codes into sliding block source codes, the basic idea of constructing a sliding block channel code is to use a punctuation sequence to stationarize a block code and to use sync words to locate the blocks in the decoded sequence. The sync word can be used to mark the beginning of a codeword and it will rarely be falsely detected during a codeword. Unfortunately, however, an r-tuple consisting of a segment of a sync and a segment of a codeword may be erroneously detected as a sync with nonnegligible probability. To resolve this confusion we look at the relative frequency of sync-detects over a sequence of blocks instead of simply trying to find a single sync. The idea is that if we look at enough blocks, the relative frequency of the sync-detects in each position should be nearly the probability of occurrence in that position and these quantities taken together give a pattern that can be used to determine the true sync location. For the ergodic theorem to apply, however, we require that blocks be ergodic and hence we first consider totally ergodic sources and channels and then generalize where possible.

Totally Ergodic Sources

Lemma 12.9.1: Let ν be a totally ergodic stationary $\bar{d}$-continuous channel. Fix $\epsilon, \delta > 0$ and assume that $\mathcal{C}_N = \{\sigma \times w_i; S \times W_i;\ i = 1, \cdots, K\}$ is a prefixed codebook satisfying (12.8.1)–(12.8.3). Let $\gamma_n : G^N \to \mathcal{C}_N$ assign an N-tuple in the prefixed codebook to each N-tuple in G^N and let $[G, \mu, U]$ be an N-stationary, N-ergodic source. Let $c(a^n)$ denote the cylinder set or rectangle of all sequences $u = (\cdots, u_{-1}, u_0, u_1, \cdots)$ for which $u^n = a^n$. There exists for sufficiently large L (which depends on the source) a sync locating function $s : B^{LN} \to \{0, 1, \cdots, N-1\}$ and a set $\Phi \in \mathcal{B}_G^m$, $m = (L+1)N$, such that if $u^m \in \Phi$ and $\gamma_N(U_{LN}^N) = \sigma \times w_i$, then

$$\inf_{x \in c(\gamma_m(u^m))} \nu_x(y : s(y^{LN}) = \theta, \theta = 0, \cdots, N-1; y_{LN} \in S \times W_i) \geq 1 - 3\epsilon. \tag{12.9.1}$$

Comments: The lemma can be interpreted as follows. The source is block encoded using γ_N. The decoder observes a possible sync word and then looks "back" in time at previous channel outputs and calculates $s(y^{LN})$ to obtain the exact sync location, which is correct with high probability. The sync locator function is constructed roughly as follows: Since μ and ν are N-stationary and N-ergodic, if $\bar{\gamma} : A^\infty \to B^\infty$ is the sequence encoder induced by the length N block code γ_N, then the encoded source $\mu\bar{\gamma}^{-1}$ and the induced channel output process η are all N-stationary and N-ergodic. The sequence $z_j = \eta(T^j c(S)))$; $j = \cdots, -1, 0, 1, \cdots$ is therefore periodic with period N. Furthermore, z_j can have no smaller period than N since from (12.8.1)–(12.8.3) $\eta(T^j c(S)) \leq \epsilon$, $j = r+1, \cdots, n-r$ and $\eta(c(S)) \geq 1-\epsilon$. Thus defining the sync pattern $\{z_j;\ j = 0, 1, \cdots, N-1\}$, the pattern is distinct from any cyclic shift of itself of the form $\{z_k, \cdots, z_{N-1}, z_0, \cdots, x_{k-1}\}$, where $k \leq N-1$. The sync locator computes the relative frequencies of the occurrence of S at intervals of length N for each of N possible starting points to obtain, say, a vector $\hat{z}^N = (\hat{z}_0, \hat{z}_1, \cdots, \hat{z}_{N-1})$. The ergodic theorem implies that the $\hat{z}_i$ will be near their expectation and hence with high probability $(\hat{z}_0, \cdots, \hat{z}_{N-1}) = (z_\theta, z_{\theta+1}, \cdots, z_{N-1}, z_0, \cdots, z_{\theta-1})$, determining θ. Another way of looking at the result is to observe that the sources ηT^j; $j = 0, \cdots, N-1$ are each N-ergodic and N-stationary and hence any two are either identical or orthogonal in the sense that they place all of their measure on disjoint N-invariant sets. (See, e.g., Exercise 1, Chapter 6 of [50].) No two can be identical, however, since if $\eta T^i = \eta T^j$ for $i \neq j$; $0 \leq i, j \leq N-1$, then η would be periodic with period $|i-j|$ strictly less than N, yielding a contradiction. Since membership in any set can be determined with high probability by observing the sequence for a long enough time, the sync locator attempts to determine which of the N distinct sources ηT^j is being observed. In fact, synchronizing the output is exactly equivalent to forcing the N sources ηT^j; $j = 0, 1, \cdots, N-1$ to be distinct N-ergodic sources. After this is accomplished, the remainder of the proof is devoted to using the properties of $\bar{d}$-continuous channels to show that synchronization of the output source when driven by μ implies that with high probability the channel output can be synchronized for all fixed input sequences in a set of high μ probability.

The lemma is stronger (and more general) than the similar results of Nedoma [107] and Vajda [141], but the extra structure is required for application to sliding block decoding.

Proof: Choose $\zeta > 0$ so that $\zeta < \epsilon/2$ and

$$\zeta < \frac{1}{8} \min_{i,j: z_i \neq z_j} |z_i - z_j|. \tag{12.9.2}$$

For $\alpha > 0$ and $\theta = 0, 1, \cdots, N-1$ define the sets $\psi(\theta, \alpha) \in \mathcal{B}_B^{LN}$ and $\tilde{\psi}(\theta, \alpha) \in \mathcal{B}_B^m$, $m = (L+1)N$ by

$$\psi(\theta, \alpha) = \{y^{LN} : |\frac{1}{L-1} \sum_{i=0}^{L-2} 1_S(y_{j+iN}^r) - z_{\theta+j}| \leq \alpha;\ j = 0, 1, \cdots, N-1\}$$

$$\tilde{\psi}(\theta, \alpha) = B^\theta \times \psi(\theta, \alpha) \times B^{N-\theta}.$$

From the ergodic theorem L can be chosen large enough so that

$$\eta(\bigcap_{\theta=0}^{N-1} T^{-\theta} c(\psi(\theta, \zeta))) = \eta^m(\bigcap_{\theta=0}^{N-1} \tilde{\psi}(\theta, \zeta)) \geq 1 - \zeta^2. \tag{12.9.3}$$

Assume also that L is large enough so that if $x_i = x_i'$, $i = 0, \cdots, m-1$ then

$$\bar{d}_m(\nu_x^m, \nu_{x'}^m) \leq (\frac{\zeta}{N})^2. \tag{12.9.4}$$

From (12.9.3)

$$\zeta^2 \geq \eta^m((\bigcap_{\theta=0}^{N-1} \tilde{\psi}(\theta, \zeta))^c) = \sum_{a^m \in G^m} \int_{c(a^m)} d\mu(u) \nu_{\bar{\gamma}(u)}^m((\bigcap_{\theta=0}^{N-1} \tilde{\psi}(\theta, \zeta)^c))$$

$$= \sum_{a^m \in G^m} \mu^m(a^m) \hat{\nu}((\bigcap_{\theta=0}^{N-1} \tilde{\psi}(\theta, \zeta))^c | \gamma_m(a^m))$$

and hence there must be a set $\Phi \in \mathcal{B}_B^m$ such that

$$\hat{\nu}^m((\bigcap_{\theta=0}^{N-1} \tilde{\psi}(\theta, \zeta))^c | \gamma_m(a^m)) \leq \zeta, a^m \in \Phi, \tag{12.9.5}$$

$$\mu^m(\Phi) \leq \zeta. \tag{12.9.6}$$

Define the sync locating function $s : B^{LN} \to \{0, 1, \cdots, N-1\}$ as follows: Define the set $\psi(\theta) = \{y^{LN} \in (\psi(\theta, \zeta))_{2\zeta/N}\}$ and then define

$$s(y^{LN}) = \begin{cases} \theta & y^{LN} \in \psi(\theta) \\ 1 & \text{otherwise} \end{cases}$$

We show that s is well defined by showing that $\psi(\theta) \subset \psi(\theta, 4\zeta)$, which sets are disjoint for $\theta = 0, 1, \cdots, N-1$ from (12.9.2). If $y^{LN} \in \psi(\theta)$, there is a $b^{LN} \in \psi(\theta, \zeta)$ for which $d_{LN}(y^{LN}, b^{LN}) \leq 2\zeta/N$ and hence for any $j \in \{0, 1, \cdots, N-1\}$ at most $LN(2\zeta/N) = 2\zeta L$ of the consecutive nonoverlapping N-tuples y^N_{j+iN}, $i = 0, 1, \cdots, L-2$, can differ from the corresponding b^N_{j+iN} and therefore

$$|\frac{1}{L-1}\sum_{i=0}^{L-2} 1_S(y^r_{j+iN}) - z_{\theta+j}|$$

$$\leq |\frac{1}{L-1}\sum_{i=0}^{L-2} 1_S(b^r_{j+iN}) - z_{\theta+j}| + 2\zeta \leq 3\zeta$$

and hence $y^{LN} \in \psi(\theta, 4\zeta)$. If $\tilde{\psi}(\theta)$ is defined to be $B^\theta \times \psi(\theta) \times B^{N-\theta} \in \mathcal{B}^m_B$, then we also have that

$$(\bigcap_{\theta=0}^{N-1} \tilde{\psi}(\theta, \zeta))_{\zeta/N} \subset \bigcap_{\theta=0}^{N-1} \tilde{\psi}(\theta)$$

since if $y^n \in (\bigcap_{\theta=0}^{N-1} \tilde{\psi}(\theta, \zeta))_{\zeta/N}$, then there is a b^m such that $b^{LN}_\theta \in \psi(\theta, \zeta)$; $\theta = 0, 1, \cdots, N-1$ and $d_m(y^m, b^m) \leq \zeta/N$ for $\theta = 0, 1, \cdots, N-1$. This implies from Lemma 12.6.1 and (12.9.4)–(12.9.6) that if $x \in \gamma^m(a^m)$ and $a^m \in \Phi$, then

$$\nu^m_x(\bigcap_{\theta=0}^{N-1} \tilde{\psi}(\theta)) \geq \nu^m_x((\bigcap_{\theta=0}^{N-1} \tilde{\psi}(\theta, \zeta))_{\zeta/N})$$

$$\geq \hat{\nu}(\bigcap_{\theta=0}^{N-1} \tilde{\psi}(\theta, \zeta)|\gamma^m(a^m)) - \frac{\zeta}{N} \geq 1 - \zeta - \frac{\zeta}{N} \geq 1 - \epsilon. \tag{12.9.7}$$

To complete the proof, we use (12.8.1)–(12.8.3) and (12.9.7) to obtain for $a^m \in \Phi$ and $\gamma_m(a_N L^N) = \sigma \times w_i$ that

$$\nu_x(y : s(y^{LN}_\theta) = \theta, \theta = 0, 1, \cdots, N-1; y^N_{LN} \in S \times W_i)$$

$$\geq \nu^m_x(\bigcap_{\theta=0}^{N-1} \psi(\theta)) - \nu^N_{T^{-NL}x}(S \times W^c_i) \geq 1 - \epsilon - 2\epsilon. \square$$

Next the prefixed block code and the sync locator function are combined with a random punctuation sequence of Lemma 9.5.2 to construct a good sliding block code for a totally ergodic source with entropy less than capacity.

Lemma 12.9.2: Given a $\bar{d}$-continuous totally ergodic stationary channel ν with Shannon capacity C, a stationary totally ergodic source $[G, \mu, U]$ with entropy rate $H(\mu) < C$, and $\delta > 0$, there exists for sufficiently large n, m a sliding block encoder $f : G^n \to A$ and decoder $g : B^m \to G$ such that $P_e(\mu, \nu, f, g) \leq \delta$.

Proof: Choose R, $\bar{H} < R < C$, and fix $\epsilon > 0$ so that $\epsilon \leq \delta/5$ and $\epsilon \leq (R - \bar{H})/2$. Choose N large enough so that the conditions and conclusions of Corollary 12.8.1 hold. Construct first a joint source and channel block encoder γ_N as follows: From the asymptotic equipartition property (Lemma 3.2.1 or Section 3.5), there is an n_0 large enough to ensure that for $N \geq n_0$ the set

$$G_N = \{u^N : |N^{-1}h_N(u) - \bar{H}| \geq \epsilon\}$$

$$= \{u^N : e^{-N(\bar{H}+\epsilon)} \leq \mu(u^N) \leq e^{-N(\bar{H}-\epsilon)}\} \tag{12.9.8}$$

has probability

$$\mu_{U^N}(G_N) \geq 1 - \epsilon. \tag{12.9.9}$$

Observe that if $M' = ||G_N||$, then

$$2^{N(\bar{H}-\epsilon)} \leq M' \leq 2^{N(\bar{H}+\epsilon)} \leq 2^{N(R-\epsilon)}. \tag{12.9.10}$$

Index the members of G_N as β_i; $i = 1, \cdots, M'$. If $u_N = \beta_i$, set $\gamma_N(u_N) = \sigma \times w_i$. Otherwise set $\gamma_N(u_N) = \sigma \times w_{M'+1}$. Since for large N, $2^{N(R-\epsilon)} + 1 \leq 2^{NR}$, γ_N is well defined. γ_N can be viewed as a synchronized extension of the almost noiseless code of Section 3.5. Define also the block decoder $\psi_N(y^N) = \beta_i$ if $y^N \in S \times W_i$; $i = 1, \cdots, M'$. Otherwise set $\psi_N(y^N) = \beta^*$, an arbitrary reference vector. Choose L so large that the conditions and conclusions of Lemma 12.9.1 hold for $\mathcal{C}$ and γ_N. The sliding block decoder $g_m : B^m \to G$, $m = (L+1)N$, yielding decoded process $\hat{U}_k = g_m(Y^m_{k-NL})$ is defined as follows: If $s(y_{k-NL}, \cdots, y_k - 1) = \theta$, form $b^N = \psi_N(y_{k-\theta}, \cdots, y_{k-\theta-N})$ and set $\hat{U}_k(y) = g_m(y_{k-NL}, \cdots, y_{k+N}) = b_\theta$, the appropriate symbol of the appropriate block.

The sliding block encoder f will send very long sequences of block words with random spacing to make the code stationary. Let K be a large number satisfying $K\epsilon \geq L + 1$ so that $m \leq \epsilon KN$ and recall that $N \geq 3$ and $L \geq 1$. We then have that

$$\frac{1}{KN} \leq \frac{1}{3K} \leq \frac{\epsilon}{6}. \tag{12.9.11}$$

Use Corollary 9.4.2 to produce a (KN, ϵ) punctuation sequence Z_n using a finite length sliding block code of the input sequence. The punctuation process is stationary and ergodic, has a ternary output and can produce only isolated 0's followed by KN 1's or individual 2's. The punctuation sequence is then used to convert the block encoder γ_N into a sliding block coder: Suppose that the encoder views an input sequence $u = \cdots, u_{-1}, u_0, u_1, \cdots$ and is to produce a single encoded symbol x_0. If u_0 is a 2, then the encoder produces an arbitrary channel symbol, say a^*. If x_0 is not a 2, then the encoder inspects u_0, u_{-1}, u_{-2} and so on into the past until it locates the first 0. This must happen within KN input symbols by construction of the punctuation sequence. Given that the first 1 occurs at, say, $Z_l = 1$,, the encoder then uses the block code γ_N to encode successive blocks of input N-tuples until the block including the symbol at time 0 is encoded. The sliding block encoder than produces the corresponding channel symbol x_0. Thus if $Z_l = 1$, then for some $J < Kx_0 = (\gamma_N(u_{l+JN}))_{l \bmod N}$ where the subscript denotes that the (l mod N)th coordinate of the block codeword is put out. The final sliding block code has a finite length given by the maximum of the lengths of the code producing the punctuation sequence and the code imbedding the block code γ_N into the sliding block code.

We now proceed to compute the probability of the error event $\{u, y : \hat{U}_0(y) \neq U_0(u)\} = E$. Let E_u denote the section $\{y : \hat{U}_0(y) \neq U_0(u)\}$, $\bar{f}$ be the sequence coder induced by f, and $F = \{u : Z_0(u) = 0\}$. Note that if $u \in T^{-1}F$, then $Tu \in F$ and hence $Z_0(Tu) = Z_1(u)$ since the coding is stationary. More generally, if $uT^{-i}F$, then $Z_i = 0$. By construction any 1 must be followed by KN 1's and hence the sets $T^{-i}F$ are disjoint for $i = 0, 1, \cdots, KN - 1$ and hence we can write

$$\begin{aligned}
P_e = \Pr(U_0 \neq \hat{U}_0) &= \mu\nu(E) \\
&= \int d\mu(u)\nu_{\bar{f}(u)}(E_u) \\
\leq \sum_{i=0}^{LN-1} \int_{T^{-i}F} d\mu(u)\nu_{\bar{f}(u)}(E_u) &+ \sum_{i=LN}^{KN-1} \int_{T^{-i}F} d\mu(u)\nu_{\bar{f}(u)}(E_u) \\
&+ \int_{(\bigcup_{i=0}^{KN-1} T^{-i}F)^c} d\mu(u) \\
= LN\mu(F) + \sum_{i=LN}^{KN-1} &\int_{T^{-i}F} d\mu(u)\nu_{\bar{f}(u)}(E_u) + \epsilon a \leq 2\epsilon \\
+ \sum_{i=LN}^{KN-1} \sum_{a^{kN} \in G^{kN}} \int_{u' \in T^{-i}(F \bigcap c(a^K N))} &d\mu(u')\nu_{\bar{f}(u')}(y' : U_0(u') \neq \hat{U}_0(u')),
\end{aligned} \tag{12.9.12}$$

where we have used the fact that $\mu(F) \leq (KN)^{-1}$ (from Corollary 9.4.2) and hence $LN\mu(F) \leq L/K \leq \epsilon$. Fix $i = kN + j$; $0 \leq j \leq N-1$ and define $u = T^{j+LN}u'$ and $y = T^{j+LN}y'$, and the integrals become

$$\int_{u' \in T^{-i}(F \bigcap c(a^{KN}))} d\mu(u')\nu_{\bar{f}(u')}(y' : U_0(u') \neq g_m(Y^m_{-NL}(y'))$$

$$= \int_{u \in T^{-(k-L)N}(F \bigcap c(a^{KN}))} d\mu(u')\nu_{\bar{f}(T^{-(j+LN)}u)}(y :$$

$$U_0(T^{j+LN}u) \neq g_m(Y\_NL^m(T^{j+NL}y)))$$

$$= \int_{u \in T^{-(k-L)N}(F \bigcap c(a^{KN}))} d\mu(u')\nu_{\bar{f}(T^{-(j+LN)}u)}(y : u_{j+LN}$$

$$\neq g_m(y_j^m)) = \int_{u \in T^{-(k-L)N}(F \bigcap c(a^{KN}))} d\mu(u')$$

$$\times \nu_{\bar{f}(T^{-(j+LN)}u)}(y : u^N_{LN} = \psi_N(y^N_{LN}) \text{ or } s(y_j^{LN} \neq j)). \tag{12.9.13}$$

If $u^N_{LN} = \beta_j \in G_N$, then $u^N_{LN} = \psi_N(y^N_{LN})$ if $y^N_{LN} \in S \times W_i$. If $u \in T^{-(k-L)N}c(a^{KN})$, then $u^m = a^m_{(k-L)N}$ and hence from Lemma 12.9.1 and stationarity we have for $i = kN + j$ that

$$\sum_{a^{KN} \in G^{KN}} \int_{T^{-i}(c(a^{KN}) \bigcap F)} d\mu(u)\nu_{\bar{f}(u)}(E_u)$$

$$\leq 3\epsilon \sum_{\substack{a^{KN} \in G^{KN} \\ a^m_{(k-L)N} \in \Phi \bigcap (G^{LN} \times G_N)}} \mu(T^{-(k-L)N}(c(a^{KN}) \bigcap F))$$

$$+ \sum_{\substack{a^{KN} \in G^{KN} \\ a^m_{(k-L)N} \notin \Phi \bigcap (G^{LN} \times G_N)}} \mu(T^{-(k-L)N}(c(a^{KN}) \bigcap F))$$

$$\leq 3\epsilon \sum_{a^{KN} \in G^{KN}} \mu(c(a^{KN}) \bigcap F))$$

$$+ \sum_{a^m_{(k-L)N} \in \Phi^c \bigcup (G^{LN} \times G_N)^c} \mu(c(a^{KN}) \bigcap F))$$

$$\leq 3\epsilon\mu(F) + \mu(c(\Phi^c) \bigcap F) + \mu(c(G_N) \bigcap F). \tag{12.9.14}$$

Choose the partition in Lemmas 9.5.1–9.5.2 to be that generated by the sets $c(\Phi^c)$ and $c(G^N)$ (the partition with all four possible intersections of these sets or their complements). Then the above expression is bounded above by

$$\frac{3\epsilon}{NK} + \frac{\epsilon}{NK} + \frac{\epsilon}{NK} \leq 5\frac{\epsilon}{NK}$$

and hence from (12.9.12)

$$P_e \leq 5\epsilon \leq \delta \tag{12.9.15}$$

which completes the proof. □

The lemma immediately yields the following corollary.

Corollary 12.9.1: If ν is a stationary $\bar{d}$-continuous totally ergodic channel with Shannon capacity C, then any totally ergodic source $[G, \mu, U]$ with $H(\mu) < C$ is admissible.

Ergodic Sources

If a prefixed blocklength N block code of Corollary 12.9.1 is used to block encode a general ergodic source $[G, \mu, U]$, then successive N-tuples from μ may not be ergodic, and hence the previous analysis does not apply. From the Nedoma ergodic decomposition [106] (see, e.g., [50], p. 232), any ergodic source μ can be represented as a mixture of N-ergodic sources, all of which are shifted versions of each other. Given an ergodic measure μ and an integer N, then there exists a decomposition of μ into M N-ergodic, N-stationary components where M divides N, that is, there is a set $\Pi \in \mathcal{B}_G^\infty$ such that

$$T^M \Pi = \Pi \tag{12.9.16}$$

$$\mu(T^i \Pi \bigcap T^j \Pi) = 0;\; i, j \leq M, i \neq j \tag{12.9.17}$$

$$\mu(\bigcup_{i=0}^{M-1} T^i \Pi) = 1$$

$$\mu(\Pi) = \frac{1}{M},$$

such that the sources $[G, \mu_i, U]$, where $\pi_i(W) = \mu(W | T^i \Pi) = M\mu(W \bigcap T^i \Pi)$ are N-ergodic and N-stationary and

$$\mu(W) = \frac{1}{M} \sum_{i=0}^{M-1} \pi_i(W) = \frac{1}{M} \sum_{i=0}^{M-1} \mu(W \bigcap T^i \Pi). \tag{12.9.18}$$

This decomposition provides a method of generalizing the results for totally ergodic sources to ergodic sources. Since $\mu(\cdot|\Pi)$ is N-ergodic, Lemma 12.9.2 is valid if μ is replaced by $\mu(\cdot|\Pi)$. If an infinite length sliding block encoder f is used, it can determine the ergodic component in effect by testing for

$T^{-i}\Pi$ in the base of the tower and insert i dummy symbols and then encode using the length N prefixed block code. In other words, the encoder can line up the block code with a prespecified one of the N-possible N-ergodic modes. A finite length encoder can then be obtained by approximating the infinite length encoder by a finite length encoder. Making these ideas precise yields the following result.

Theorem 12.9.1: If ν is a stationary $\bar{d}$-continuous totally ergodic channel with Shannon capacity C, then any ergodic source $[G, \mu, U]$ with $H(\mu) < C$ is admissible.

Proof: Assume that N is large enough for Corollary 12.8.1 and (12.9.8)–(12.9.10) to hold. From the Nedoma decomposition

$$\frac{1}{M}\sum_{i=0}^{M-1} \mu^N(G_N|T^i\Pi) = \mu^N(G_N) \geq 1-\epsilon.$$

and hence there exists at least one i for which

$$\mu^N(G_N|T^i\Pi) \geq 1-\epsilon;$$

that is, at least one N-ergodic mode must put high probability on the set G_N of typical N-tuples for μ. For convenience relabel the indices so that this good mode is $\mu(\cdot|\Pi)$ and call it the design mode. Since $\mu(\cdot|\Pi)$ is N-ergodic and N-stationary, Lemma 12.9.1 holds with μ replaced by $\mu(\cdot|\Pi)$; that is, there is a source/channel block code (γ_N, ψ_N) and a sync locating function $s : B^{LN} \to \{0, 1, \cdots, M-1\}$ such that there is a set $\Phi \in G_m$; $m = (L+1)N$, for which (12.9.1) holds and

$$\mu^m(\Phi|\Pi) \geq 1-\epsilon.$$

The sliding block decoder is exacted exactly as in Lemma 12.9.1. The sliding block encoder, however, is somewhat different. Consider a punctuation sequence or tower as in Lemma 9.5.2, but now consider the partition generated by Φ, G_N, and $T^i\Pi$, $i = 0, 1, \cdots, M-1$. The infinite length sliding block code is defined as follows: If $u \notin \bigcup_{k=0}^{NK-1} T^k F$, then $f(u) = a^*$, an arbitrary channel symbol. If $u \in T^i(F \bigcap T^{-j}\Pi)$ and if $i < j$, set $f(u) = a^*$ (these are spacing symbols to force alignment with the proper N-ergodic mode). If $j \leq i \leq KN - (M-j)$, then $i = j + kN + r$ for some $0 \leq k \leq (K-1)N$, $r \leq N-1$. Form $G_N(u^N_{j+kN}) = a^N$ and set $f(u) = a_r$. This is the same encoder as before, except that if $u \in T^j\Pi$, then block encoding is postponed for j symbols (at which time $u \in \Pi$). Lastly, if $KN - (M-j) \leq i \leq KN-1$, then $f(u) = a^*$.

As in the proof of Lemma 12.9.2

$$P_e(\mu, \nu, f, g_m) = \int d\mu(u)\nu_{f(u)}(y : U_0(u) \neq g_m(Y^m_{-LN}(y)))$$

$$\leq 2\epsilon + \sum_{i=LN}^{KN-1} \int u \in T^i F d\mu(u)\nu_{f(u)}(y : U_0(u) \neq \hat{U}_0(y))$$

$$= 2\epsilon + \sum_{i=LN}^{KN-1} \sum_{j=0}^{M-1} \sum_{a^{KN} \in G^{KN}}$$

$$\int_{u \in T^i(c(a^{KN}) \bigcap F \bigcap T^{-j}\Pi)} d\mu(u)\nu_{f(u)}(y : U_0(u) \neq \hat{U}_0(y))$$

$$\leq 2\epsilon + \sum_{j=0}^{M-1} \sum_{i=LN+j}^{KN-(M-j)} \sum_{a^{KN} \in G^{KN}}$$

$$\int_{u \in T^i(c(a^{KN}) \bigcap F \bigcap T^{-j}\Pi)} d\mu(u)\nu_{f(u)}(y : U_0(u) \neq \hat{U}_0(y))$$

$$+ \sum_{j=0}^{M-1} M\mu(F \bigcap T^{-j}\Pi), \tag{12.9.19}$$

where the rightmost term is

$$M \sum_{j=0}^{M-1} \mu(F \bigcap T^{-j}\Pi) \leq \frac{M}{KN} \leq \frac{1}{K} \leq \epsilon.$$

Thus

$$P_e(\mu, \nu, f, g_m) \leq 3\epsilon + \sum_{j=0}^{M-1} \sum_{i=LN+j}^{KN-(M-j)} \sum_{a^{KN} \in G^{KN}}$$

$$\int_{u \in T^i(c(a^{KN}) \bigcap F \bigcap T^{-j}\Pi)} d\mu(u)\nu_{f(u)}(y : U_0(u) \neq \hat{U}_0(y)).$$

Analogous to (12.9.13) (except that here $i = j + kN + r$, $u = T^{-(LN+r)}u'$)

$$\int_{u' \in T^i(c(a^{KN}) \bigcap F \bigcap T^{-j}\Pi)} d\mu(u')\nu_{f(u')}(y' : U_0(u') = g_m(Y^m_{-LN}(y')))$$

$$\leq \int_{T^{j+(k-L)N}(c(a^{KN}) \bigcap F \bigcap T^{-j}\Pi)} d\mu(u)$$

$$\nu_{f(T^i+LNu)}(y : u^N_{LN} \neq \psi_N(y^N_{LN}) \mathrm{or} s(y^{LN}_r) \neq r).$$

Thus since $u \in T^{j+(k-L)N}(c(a^{KN})\bigcap F\bigcap T^{-j}\Pi$ implies $u^m = a^m_{j+(k-L)N}$, analogous to (12.9.14) we have that for $i = j + kN + r$

$$\sum_{a^{KN}\in G^{KN}} \int_{T^i(c(a^{KN})\bigcap F\bigcap T^{-j}\Pi)} d\mu(u)\nu_{f(u)}(y : U_0(u) \neq g_m(Y\_LN^m(y)))$$

$$= \epsilon \sum_{a^{KN}:a^m_{j+(k-L)N}\in\Phi} \mu(T^{j+(k-L)N}(c(a^{KN})\bigcap F\bigcap T^{-j}\Pi))$$

$$+ \sum_{a^{KN}:a^m_{j+(k-L)N}\notin\Phi} \mu(T^{j+(k-L)N}(c(a^{KN})\bigcap F\bigcap T^{-j}\Pi))$$

$$= \epsilon \sum_{a^{KN}:a^m_{j+(k-L)N}\in\Phi} \mu(c(a^{KN})\bigcap F\bigcap T^{-j}\Pi)$$

$$+ \sum_{a^{KN}:a^m_{j+(k-L)N}\notin\Phi} \mu(c(a^{KN})\bigcap F\bigcap T^{-j}\Pi)$$

$$= \epsilon\mu(T^{-(j+(k-L)N)}c(\Phi)\bigcap F\bigcap T^{-j}\Pi)$$

$$+\mu(T^{-(j+(k-L)N)}c(\Phi)^c\bigcap F\bigcap T^{-j}\Pi).$$

From Lemma 9.5.2 (the Rohlin-Kakutani theorem), this is bounded above by

$$\epsilon\frac{\mu(T^{-(j+(k-L)N)}c(\Phi)\bigcap T^{-j}\Pi)}{KN} + \frac{\mu(T^{-(j+(k-L)N)}c(\Phi)^c\bigcap T^{-j}\Pi)}{KN}$$

$$= \epsilon\frac{\mu(T^{-(j+(k-L)N)}c(\Phi)|T^{-j}\Pi)\mu(\Pi)}{KN} + \frac{\mu(T^{-(j+(k-L)N)}c(\Phi)^c|T^{-j}\Pi)\mu(\Pi)}{KN}$$

$$= \epsilon\mu(c(\Phi)|\Pi)\frac{\mu(\Pi)}{KN}\mu(c(\Phi)^c|\Pi)\frac{\mu(\Pi)}{KN} + \leq \frac{2\epsilon}{MKN}.$$

With (12.9.18)–(12.9.19) this yields

$$P_e(\mu,\nu,f,g_m) \leq 3\epsilon + \frac{MKN2\epsilon}{MKN} \leq 5\epsilon, \tag{12.9.20}$$

which completes the result for an infinite sliding block code.

The proof is completed by applying Corollary 10.5.1, which shows that by choosing a finite length sliding block code f_0 from Lemma 4.2.4 so that $\Pr(f \neq f_0)$ is sufficiently small, then the resulting P_e is close to that for the infinite length sliding block code. □

In closing we note that the theorem can be combined with the sliding block source coding theorem to prove a joint source and channel coding theorem similar to Theorem 12.7.1, that is, one can show that given a source with distortion rate function $D(R)$ and a channel with capacity C, then sliding block codes exist with average distortion approximately $D(C)$.

Bibliography

[1] N. M. Abramson. *Information Theory and Coding.* McGraw-Hill, New York, 1963.

[2] R. Adler. Ergodic and mixing properties of infinite memory channels. *Proc. Amer. Math. Soc.*, 12:924–930, 1961.

[3] R. L. Adler, D. Coppersmith, and M. Hassner. Algorithms for sliding-block codes–an application of symbolic dynamics to information theory. *IEEE Trans. Inform. Theory*, IT-29:5–22, 1983.

[4] R. Ahlswede and P. Gács. Two contributions to information theory. In *Topics in Information Theory*, pages 17–40, Keszthely, Hungary, 1975.

[5] R. Ahlswede and J. Wolfowitz. Channels without synchronization. *Adv. in Appl. Probab.*, 3:383–403, 1971.

[6] P. Algoet. *Log-Optimal Investment.* PhD thesis, Stanford University, 1985.

[7] P. Algoet and T. Cover. A sandwich proof of the shannon-mcmillan-breiman theorem. *Ann. Probab.*, 16:899–909, 1988.

[8] E. Ayanŏglu and R. M. Gray. The design of joint source and channel trellis waveform coders. *IEEE Trans. Inform. Theory*, IT-33:855–865, November 1987.

[9] A. R. Barron. The strong ergodic theorem for densities: generalized shannon-mcmillan-breiman theorem. *Ann. Probab.*, 13:1292–1303, 1985.

[10] T. Berger. Rate distortion theory for sources with abstract alphabets and memory. *Inform. and Control*, 13:254–273, 1968.

[11] T. Berger. *Rate Distortion Theory*. Prentice-Hall Inc., Englewood Cliffs, New Jersey, 1971.

[12] T. Berger. Multiterminal source coding. In G. Longo, editor, *The Information Theory Approach to Communications*, volume 229 of *CISM Courses and Lectures*, pages 171–231. Springer-Verlag, Vienna and New York, 1978.

[13] E. Berlekamp. *Algebraic Coding Theory*. McGraw-Hill, New York, 1968.

[14] E. Berlekamp, editor. *Key Papers in the Development of Coding Theory*. IEEE Press, New York, 1974.

[15] P. Billingsley. *Ergodic Theory and Information*. Wiley, New York, 1965.

[16] G. D. Birkhoff. Proof of the ergodic theorem. *Proc. Nat. Acad. Sci.*, 17:656–660, 1931.

[17] R. E. Blahut. Computation of channel capacity and rate-distortion functions. *IEEE Trans. Inform. Theory*, IT-18:460–473, 1972.

[18] R. E. Blahut. *Theory and Practice of Error Control Codes*. Addison Wesley, Reading, Mass., 1987.

[19] L. Breiman. The individual ergodic theorem of information theory. *Ann. of Math. Statist.*, 28:809–811, 1957.

[20] L. Breiman. A correction to 'The individual ergodic theorem of information theory'. *Ann. of Math. Statist.*, 31:809–810, 1960.

[21] J. R. Brown. *Ergodic Theory and Topological Dynamics*. Academic Press, New York, 1976.

[22] J. A. Bucklew. A large deviation theory proof of the abstract alphabet source coding theorem. *IEEE Trans. Inform. Theory*, IT-34:1081–1083, 1988.

[23] T. M. Cover, P. Gacs, and R. M. Gray. Kolmogorov's contributions to information theory and algorithmic complexity. *Ann. Probab.*, 17:840–865, 1989.

[24] I. Csiszár. Information-type measures of difference of probability distributions and indirect observations. *Studia Scientiarum Mathematicarum Hungarica*, 2:299–318, 1967.

[25] I. Csiszár. I-divergence geometry of probability distributions and minimization problems. *Ann. Probab.*, 3(1):146–158, 1975.

[26] I. Csiszár and J. Körner. *Coding Theorems of Information Theory.* Academic Press/Hungarian Academy of Sciences, Budapest, 1981.

[27] L. D. Davisson and R.M. Gray. A simplified proof of the sliding-block source coding theorem and its universal extension. In *Conf. Record 1978 Int'l. Conf. on Comm. 2*, pages 34.4.1–34.4.5, Toronto, 1978.

[28] L. D. Davisson, R. J. McEliece, M. B. Pursley, and M. S. Wallace. Efficient universal noiseless source codes. *IEEE Trans. Inform. Theory*, IT-27:269–279, 1981.

[29] L. D. Davisson and M. B. Pursley. An alternate proof of the coding theorem for stationary ergodic sources. In *Proceedings of the Eighth Annual Princeton Conference on Information Sciences and Systems*, 1974.

[30] M. Denker, C. Grillenberger, and K. Sigmund. *Ergodic Theory on Compact Spaces*, volume 57 of *Lecture Notes in Mathematics.* Springer-Verlag, New York, 1970.

[31] J.-D. Deushcel and D. W. Stroock. *Large Deviations*, volume 137 of *Pure and Applied Mathematics.* Academic Press, Boston, 1989.

[32] R. L. Dobrushin. A general formulation of the fundamental Shannon theorem in information theory. *Uspehi Mat. Akad. Nauk. SSSR*, 14:3–104, 1959. Translation in *Transactions Amer. Math. Soc*, series 2, vol. 33, 323–438.

[33] R. L. Dobrushin. Shannon's theorems for channels with synchronization errors. *Problemy Peredaci Informatsii*, 3:18–36, 1967. Translated in *Problems of Information Transmission*, vol., 3, 11–36 (1967), Plenum Publishing Corporation.

[34] M. D. Donsker and S. R. S. Varadhan. Asymptotic evaluation of certain markov process expectations for large time. *J. Comm. Pure Appl. Math.*, 28:1–47, 1975.

[35] J. G. Dunham. A note on the abstract alphabet block source coding with a fidelity criterion theorem. *IEEE Trans. Inform. Theory*, IT-24:760, November 1978.

[36] P. Elias. Two famous papers. *IRE Transactions on Information Theory*, page 99, 1958.

[37] R. M. Fano. *Transmission of Information.* Wiley, New York, 1961.

[38] A. Feinstein. A new basic theorem of information theory. *IRE Transactions on Information Theory*, pages 2–20, 1954.

[39] A. Feinstein. *Foundations of Information Theory.* McGraw-Hill, New York, 1958.

[40] A. Feinstein. On the coding theorem and its converse for finite-memory channels. *Inform. and Control*, 2:25–44, 1959.

[41] G. D. Forney. The viterbi algorithm. *Proc. IEEE*, 61:268–278, March 1973.

[42] N. A. Friedman. *Introduction to Ergodic Theory.* Van Nostrand Reinhold Company, New York, 1970.

[43] R. G. Gallager. *Information Theory and Reliable Communication.* John Wiley & Sons, New York, 1968.

[44] A. El Gamal and T. Cover. Multiple user information theory. *Proc. IEEE*, 68:1466–1483, 1980.

[45] I. M. Gelfand, A. N. Kolmogorov, and A. M. Yaglom. On the general definitions of the quantity of information. *Dokl. Akad. Nauk*, 111:745–748, 1956. (In Russian.).

[46] A. Gersho and V. Cuperman. Vector quantization: A pattern-matching technique for speech coding. *IEEE Communications Magazine*, 21:15–21, December 1983.

[47] A. Gersho and R. M. Gray. *Vector Quantization and Signal Compression.* Kluwer Academic Press, 1990. To appear.

[48] R. M. Gray. Tree-searched block source codes. In *Proceedings of the 1980 Allerton Conference*, Allerton IL, Oct. 1980.

[49] R. M. Gray. Vector quantization. *IEEE ASSP Magazine*, 1, No. 2:4–29, April 1984.

[50] R. M. Gray. *Probability, Random Processes, and Ergodic Properties.* Springer-Verlag, New York, 1988.

[51] R. M. Gray. Spectral analysis of quantization noise in a single-loop sigma-delta modulator with dc input. *IEEE Trans. Comm.*, COM-37:588–599, 1989.

[52] R. M. Gray. *Source Coding Theory.* Kluwer Academic Press, Boston, 1990.

[53] R. M. Gray and L. D. Davisson. Source coding without the ergodic assumption. *IEEE Trans. Inform. Theory*, IT-20:502–516, 1974.

[54] R. M. Gray and J. C. Kieffer. Asymptotically mean stationary measures. *Ann. Probab.*, 8:962–973, 1980.

[55] R. M. Gray, D. L. Neuhoff, and J. K. Omura. Process definitions of distortion rate functions and source coding theorems. *IEEE Trans. Inform. Theory*, IT-21:524–532, 1975.

[56] R. M. Gray, D. L. Neuhoff, and D. Ornstein. Nonblock source coding with a fidelity criterion. *Ann. Probab.*, 3:478–491, 1975.

[57] R. M. Gray, D. L. Neuhoff, and P. C. Shields. A generalization of ornstein's d-bar distance with applications to information theory. *Ann. Probab.*, 3:315–328, April 1975.

[58] R. M. Gray and D. S. Ornstein. Sliding-block joint source/noisy-channel coding theorems. *IEEE Trans. Inform. Theory*, IT-22:682–690, 1976.

[59] R. M. Gray, D. S. Ornstein, and R. L. Dobrushin. Block synchronization, sliding-block coding, invulnerable sources and zero error codes for discrete noisy channels. *Ann. Probab.*, 8:639–674, 1980.

[60] R. M. Gray, M. Ostendorf, and R. Gobbi. Ergodicity of markov channels. *IEEE Trans. Inform. Theory*, 33:656–664, September 1987.

[61] R. M. Gray and F. Saadat. Block source coding theory for asymptotically mean stationary sources. *IEEE Trans. Inform. Theory*, 30:64–67, 1984.

[62] P. R. Halmos. *Lectures on Ergodic Theory.* Chelsea, New York, 1956.

[63] G. H. Hardy, J. E. Littlewood, and G. Polya. *Inequalities.* Cambridge Univ. Press, London, 1952. Second Edition, 1959.

[64] R. V. L. Hartley. Transmission of information. *Bell System Tech. J.*, 7:535–563, 1928.

[65] E. Hoph. *Ergodentheorie.* Springer-Verlag, Berlin, 1937.

[66] K. Jacobs. Die Übertragung diskreter Informationen durch periodishce und fastperiodische Kanale. *Math. Annalen*, 137:125–135, 1959.

[67] K. Jacobs. Über die Struktur der mittleren Entropie. *Math. Z.*, 78:33–43, 1962.

[68] K. Jacobs. The ergodic decomposition of the Kolmogorov-Sinai invariant. In F. B. Wright and F. B. Wright, editors, *Ergodic Theory*. Academic Press, New York, 1963.

[69] N. S. Jayant and P. Noll. *Digital Coding of Waveforms*. Prentice-Hall, Englewood Cliffs, New Jersey, 1984.

[70] T. Kadota. Generalization of feinstein's fundamental lemma. *IEEE Trans. Inform. Theory*, IT-16:791–792, 1970.

[71] S. Kakutani. Induced measure preserving transformations. In *Proceedings of the Imperial Academy of Tokyo*, volume 19, pages 635–641, 1943.

[72] A. J. Khinchine. The entropy concept in probability theory. *Uspekhi Matematicheskikh Nauk.*, 8:3–20, 1953. Translated in *Mathematical Foundations of Information Theory*, Dover New York (1957).

[73] A. J. Khinchine. On the fundamental theorems of information theory. *Uspekhi Matematicheskikh Nauk.*, 11:17–75, 1957. Translated in *Mathematical Foundations of Information Theory*, Dover New York (1957).

[74] J. C. Kieffer. A counterexample to Perez's generalization of the Shannon-McMillan theorem. *Ann. Probab.*, 1:362–364, 1973.

[75] J. C. Kieffer. A general formula for the capacity of stationary nonanticipatory channels. *Inform. and Control*, 26:381–391, 1974.

[76] J. C. Kieffer. On the optimum average distortion attainable by fixed-rate coding of a nonergodic source. *IEEE Trans. Inform. Theory*, IT-21:190–193, March 1975.

[77] J. C. Kieffer. A generalization of the pursley-davisson-mackenthun universal variable-rate coding theorem. *IEEE Trans. Inform. Theory*, IT-23:694–697, 1977.

[78] J. C. Kieffer. A unified approach to weak universal source coding. *IEEE Trans. Inform. Theory*, IT-24:674–682, 1978.

[79] J. C. Kieffer. Extension of source coding theorems for block codes to sliding block codes. *IEEE Trans. Inform. Theory*, IT-26:679–692, 1980.

[80] J. C. Kieffer. Block coding for weakly continuous channels. *IEEE Trans. Inform. Theory*, IT-27:721–727, 1981.

[81] J. C. Kieffer. Sliding-block coding for weakly continuous channels. *IEEE Trans. Inform. Theory*, IT-28:2–10, 1982.

[82] J. C. Kieffer. Coding theorem with strong converse for block source coding subject to a fidelity constraint, 1989. Preprint.

[83] J. C. Kieffer. An ergodic theorem for constrained sequences of functions. *Bulletin American Math Society*, 1989.

[84] J. C. Kieffer. Sample converses in source coding theory, 1989. Preprint.

[85] J. C. Kieffer. *Elementary Information Theory*. 1990. Book in Progress.

[86] J. C. Kieffer and M. Rahe. Markov channels are asymptotically mean stationary. *Siam Journal of Mathematical Analysis*, 12:293–305, 1980.

[87] A. N. Kolmogorov. On the Shannon theory of information in the case of continuous signals. *IRE Transactions Inform. Theory*, IT-2:102–108, 1956.

[88] A. N. Kolmogorov. A new metric invariant of transitive dynamic systems and automorphisms in lebesgue spaces. *Dokl. Akad. Nauk SSR*, 119:861–864, 1958. (In Russian.).

[89] A. N. Kolmogorov. On the entropy per unit time as a metric invariant of automorphisms. *Dokl. Akad. Naud SSSR*, 124:768–771, 1959. (In Russian.).

[90] A. N. Kolmogorov, A. M. Yaglom, and I. M. Gelfand. Quantity of information and entropy for continuous distributions. In *Proceedings 3rd All-Union Mat. Conf.*, volume 3, pages 300–320. Izd. Akad. Nauk. SSSR, 1956.

[91] S. Kullback. A lower bound for discrimination in terms of variation. *IEEE Trans. Inform. Theory*, IT-13:126–127, 1967.

[92] S. Kullback. *Information Theory and Statistics*. Dover, New York, 1968. Reprint of 1959 edition published by Wiley.

[93] B. M. Leiner and R. M. Gray. Bounds on rate-distortion functions for stationary sources and context-dependent fidelity criteria. *IEEE Trans. Inform. Theory*, IT-19:706–708, Sept. 1973.

[94] V. I. Levenshtein. Binary codes capable of correcting deletions, insertions, and reversals. *Sov. Phys.-Dokl.*, 10:707–710, 1966.

[95] S. Lin. *Introduction to Error Correcting Codes.* Prentice-Hall, Englewood Cliffs,NJ, 1970.

[96] K. M. Mackenthun and M. B. Pursley. Strongly and weakly universal source coding. In *Proceedings of the 1977 Conference on Information Science and Systems*, pages 286–291, Johns Hopkins University, 1977.

[97] F. J. MacWilliams and N. J. A. Sloane. *The Theory of Error-Correcting Codes.* North-Holland, New York, 1977.

[98] A. Maitra. Integral representations of invariant measures. *Transactions of the American Mathematical Society*, 228:209–235, 1977.

[99] J. Makhoul, S. Roucos, and H. Gish. Vector quantization in speech coding. *Proc. IEEE*, 73. No. 11:1551–1587, November 1985.

[100] B. Marcus. Sophic systems and encoding data. *IEEE Trans. Inform. Theory*, IT-31:366–377, 1985.

[101] K. Marton. On the rate distortion function of stationary sources. *Problems of Control and Information Theory*, 4:289–297, 1975.

[102] R. McEliece. *The theory of information and coding.* Cambridge University Press, New York, NY, 1984.

[103] B. McMillan. The basic theorems of information theory. *Ann. of Math. Statist.*, 24:196–219, 1953.

[104] L. D. Meshalkin. A case of isomorphisms of bernoulli scheme. *Dokl. Akad. Nauk SSSR*, 128:41–44, 1959. (In Russian.).

[105] Shu-Teh C. Moy. Generalizations of shannon-mcmillan theorem. *Pacific Journal Math.*, 11:705–714, 1961.

[106] J. Nedoma. On the ergodicity and r-ergodicity of stationary probability measures. *Z. Wahrsch. Verw. Gebiete*, 2:90–97, 1963.

[107] J. Nedoma. The synchronization for ergodic channels. *Transactions Third Prague Conf. Information Theory, Stat. Decision Functions, and Random Processes*, pages 529–539, 1964.

[108] D. L. Neuhoff and R. K. Gilbert. Causal source codes. *IEEE Trans. Inform. Theory*, IT-28:701–713, 1982.

[109] D. L. Neuhoff, R. M. Gray, and L. D. Davisson. Fixed rate universal block source coding with a fidelity criterion. *IEEE Trans. Inform. Theory*, 21:511–523, 1975.

[110] D. L. Neuhoff and P. C. Shields. Channels with almost finite memory. *IEEE Trans. Inform. Theory*, pages 440–447, 1979.

[111] D. L. Neuhoff and P. C. Shields. Channel distances and exact representation. *Inform. and Control*, 55(1), 1982.

[112] D. L. Neuhoff and P. C. Shields. Channel entropy and primitive approximation. *Ann. Probab.*, 10(1):188–198, 1982.

[113] D. L. Neuhoff and P. C. Shields. Indecomposable finite state channels and primitive approximation. *IEEE Trans. Inform. Theory*, IT-28:11–19, 1982.

[114] D. Ornstein. Bernoulli shifts with the same entropy are isomorphic. *Advances in Math.*, 4:337–352, 1970.

[115] D. Ornstein. An application of ergodic theory to probability theory. *Ann. Probab.*, 1:43–58, 1973.

[116] D. Ornstein. *Ergodic Theory, Randomness, and Dynamical Systems*. Yale University Press, New Haven, 1975.

[117] D. Ornstein and B. Weiss. The Shannon-McMillan-Breiman theorem for a class of amenable groups. *Israel J. of Math*, 44:53–60, 1983.

[118] D. O'Shaughnessy. *Speech Communication*. Addison-Wesley, Reading, Mass., 1987.

[119] P. Papantoni-Kazakos and R. M. Gray. Robustness of estimators on stationary observations. *Ann. Probab.*, 7:989–1002, Dec. 1979.

[120] A. Perez. Notions généralisées d'incertitude, d'entropie et d'information du point de vue de la théorie des martingales. In *Transactions First Prague Conf. on Information Theory, Stat. Decision Functions, and Random Processes*, pages 183–208. Czech. Acad. Sci. Publishing House, 1957.

[121] A. Perez. Sur la convergence des incertitudes, entropies et informations échantillon vers leurs valeurs vraies. In *Transactions First Prague Conf. on Information Theory, Stat. Decision Functions, and Random Processes*, pages 245–252. Czech. Acad. Sci. Publishing House, 1957.

[122] A. Perez. Sur la théorie de l'information dans le cas d'un alphabet abstrait. In *Transactions First Prague Conf. on Information Theory, Stat. Decision Functions, Random Processes*, pages 209–244. Czech. Acad. Sci. Publishing House, 1957.

[123] A. Perez. Extensions of shannon-mcmillan's limit theorem to more general stochastic processes processes. In *Third Prague Conf. on Inform. Theory, Decision Functions, and Random Processes*, pages 545–574, Prague and New York, 1964. Publishing House Czech. Akad. Sci. and Academic Press.

[124] K. Petersen. *Ergodic Theory.* Cambridge University Press, Cambridge, 1983.

[125] M. S. Pinsker. Dynamical systems with completely positive or zero entropy. *Soviet Math. Dokl.*, 1:937–938, 1960.

[126] D. Ramachandran. *Perfect Measures.* ISI Lecture Notes, No. 6 and 7. Indian Statistical Institute, Calcutta, India, 1979.

[127] V. A. Rohlin and Ya. G. Sinai. Construction and properties of invariant measurable partitions. *Soviet Math. Dokl.*, 2:1611–1614, 1962.

[128] V. V. Sazanov. On perfect measures. *Izv. Akad. Nauk SSSR*, 26:391–414, 1962. American Math. Soc. Translations, Series 2, No. 48, pp. 229-254, 1965.

[129] C. E. Shannon. A mathematical theory of communication. *Bell Syst. Tech. J.*, 27:379–423,623–656, 1948.

[130] C. E. Shannon. Coding theorems for a discrete source with a fidelity criterion. In *IRE National Convention Record, Part 4*, pages 142–163, 1959.

[131] P. C. Shields. *The Theory of Bernoulli Shifts.* The University of Chicago Press, Chicago, Ill., 1973.

[132] P. C. Shields. The ergodic and entropy theorems revisited. *IEEE Trans. Inform. Theory*, IT-33:263–266, 1987.

[133] P. C. Shields and D. L. Neuhoff. Block and sliding-block source coding. *IEEE Trans. Inform. Theory*, IT-23:211–215, 1977.

[134] Ya. G. Sinai. On the concept of entropy of a dynamical system. *Dokl. Akad. Nauk. SSSR*, 124:768–771, 1959. (In Russian.).

[135] Ya. G. Sinai. Weak isomorphism of transformations with an invariant measure. *Soviet Math. Dokl.*, 3:1725–1729, 1962.

[136] Ya. G. Sinai. *Introduction to Ergodic Theory*. Mathematical Notes, Princeton University Press, Princeton, 1976.

[137] D. Slepian. A class of binary signaling alphabets. *Bell Syst. Tech. J.*, 35:203–234, 1956.

[138] D. Slepian, editor. *Key Papers in the Development of Information Theory*. IEEE Press, New York, 1973.

[139] A. D. Sokai. Existence of compatible families of proper regular conditional probabilities. *Z. Wahrsch. Verw. Gebiete*, 56:537–548, 1981.

[140] J. Storer. *Data Compression*. Computer Science Press, Rockville, Maryland, 1988.

[141] I. Vajda. A synchronization method for totally ergodic channels. In *Transactions of the Fourth Prague Conf. on Information Theory, Decision Functions, and Random Processes*, pages 611–625, Prague, 1965.

[142] E. van der Meulen. A survey of multi-way channels in information theory: 1961–1976. *IEEE Trans. Inform. Theory*, IT-23:1–37, 1977.

[143] S. R. S. Varadhan. *Large Deviations and Applications*. Society for Industrial and Applied Mathematics, Philadelphia, 1984.

[144] L. N. Vasershtein. Markov processes on countable product space describing large systems of automata. *Problemy Peredachi Informatsii*, 5:64–73, 1969.

[145] A. J. Viterbi and J. K. Omura. *Principles of Digital Communication and Coding*. McGraw-Hill, New York, 1979.

[146] J. von Neumann. Zur operatorenmethode in der klassischen mechanik. *Ann. of Math.*, 33:587–642, 1932.

[147] P. Walters. *Ergodic Theory-Introductory Lectures*. Lecture Notes in Mathematics No. 458. Springer-Verlag, New York, 1975.

[148] E. J. Weldon, Jr. and W. W. Peterson. *Error Correcting Codes.* MIT Press, Cambridge, Mass., 1971. Second Ed.

[149] K. Winkelbauer. Communication channels with finite past history. *Transactions of the Second Prague Conf. on Information Theory, Decision Functions, and Random Processes*, pages 685–831, 1960.

[150] J. Wolfowitz. Strong converse of the coding theorem for the general discrete finite-memory channel. *Inform. and Control*, 3:89–93, 1960.

[151] J. Wolfowitz. *Coding Theorems of Information Theory.* Springer-Verlag, New York, 1978. Third edition.

[152] A. Wyner. A definition of conditional mutual information for arbitrary ensembles. *Inform. and Control*, pages 51–59, 1978.

Index

www.ingramcontent.com/pod-product-compliance
Ingram Content Group UK Ltd.
Pitfield, Milton Keynes, MK11 3LW, UK
UKHW012153240726
13966UKWH00002B/313